Astrophysics of the Interstellar Medium

Walter J. Maciel

Astrophysics of the Interstellar Medium

Translated by Margarida Serote Roos

Walter J. Maciel
Departamento de Astronomia IAG/USP
Cidade Universitaria
São Paulo, Brazil

ISBN 978-1-4899-9691-6 ISBN 978-1-4614-3767-3 (eBook)
DOI 10.1007/978-1-4614-3767-3
Springer New York Heidelberg Dordrecht London

Printed on acid-free paper

Springer is part of Springer Science+Business Media (www.springer.com)

qui dat nivem sicut lanam
nebulam sicut cinerem spargit

Psalm 147, 16

"He spreads the snow like wool and scatters the frost like ashes"

To the memory of my father

Preface

In the space between the stars, there is a large diversity of objects, where fundamental physical processes for galaxy structure and evolution are taking place. This book aims to describe the main processes and to give, besides a simple description, numerical estimates of the most relevant quantities for interstellar astrophysics.

The principal objects that permeate the interstellar regions are described and analyzed according to their physical properties, but the main focus of the book are the physical processes taking place within these objects. These processes can also occur in other astrophysical contexts, such as stellar physics, galactic astronomy, and in regions like active galactic nuclei. For this reason, the book, besides being an introduction course on astrophysics of the interstellar medium, may also be a useful tool for students of astrophysics and researchers interested in these fields of knowledge.

The text of this book is based on lecture notes of the postgraduation course on interstellar medium that has been lectured by the Astronomy Department of the Astronomical and Geophysical Institute, São Paulo University, for about twenty years. The main part of the course is found in Chaps. 2 to 11, where the principal physical processes involved are described. There is not—yet—a general theory for the interstellar medium; thus, the various processes still represent open problems in this area of expertise, for which answers are presently being sought.

The units used in astrophysics are deeply embedded in everyday life, making the full use of a system such as the International System of Units (SI) virtually impossible. Besides some unusual units, like the parsec, the stellar magnitudes, and the solar units, we mainly use the cgs system complemented with atomic units, such as electron volt (eV) and ångström (Å). In this book, we maintain the traditional units, the ones the reader will find in every specialized paper. However, in the end, we provide a table with the principal physical and astronomical constants in their usual units and also converted to SI units. For a better memorization of the numerical values of the most important quantities, each chapter includes some exercises, which must be taken as integral parts of the text. There is also basic bibliography for each chapter, where the reader will find a wide variety of materials for a more detailed study.

This work was made possible thanks to the collaboration and support of many people to whom I am greatly indebted: Professors José Antônio de Freitas Pacheco, Sayd Codina, Jean Lefèvre, and Stuart Pottasch, for the postgraduation courses that were the heart of this manuscript; Beatriz Barbuy and Jacques Lépine who, as heads of department, supported the distribution of the first versions of this work, which is an essential step for the establishment of the text presented in this manuscript; Roberto Costa, who provided the spectrum of NGC 2346; Rodrigo Prates Campos, for images of NGC 4594, and Orion Nebula; Sílvia Lorenz, for the data and model of star AFGL 1141; Gilberto Sanzovo who commented on Chapter 10; and finally Ruth Gruenwald, for her careful reading of the whole manuscript and her helpful suggestions and corrections. Naturally, all errors and omissions that may have remained are my own responsibility.

São Paulo, Brazil Walter J. Maciel

About the Author

Walter Junqueira Maciel was born in Cruzília, MG, Brazil. He graduated in Physics at UFMG (Minas Gerais Federal University), in Belo Horizonte, and obtained a master's degree at ITA (Aeronautics Technological Institute), São José dos Campos, and a PhD at São Paulo University. He did several internships in Groningen, the Netherlands, and in Heidelberg, Germany. He is a full professor in the Astronomy Department at the Astronomical and Geophysical Institute, São Paulo University, where he has been working since 1974. He was the head of department between 1992 and 1994. He has published over a hundred scientific papers in international journals and around fifty papers (science, education, and outreach) in national journals. He is the author of the book "Introduction to Stellar Structure and Evolution" (Edusp, 1999), which won the Jabuti Prize in 2000 in the field of exact sciences, technology, and computer sciences and of the book Hydrodynamics and stellar winds: an introduction (Edusp, 2005).

Contents

Chapter 1
General Overview of the Interstellar Medium

1.1 Introduction

The night sky is endowed with stars. It is therefore natural to accept the existence of a *stellar medium*. However, this seems to indicate that the space between the stars is empty. Yet the most astute observers might suspect that this is not entirely true, since some regions are seen to be darker than others, while some regions are bright and diffuse. The idea of an interstellar medium, as opposed to the stellar medium, came from the clash between some observations and known stellar properties.

We can initially characterize the interstellar medium by means of a scale factor. If we consider for instance the Local Group of Galaxies, which is a grouping of galaxies containing the Milky Way, M31, and about 40 smaller galaxies, we can distinguish three different regions: the *intergalactic medium*, with dimensions that span from several hundred kpc to about one Mpc; the *interstellar* galactic medium, spanning some tens of kpc; and the *interplanetary medium*, characterized by the dimensions of the Solar System, some thousandth of a pc.

In a more detailed manner, we can also distinguish a circumstellar medium, characterized by regions around young or evolved stars, with dimensions of the order or even bigger than the ones of the Solar System. This is the case, for instance, of the circumstellar regions around giant stars with intermediate mass, where infrared emission from solid particles and maser emission from some simple molecules are observed. In this case, the limits between these different "media" are less precise, without a clear separation between them.

The *galactic disk* has a 30 kpc diameter and is 1 kpc thick. The connection between the interstellar medium and the disk region can easily be identified in pictures of galaxies with disks or spirals. In these objects, a dark lane is normally observed composed mainly of diffuse gas and dust, with a high enough concentration to obscure the light coming from the stars in the same line of sight, as shown in Fig. 1.1 for galaxy NGC 4594. Our own galaxy, named Galaxy or Milky Way, shows the same evidence. Further away from the disk, interstellar material can also be found, though its concentration is generally lower than in the disk. This is also true for elliptical galaxies.

W.J. Maciel, *Astrophysics of the Interstellar Medium*,
DOI 10.1007/978-1-4614-3767-3_1, © Springer Science+Business Media New York 2013

Fig. 1.1 The galaxy NGC 4594 (credits: Rodrigo Prates Campos, LNA)

In spite of being mainly concentrated in the disk, the interstellar material has a heterogeneous and fragmentary distribution. Several regions of the disk are filled by a diffuse gas, with some areas denser than others. Some of these denser areas are associated with young stars and with the spiral arms, showing the connection between the interstellar medium and star formation. Others occur in the neighborhood of evolved stars, both massive and intermediate mass, showing the final stage of their evolution.

The observed fragmentation in the interstellar medium makes its study more complex, and different interstellar "objects" are identified, depending on the observational method used.

In this book, we emphasize the *physical processes* that occur in the interstellar medium, particularly in our own Galaxy, bringing up the phenomenological aspects when deemed necessary. Many of these processes also occur in other astrophysical situations, not necessarily related to the interstellar medium, such as some important processes taking place in the interior of stars, in active galactic nuclei, and in the intergalactic medium, meaning that the basic principles studied here will also be useful in these areas of research.

1.2 Density of the Interstellar Medium

The most important characteristic of the interstellar medium is probably its density. Though easily observed from afar, as seen in Fig. 1.1, the material that fills the space between the stars is extremely tenuous when compared with the normal

Table 1.1 Typical densities in different environments

Region	n (cm^{-3})	Particles/cup
Solar interior	7×10^{26}	1×10^{29}
Water	3×10^{22}	8×10^{24}
Air	2×10^{19}	6×10^{21}
Solar photosphere	1×10^{17}	3×10^{19}
Atmosphere (M5 III)	2×10^{15}	6×10^{17}
Circumstellar envelope (M)	1×10^{8}	2×10^{10}
Dense interstellar region	1×10^{4}	2×10^{6}
Diffuse interstellar region	10	2×10^{3}
Intercloud medium	0.1	20
Coronal gas/IGM	10^{-4}	2×10^{-2}

densities in the laboratory and even with other astronomical objects. At first glance, we can consider the interstellar medium composed of dense and diffuse regions, as well as an even more rarefied medium permeating these regions. In Table 1.1, we show the typical density, n, measured in particles per cm^3 and the total number of particles in a "cup" of interstellar material. For comparison, the table shows these quantities in different astronomical situations, as well as two everyday examples (water, air).

As we can see from Table 1.1, even dense interstellar (IS) regions are many orders of magnitude more diluted than a typical stellar atmosphere or the extended envelopes surrounding red giants of spectral type M (luminosity class III). Giant molecular clouds can eventually reach densities of one or two orders of magnitude higher than the value of the interstellar regions indicated in Table 1.1, though in very localized areas. The lower limit is reached by coronal gas that surrounds, not just the disk, but the whole Galaxy, and whose density is similar to that of the intergalactic medium (IGM). Note that the best vacuum obtained in the laboratory corresponds to pressures of the order of 10^{-9} Torr ~10^{-12} atm ~10^{-6} dyne cm^{-2}, to densities of around 10^7 particles per cm^3. Thus, a cup of *vacuum* contains about 2×10^9 particles, much more than in any normal situation found in the interstellar medium!

1.3 The Interstellar Zoo

Stars in our Galaxy fill a spherical volume bigger than 10^{68} cm^3, where a thin disk with a volume of about 10^{67} cm^3 is immersed. This volume is also occupied by the interstellar medium, but its main mass is concentrated within an even thinner disk, as observed in external galaxies, with a 10^{66} cm^3 volume and a 300 pc thickness. In this region, there are many bright stars, in particular OB stars. So, the most natural component of the interstellar medium is the *photon*, mainly the ultraviolet photon emitted by these stars. That is why there is, in general, a *radiation field* associated with the interstellar medium that can heat up and ionize the gas while interacting with the other components of the interstellar space at the same time.

Interstellar gas is generally associated with a solid component, the interstellar dust particles. If a cloud containing gas and dust is not associated with bright stars, dust will absorb the interstellar radiation field and we will see it as a *dark nebula*. That is the case of the Coal Sack or the Horsehead Nebula in the Orion constellation. This kind of nebula was not always recognized as such. For instance, when observing the contrast between dark and bright regions in the Milky Way, William Herschel thought they were "holes in the sky." The basics for the study of interstellar absorption by dust were already laid down in the eighteenth century, but real evidence about *interstellar extinction* was only brought to light by Robert Trumpler's work in 1930. Since then, detailed studies about the nature of the interstellar dust have been made based on observations, theoretical studies, and even laboratory experiments, the latter being quite unusual for astrophysics.

If the dust cloud is associated with moderately hot stars, with effective temperatures of about $T_{\rm eff} \simeq 25{,}000$ K, the dust particles may scatter the stellar radiation, forming a *reflection nebula*. These nebulae are important to determine the physical properties of interstellar dust particles because the observed radiation comes from a known source (the star itself), and thus, its analysis gives us direct information about the nature of the particles.

When the nebula is associated with very hot stars ($T_{\rm eff} \gtrsim 25{,}000$ K), the gas surrounding it is photoionized, and the nebula is called a *diffuse nebula or H II region*. Since Bengt Strömgren's work in 1939, we know that ionized gas occupies a well-defined region in space, with a clear separation between the hot and ionized H II region, with an electron temperature $T_{\rm e} \sim 10^4$ K, and the cold and neutral H I region, with a kinetic temperature $T_{\rm k} \sim 10^2$ K. Several physical processes occur in these regions, which present strong emission of the ultraviolet and optical lines of H, He, and heavier elements and also radio and infrared emission. Connected with star-forming regions, H II regions play an important role in the study of the chemical evolution of the Galaxy and in the definition of spiral arms, for which they are the principal tracers.

Molecular clouds are generally associated with H II regions but are undetected in optical images. These objects are essentially gas and dust clouds with relatively high densities ($n \gtrsim 10^4$ cm^{-3}) and relatively low kinetic temperatures, $T_{\rm k} \simeq 10-100$ K. The detection of these clouds was only made possible due to the development of radio-astronomy techniques, such as microwave detection of CO, OH, and NH_3 molecules.

There is the general belief concerning stellar evolution that stars with masses of the order of one solar mass or a bit more die after expelling their external layers. The star becomes a white dwarf and later a black dwarf, and its former atmosphere becomes a *planetary nebula*. The central stars of these nebulae are very hot objects ($T_{\rm eff} \gtrsim 30{,}000$ K), in fact the hottest stars in the Universe. The nebula, illuminated by the star, is also photoionized presenting properties similar to ones of an H II region. Both have electron temperatures of the order of 10^4 K and densities of 10^2–10^4 cm^{-3}, though planetary nebulae are denser and much smaller.

Reflection nebulae, H II regions, and planetary nebulae are all gaseous bright nebulae, as opposed to objects such as the Andromeda "Nebula," which is a galaxy like our own, meaning that it is a big ensemble of stars and interstellar gas. This distinction was not always clear because these objects appear as diffuse bright nebulae in photographs. The stellar nature of galaxies and the gaseous nature of nebulae was only discovered after the development of astronomical spectroscopy in the last half of the nineteenth century. The very first spectra were obtained by Huggins and showed that galaxies had continuum spectra typical of the stars they were composed of, and nebulae had spectra with bright lines, according to Kirchhoff's second law.

When an interstellar cloud has a very low density (Table 1.1), it can be detected by radio emission of the neutral H 21 cm line or by absorption of radiation emitted by distant stars in the direction of the cloud. This latter was the first technique used to study this type of clouds and, in fact, the only way for a long time. With the development of radio-astronomy techniques, the H 21 cm line was mapped in the Galaxy. More recently, the ultraviolet extension of the observed spectrum allowed the analysis of many important lines of the interstellar medium, unaccessible until then.

Though the definition of interstellar "cloud" is not very precise, there certainly exist density contrasts in the interstellar medium. Hot ($T_k \sim 10^4$ K) and less dense regions, or *intercloud medium* are predicted by the theory, although their detection is difficult. There is also a much diluted hot gas ($T_k \sim 10^5$ K) known as *coronal gas* detected by observations of some ion absorption lines such as O VI.

Stars more massive than nine solar masses end their lives in supernova explosions, ejecting all or part of their mass. The stellar material spreads across the interstellar space forming a bright gaseous nebula, known as a *supernova remnant*. Gas in this region is ionized by collisions between the ejected material and the interstellar medium. Emission of radiation occurs mainly due to nonthermal processes, such as synchrotron emission.

Interstellar dust grains originate stellar radiation polarization, but only if they show some anisotropy and are aligned along a preferential direction. This may be accomplished by a *magnetic field*, so dust grains indicate the presence of this field in the interstellar medium. It is a weak field with intensity $B \sim 10^{-6}$ Gauss, associated with the galactic disk. It interacts with the other components and plays an important role on interstellar medium dynamics and star formation.

Finally, the interstellar space contains *cosmic rays*, high-energy particles such as protons, electrons, and heavy elements nuclei that cross the disk at almost the speed of light. The detection and analysis of cosmic rays enable us to study the acceleration processes they have gone through and thus the physical conditions of their birthplaces.

These are the best known objects in the interstellar "zoo." Many others may be observed, particularly from high-resolution observations, such as jets, "elephant-trunk" and "comet-tail" structures, high-speed clouds, compact H II regions, and Bok globules.

1.4 Historical Sketch

Let us briefly summarize the history of research development of the interstellar medium, particularly until the 1950s. After the 1960s, spacecrafts and satellites started operating in the ultraviolet, radically advancing our knowledge of interstellar regions.

Research development of the interstellar medium occurred, and still does, in a continuous way along every frontier of knowledge. For didactical purposes only, we will consider in this sketch the traditional three-part division of the interstellar medium, that is, bright nebulae, dark nebulae, and diffuse interstellar gas.

1.4.1 Bright Nebulae

As previously explained, the word "nebula" had a frequently ambiguous meaning, standing for "gas cloud," "star cluster," or even "galaxy," because these objects' characteristics had just started to be known from the second half of the nineteenth century onward. Diffuse objects such as the Magellanic Clouds can be observed by the naked eye on clear nights, which means that observation of bright "nebulae" or "clouds" in the space between the stars is as ancient as humanity.

Regular scientific observations, however, date from telescopic times, such as observations of "nebulae," "nebular stars," and "clusters" made by Nicolas Louis de Lacaille (1713–1762) during a scientific expedition to the Cape of Good Hope in South Africa (1750–1754). Lacaille, as he is frequently named, observed several tens of these objects before dying in 1762 of work excess.

Another French astronomer, Charles Messier (1730–1817), compiled a catalogue (1781) with positions for more than a hundred "nebulae" in order to avoid confusion between these and the comets he was interested in. We know today that most of Messier "nebulae" are globular clusters, like M 15 (or NGC 7078, see below), galactic clusters, like M 23 = NGC 6494 or even galaxies like Andromeda (M 31 = NGC 224). However, about ten of these objects are real nebulae as we know them today, that is, gas clouds of nonstellar nature. They can be diffuse emission nebulae, such as the Orion Nebula (M 42 = NGC 1976), or planetary nebulae like the Ring Nebula, in the Lyra constellation (M 57 = NGC 6720).

The first systematic studies concerning nebulae were done by Frederick William Herschel (1738–1822), astronomer and musician, who had gained notoriety among scientists because of the discovery of Uranus (1781), the first planet to be discovered after the advent of the telescope. Herschel became famous for his work on perfecting reflector telescopes that were becoming better and better. His observation of the Orion Nebula (1774) can be considered the initial milestone of his astronomical career, even though he continued to perform as a musician for some time afterward.

Around a hundred nebulae were known before Herschel's observations, essentially those included in Messier's catalogue. This number increased to two thousand in three subsequent catalogues. From his observations, Herschel determined positions, morphological characteristics, and later on, classification and spatial distribution. As for nebulae classification, the most important problem was to discern between "stellar" nebulae and "nonstellar" ones. Herschel had already noted that several objects—such as the Pleiades—can appear as a faint haziness to some people, though six stars are normally identified. Moreover, several objects classified as "nebulae" in Messier's catalogue were resolved into stars by Herschel himself. So, for some time Herschel was led to conclude that the difference between stellar and nonstellar nebulae was solely due to telescope power. However, around 1791, new observations led him to propose the existence of nonstellar nebulae, "bright fluids" whose properties were completely unknown. In 1811–1814, Herschel finally published a complete theory of star clusters condensation from this bright fluid that was the nebulae. Each phase of the process was illustrated with real nebulae observations.

From a theoretical point of view or rather in a mere speculative way, the idea of star formation due to cloud contraction had already been proposed in 1755 by philosopher Immanuel Kant (1724–1804), for whom nebulae were real island universes, similar to the Milky Way system.

A more elaborate theory along Kant's lines was presented by mathematician and astronomer Pierre Simon Laplace (1749–1827), well known for his contributions to celestial mechanics. Taking into account the planets' and satellites' regularity of motion, Laplace suggested his famous *nebular hypothesis* in 1796, according to which the bodies of the Solar System had the same origin, coming from a large fluid material that had evolved into a central condensation from where the Sun had then formed. The nebular hypothesis was seemingly supported by Herschel's observations and it referred both to the Solar System and to fixed stars with their eventual planets.

Following his father's work, John Frederick William Herschel (1792–1871) made a survey of nebulae in 1833, presenting a catalogue with 2,500 objects, 500 of which discovered by himself. On a trip to the Cape of Good Hope (1833–1838), he discovered more than a thousand nebulae, having observed another 500. His observations of the η Carinae Nebula (former part of the Argus constellation) are of particular interest, showing the contrast between bright regions and dark zones, known as "Coal Sacks." Returning to England, he published a large catalogue of all known nebulae and clusters with a total of 5,079 objects (1864). This catalogue was later revised and extended. By the end of the nineteenth century, more than 9,000 "nebulae" were known, half of which discovered by the Herschels, both father and son. The organization of nebulae catalogues culminated with the publication of the New General Catalogue of Nebulae and Clusters of Stars (NGC) between 1888 and 1908, and its supplements by Johan Ludvig Emil Dreyer (1852–1926).

Great progress accomplished in telescope construction led William Parsons, Lord Rosse (1800–1867), to identify the stars that composed some of the "nebulae" previously catalogued by Herschel as bright fluids, using his gigantic 1.80 m diameter telescope. Rosse is better known for his observations of spiral structures in some nebulae, such as M 51 in the Canes Venatici constellation. He also observed "ring" nebulae (planetary nebulae) and tried to relate them to spiral nebulae.

The final answer to the question of the nature of nebulae started to take shape still in the nineteenth century, with the application of spectral analysis by William Huggins (1824–1910). The first spectra obtained in 1864 showed three bright lines, later observed in many other nebulae. Until 1864, a third of the seventy nebulae observed by Huggins (the Orion Nebula among them) showed these features, revealing their gaseous nature and thus confirming the "bright fluid" hypothesis of Herschel. The other "nebulae" showed a continuous spectrum, revealing their stellar nature.

The stronger lines observed in gaseous nebulae were attributed to an unknown element, *nebulium*, for many years. It was only in 1927 that Ira Sprague Bowen (1898–1973) showed that these lines belonged to forbidden transitions of ions of known elements, such as oxygen and nitrogen.

In the twentieth century, the study of nebulae progressed rapidly and unabated. In 1913, Vesto Melvin Slipher (1875–1969) proved the existence of a class of nebulae, named *reflection nebulae*, that presented a continuous spectrum with absorption lines similar to those of the associated star. This was later confirmed by Otto Struve (1897–1963).

One of the greatest steps forward on the theory of nebulae associated with hot stars was taken by Bengt Georg Daniel Strömgren (1908–1987), who demonstrated the existence of ionized H regions (H II regions) separated from neutral H regions (H I regions) by a relatively thin layer.

The study of physical processes occurring in bright nebulae—both diffuse and planetary—is thus a product of "modern" astrophysics, that is, astrophysics developed during the twentieth century.

1.4.2 Dark Nebulae

The ancients had already noticed the existence of dark nebulae in the sky. Regions with large angular dimensions and apparently no stars were named "Coal Sacks," a name that persists even today for a dark region observed in the southern hemisphere near the Southern Cross constellation. Similar to the case of bright nebulae, the systematic study of dark nebulae started in the eighteenth century with the advent of telescopes.

The presence of interstellar matter that could somehow attenuate the light coming from distant stars was suggested by Edmond Halley (1656–1742), who is known for his studies of the comet that bores his name. William Herschel called these nebulae "holes in the sky," highlighting the contrast between zones without stars and bright regions in the Milky Way.

A detailed mathematical theory about the interstellar absorption phenomenon was presented in 1847 by Frederick Georg Wilhelm Struve (1793–1864), the first of a well-known family of astronomers. Using studies of stellar distribution in space based on Herschel's catalogues, Struve managed to deduce an interstellar extinction rate of 1 mag kpc^{-1}, a value similar to today's estimates.

Years later in 1877, priest Angelo Secchi (1818–1878), known for his spectral classification system, published a work about "dark masses," where he considered them regions projected on a bright background of stars, whose light beams were intercepted by the clouds. Similar ideas were proposed by Arthur Cowper Ranyard (1845–1894) and Heber Doust Curtis (1872–1942), in opposition to the interpretation of dark nebulae as holes in the sky.

The use of photographic techniques to study nebulae started in 1880 with photographs taken from the Orion Nebula by Henry Draper (1837–1882). Special highlight goes to observations made by Edward Emerson Barnard (1857–1923) in 1919. In his atlas published in 1927, a catalogue of dark nebulae is included, the only one in existence until 1960.

Max Wolf (1863–1932) carried out important research concerning dark nebulae, in parallel to Barnard's work. By comparing the number of stars per unit area in the sky as a function of apparent brightness of the stars for regions with and without obscuration, Wolf was able to roughly estimate the interstellar absorption and the distance to the cloud. The indiscriminate use of "Wolf's curves," however, was already criticized in 1937 by Bart Jan Bok (1906–1983), who had made relevant research in the field of dark nebulae and star-forming regions, studying what we call today *Bok globules*.

In spite of several observations and theoretical studies done until the first quarter of the twentieth century, conclusive evidence about interstellar extinction was only given by Robert Julius Trumpler (1886–1956) in 1930 based on galactic clusters observations. Trumpler obtained a mean value of 0.67 mag kpc^{-1} for the extinction, similar to the former determination by F.G.W. Struve.

Due to the nonuniform distribution of interstellar matter, it was quite clear from the beginning that the application of a constant extinction coefficient was not possible, especially when large regions were considered. Studies about this, as well as indications of the reddening (selective absorption) of distant stars, were made by Joel Stebbins (1878–1966) and collaborators using photoelectric measurements.

Apart from extinction, another important effect of interstellar dust is the polarization of starlight, accidentally discovered in 1949 by William Albert Hiltner (1914–1991) and John Scoville Hall (1908–1991). Polarization is correlated with extinction and requires the existence of a diffuse magnetic field to align the dust particles scattered around the interstellar space.

1.4.3 *Diffuse Interstellar Gas*

The idea of interstellar gas extending throughout the Galaxy is quite recent, when compared with the already mentioned nebulae.

First evidence dates back to Johannes Franz Hartmann (1865–1936), whose observations of δ Orionis in 1904, a spectroscopic binary, revealed the presence of Ca II absorption lines that did not show the orbital motion of the stars around each other. Hartmann concluded on the existence of a calcium cloud in the line of sight of δ Orionis, which produced the absorption and was moving away with a radial velocity of 16 km s^{-1}.

Na I followed Ca and later similar lines were observed from elements Ti, Ca, K, and Fe, as well as from molecules CH, CN, and CH^+. Analysis of these lines provided a way to determine the chemical composition of the interstellar gas. Although subject to controversy for a time, the presence of a gas layer containing the above elements was generally accepted around 1935, in addition to a layer of dust grains responsible for extinction in the interstellar space.

Theoretical studies of the physical properties of the interstellar gas were initiated in 1926 by Arthur Stanley Eddington (1882–1944) with an estimate of kinetic temperature and ionization phase of the atoms present in "diffuse matter in space."

The irregular distribution of interstellar gas—as well as dust, as noted from extinction measurements—also became obvious thanks to radial velocities measured from absorption lines. These lines show quite frequently several components, giving rise to the idea of interstellar clouds and providing a link with the already mentioned "nebulae." In this regard, it is worthwhile to mention the observations of absorption lines along the line of sight for several hundred stars by Walter Sydney Adams (1876–1956), who determined their radial velocities.

The most important contribution for the study of diffuse interstellar gas in the first half of the twentieth century was probably made by Hendrik Christoffel van de Hulst (1918–2000) who, in 1945, made a theoretical prevision of the possibility of observing the 21 cm radiation of neutral H, presumably the principal component of interstellar clouds. This radiation was detected in 1951 and its systematic study allowed the mapping of enormous regions within the Galaxy, pointing to the link between interstellar gas and spiral arms.

1.5 The Oort Limit

Several components of the Galaxy, such as highly ionized atoms, some molecules, and solid particles of large sizes, are difficult to detect through the usual astrophysical methods, that is, using observations of some type of radiation. Thus, it is interesting to determine the limits of the contribution of this non-detectable matter, which can be done through estimates of the total gravitational attraction perpendicular to the galactic plane. As initially shown by Jan Hendrik Oort (1900–1992), this method can also produce limits for the contribution of interstellar matter in the Galaxy.

If we consider the Galaxy to be a flat disk, the determination of gravitational acceleration g_z in the z direction, perpendicular to the disk, provides an estimate of the total mass in the solar neighborhood, including gas and stars. The determination of g_z is done using measurements of velocity v_z and density gradient in the z direction of homogeneous and bright enough groups of stars.

For simplicity, we will consider a one-dimensional galactic disk, that is, we will neglect the motions of the stars in directions r and θ, considering cylindrical coordinates, which leaves us only with z direction motions, perpendicular to the plane. In this case and supposing a steady state, we can define a distribution function $f(z,p_z)\mathrm{d}p_z\mathrm{d}V$, that is, the number of stars inside volume $\mathrm{d}V$ with momentum between p_z and $p_z + \mathrm{d}p_z$.

A group of stars moving in the direction perpendicular to the galactic plane is not a "fluid" in the usual sense, like, for instance, the gas layer in a stellar atmosphere. Thus, instead of the usual hydrodynamics equations, we must apply the *Liouville Theorem*. According to this theorem, in a set of point masses moving without dissipation in a potential field, the density f of the points in the phase space is constant along a dynamical trajectory. In other words, the difference between stars moving inside a phase space element and those moving *outside* the element must be equal to the increase of f in the given element. Generally we may write

$$\frac{\mathrm{d}f}{\mathrm{d}t} = \frac{\partial f}{\partial t} + \sum \left(\frac{\partial f}{\partial q_i} \dot{q}_i + \frac{\partial f}{\partial p_i} \dot{p}_i \right) = 0, \tag{1.1}$$

where we use the generalized coordinates q_i and p_i. In our case, we have

$$\frac{\partial f}{\partial z} v_z + \frac{\partial f}{\partial p_z} m\, g_z = 0, \tag{1.2}$$

or

$$\frac{p_z}{m} \frac{\partial f}{\partial z} - m \frac{\mathrm{d}\phi}{\mathrm{d}z} \frac{\partial f}{\partial p_z} = 0, \tag{1.3}$$

where m is the mass of the stars, assumed constant, and $\phi(z)$ is the gravitational potential of the disk, which is a function of z only. In (1.3), we use the fact that

$$g_z = -\frac{\mathrm{d}\phi}{\mathrm{d}z}. \tag{1.4}$$

We may write

$$p_z \frac{\partial f}{\partial z} = -m^2 g_z \frac{\partial f}{\partial p_z}. \tag{1.5}$$

The stellar density n is

$$n = \int_{-\infty}^{\infty} f(z, p_z)\, \mathrm{d}p_z \tag{1.6}$$

and the mean quadratic momentum is

$$\langle p_z^2 \rangle = \frac{\int_{-\infty}^{\infty} f(z,p_z) p_z^2 \mathrm{d}p_z}{\int_{-\infty}^{\infty} f(z,p_z) \mathrm{d}p_z} = \frac{1}{n} \int_{-\infty}^{\infty} f(z,p_z) p_z^2 \, \mathrm{d}p_z. \tag{1.7}$$

If we multiply this equation by n and calculate the derivative with respect to z,

$$\begin{aligned} \frac{\mathrm{d}(n\langle p_z^2 \rangle)}{\mathrm{d}z} &= \int_{-\infty}^{\infty} \frac{\partial f(z,p_z)}{\partial z} p_z^2 \, \mathrm{d}p_z \\ &= -m^2 g_z \int_{-\infty}^{\infty} \frac{\partial f(z,p_z)}{\partial p_z} p_z \, \mathrm{d}p_z, \end{aligned} \tag{1.8}$$

where we used (1.5). Integrating (1.8) by parts and using (1.6), we obtain

$$\frac{1}{n} \frac{\mathrm{d}(n\langle v_z^2 \rangle)}{\mathrm{d}z} = g_z, \tag{1.9}$$

where we use the fact that $p_z = mv_z$. The total mass density $\rho(z)$ can be obtained by the Poisson equation

$$\nabla^2 \phi = \frac{\mathrm{d}^2 \phi}{\mathrm{d}z^2} = 4\pi G \rho(z). \tag{1.10}$$

Considering (1.4), we may write

$$\rho(z) = -\frac{1}{4\pi G} \frac{\mathrm{d}g_z}{\mathrm{d}z}. \tag{1.11}$$

From observations of velocities v_z and densities n of homogeneous groups of stars, assumed to have the same mass, we can determine g_z as a function of z using (1.9), and the total mass density $\rho(z)$ using (1.11). This was done by Oort for a group of K giant stars. The first analysis made by Oort dates back from 1932, but his complete work dates from 1960. The function g_z obtained by Oort for the solar neighborhood is shown in Fig. 1.2.

For the galactic plane, $z = 0$ and the total mass density is $\rho_t \simeq 10 \times 10^{-24}$ g cm^{-3} = 0.15 $M_\odot$/pc^3, where $M_\odot = 1.99 \times 10^{33}$ g is the solar mass. The portion of this material coming from the stars is known independently, being $\rho_* \simeq 4 \times 10^{-24}$ g cm^{-3} = 0.06 $M_\odot$/pc^3.Thus, the total mass density of the interstellar matter in the solar neighborhood must be $\rho_{\mathrm{im}} \lesssim 6 \times 10^{-24}$ g cm^{-3} = 0.09 $M_\odot$/pc^3, which corresponds to 2.6 H atoms per cubic centimeter, if we adopt an abundance of He/H = 0.1 per number of atoms. This limit is known as the *Oort limit*. Of course, this value is subject to uncertainty, not only due to observations but also due to mathematical approximations used during the calculation process. However, more recent works produce essentially the same results.

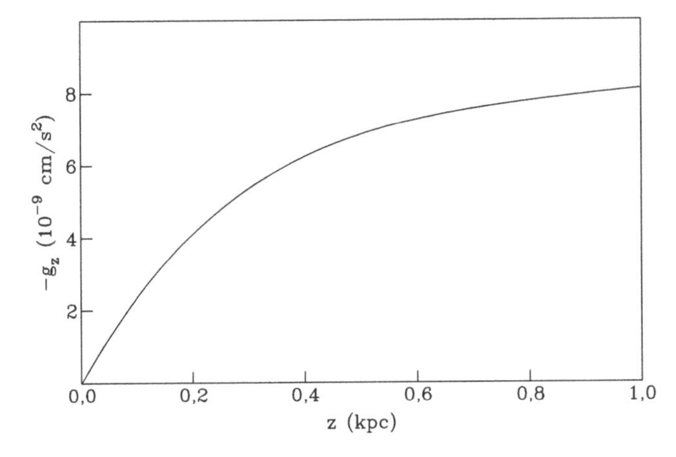

Fig. 1.2 Relation between the gravitational acceleration and height above the galactic plane for a sample of K giants

The value of the interstellar medium density obtained from direct measurements of gas and dust in the solar neighborhood is of the same order of magnitude as the Oort limit. So, the stars, the interstellar medium and the non-detectable matter seem to have comparable densities. Most of the mass of the interstellar medium, about 90%, is concentrated in the gas, which leaves approximately 10% for the interstellar dust grains.

Though apparently low, the density of the interstellar medium corresponds to a high enough total mass, since it is spread along the whole galactic disk. To illustrate this point, we will consider the disk to have a diameter of the order of $2R \simeq 30$ kpc and a thickness $h \simeq 500$ pc. With a density of $\rho_{im} \simeq 3 \times 10^{-24}$ g cm^{-3}, the mass is $M_{im} \simeq \pi R^2 h\, \rho_{im} \sim 10^{43}$ g, quite enough to form about 10^{10} stars of one solar mass each.

Exercises

1.1 Consider an interstellar cloud composed of atomic hydrogen, with a density of 10 particles per cubic centimeter and kinetic temperature of 100 K. (a) What is the cloud density in g cm^{-3}? (b) Estimate the pressure inside the cloud. Compare the result with the pressure of a typical laboratory vacuum.

1.2 Suppose that an interstellar cloud with a density of 10 particles per cubic centimeter and temperature of 100 K is in pressure equilibrium with the intercloud medium, where the density is 0.1 particles per cubic centimeter, according to Table 1.1. What would be, in order of magnitude, the temperature of the intercloud medium?

1.3 A spherical solid dust grain in an interstellar cloud has a radius of a $\simeq$1,000 Å $= 10^{-5}$ cm and an internal density of $s_d \simeq 3$ g cm^{-3}. (a) What is the grain mass? (b) Consider a typical interstellar cloud where the concentration of the dust grains is $n_d \sim 10^{-11}$ cm^{-3}. What would be the volume of the cloud occupied by a person with 70 kg, if the whole body was pulverized into interstellar grains and spread across the cloud?

1.4 By means of a theoretical treatment of the oscillatory motions perpendicular to the galactic plane, F. House and D. Kilkenny (Astronomy & Astrophysics vol. 81, p. 251, 1980) have derived an analytical expression for gravitational acceleration g_z, valid for $|z| \leq 1$ kpc:

$$g_z = A_0 \sin\left(\frac{2z}{R} + B_0\right) + C_0 \exp(-\alpha z)$$

where A_0, B_0, and C_0 are constants, R is the distance to the galactic axis, and $\alpha = 1/h$, h being the effective thickness of the layer of gas and stars above the galactic plane. The constants have been determined by radial velocity measurements of OB stars in the solar neighborhood, being $A_0 = 9.6 \times 10^{-9}$ cm s^{-2}, $B_0 = 5$ rad, and $C_0 = 9.0 \times 10^{-9}$ cm s^{-2}. Assume a mean value 2 h $\simeq$ 800 pc and $R \simeq 8.5$ kpc and calculate the total mass density in the galactic plane for the solar neighborhood. Compare the result with the value obtained by Oort, based on the analysis of K giants.

1.5 Determinations of mass density distribution in the form of stars in the solar neighborhood yield the following values: 0.038 $M_\odot$/pc^3 for dwarf stars of spectral type G, K, and M; 0.02 $M_\odot$/pc^3 for white dwarfs; and 0.006 $M_\odot$/pc^3 for the rest. What is the total mass in the form of stars, in $M_\odot$/pc^3 and g cm^{-3}? Compare the result with the Oort limit.

Bibliography

Bowers, R.L., Deeming, T.: Astrophysics II. Jones & Bartlett, Boston (1984). Basic text on astrophysics, with detailed discussion on the principal aspects of neutral and ionized interstellar regions

Dyson, J., Williams, D.A.: The Physics of the Interstellar Medium. Institute of Physics Publishing, London (1997). Introductory text, quite accessible, covering the main physical processes of the interstellar medium

Kaplan, S.A., Pikelner, S.B.: The Interstellar Medium. Harvard University Press, Cambridge (1970). Basic text, written by two Russian specialists, highlighting hydrogen ionization processes in the interstellar space

Lang, K.R., Gingerich, O. (eds.): A Source Book in Astronomy and Astrophysics 1900–1975. Harvard University Press, Cambridge (1979). Compilation of a series of classical articles published between 1900 and 1975, with the inclusion of some fundamental works on astrophysics of the interstellar medium. See also Shapley H. & Howarth H.E. (eds.). Source Book in Astronomy 1900-1950. Cambridge, Harvard University Press

Middlehurst, B.M., Aller, L.H. (eds.): Nebulae and Interstellar Matter. University of Chicago Press, Chicago (1968). Series of classical review articles about several aspects of astrophysics of the interstellar medium

Oort, J.H.: Stellar dynamics. In: Blaauw, A., Schmidt, M. (eds.) Galactic Structure, p. 455. University of Chicago Press, Chicago (1965). Article by Jan Oort, one of the principal astrophysicists of the 20th century, about density determination of the interstellar matter

Osterbrock, D.: Astrophysics of Gaseous Nebulae and Active Galactic Nuclei. University Science Books, Mill Valley (1989). Fundamental book to the study of photoionized gaseous nebulae, including a detailed study of the physical processes taking place in H II regions and planetary nebulae

Pikelner, S.B.: Physics of Interstellar Space. Foreign Languages Publishing House, Moscow (1961). Outreach introductory book written by one to the leading Russian astrophysicists showing in a simple way hydrogen ionization processes in the interstellar space

Scheffler, H., Elsässer, H.: Physics of the Galaxy and Interstellar Matter. Springer, Berlin (1988). Advanced level text covering a wide range of issues, more recent than the classical book by Spitzer, presenting a discussion on interstellar phenomena in a galactic context

Spitzer, L.: Physical Processes in the Interstellar Medium. Wiley, New York (1978). Classical text about the principal physical processes occurring in the interstellar medium and a must read for the study of these processes. The present book may be considered as an introduction to the more accurate treatment presented by Spitzer

Spitzer, L.: Searching Between the Stars. Yale University Press, New Haven (1982). Personal and accessible account of the main progress achieved in the study of the interstellar medium from ultraviolet observations, by one of the leading specialists

Strömgren, B.: The physical state of interstellar hydrogen. Astrophys J **89**, 526 (1939). Classical article by Strömgren on hydrogen ionization in H II regions

Struve, O., Zebergs, V.: Astronomy of the 20th Century. MacMillan, New York (1962). An account of astronomy history in the 20th century, focussing particularly on the development of studies concerning the interstellar medium

Verschuur, G.L.: Interstellar Matters: Essays on Curiosity and Astronomical Discovery. Springer, New York (1989). History of the interstellar matter discovery, highlighting radio astronomical data

Wynn-Williams, G.: The Fullness of Space. Cambridge University Press, Cambridge (1992). Recent introductory book focussing on the principal physical aspects of the interstellar medium with a qualitative treatment

Chapter 2
The Interstellar Radiation Field

2.1 Introduction

The distribution of radiation in the Galaxy varies greatly with the observed wavelength, reflecting the enormous differences of the physical processes responsible for the radiation field. Optical observations of the Milky Way show a stellar component similar to the one of the galaxy shown in Fig. 1.1, concentrated in a relatively thin disk and in a bright bulge. Infrared observations around 25–60 μm and 1.2–3.4 μm obtained with the *Cosmic Background Explorer* (COBE) satellite reveal a thinner layer mainly composed of gas and dust. These layers have a dramatic effect on distant objects, completely extinguishing visible light in some directions. However, at shorter wavelengths, typical of high-energy radiation, and besides scattered point sources, a diffuse radiation is observed spreading across the Galaxy.

In a more quantitative way, the main features of the interstellar radiation field are schematically described in Fig. 2.1, from radio wavelengths to high-energy radiation.

In this figure, the frequency ν(Hz) is on the x-axis and νU_ν is on the y-axis, where U_ν (erg cm^{-3} Hz^{-1}) is the energy density per frequency interval. Since we have

$$\int U_\nu \, d\nu = \int \nu U_\nu \, d \ln \nu, \tag{2.1}$$

the quantity νU_ν represents the radiation field energy density per frequency logarithmic interval. The principal spectral regions in the figure are:

A: Radio-integrated radiation

B: Cosmic background radiation, with a temperature of 2.7 K

C: Infrared radiation, including galactic and extragalactic contributions, with a major component due to thermal emission of the interstellar dust grains (Chap. 9)

D: Ultraviolet integrated stellar radiation, the main subject of this chapter, with a cut at $\nu = 3.29 \times 10^{15}$ Hz or $\lambda = 912$ Å

E: High-energy diffuse radiation, meaning X- and γ-rays of galactic and extragalactic origin

W.J. Maciel, *Astrophysics of the Interstellar Medium*,
DOI 10.1007/978-1-4614-3767-3_2, © Springer Science+Business Media New York 2013

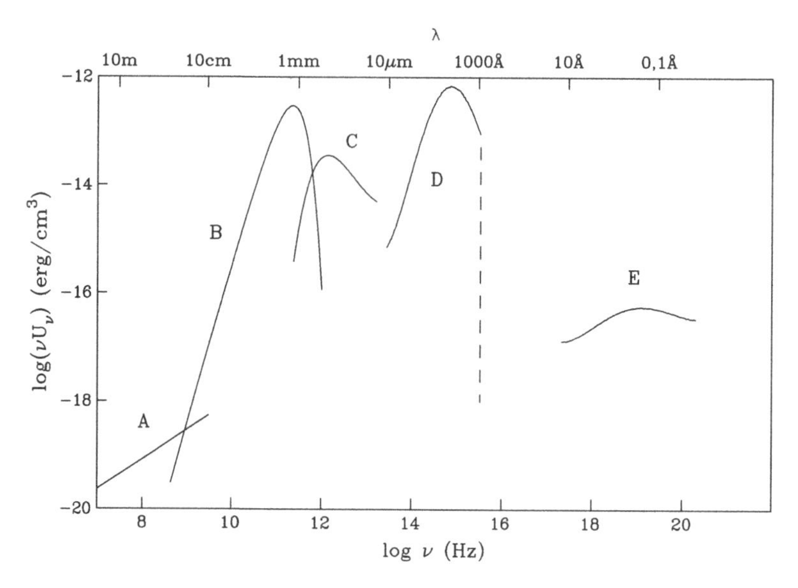

Fig. 2.1 A compilation of the interstellar radiation field from radio waves to gamma rays

2.2 Radio-Frequency Integrated Radiation

Part of the observed radio-frequency radiation comes from extragalactic sources, and we will not discuss them in this book. The major part, however, including the neutral H 21 cm (1,420 MHz) line, is of galactic origin, being produced by several relatively well-known mechanisms. We will present a summary of the main characteristics of these processes. For more details, see Rybicki and Lightman (1979).

2.2.1 *Bremsstrahlung Radiation*

This is a kind of continuum radio radiation emitted by a plasma through free–free transitions of H and He ions. This radiation is produced in H II regions with electron density $n_e \lesssim 10^{-5}$ cm^{-3}, electron temperature $T_e \sim 10^3 - 10^4$ K, and ratio density between electrons and H atoms $n_e/n_H \gg 1$.

Plasma emissivity, defined as the total power emitted per unit volume per unit solid angle per unit frequency interval between ν and $\nu + d\nu$, is given by

$$\epsilon_\nu = \frac{n_e}{4\pi} \int P(v, \nu) f(v) dv, \tag{2.2}$$

and it is generally measured in erg cm^{-3} s^{-1} sr^{-1} Hz^{-1}, where $f(v)$ is the distribution function of electron velocities and $P(v,\nu)$ is the total power emitted per unit frequency interval during the collision between an electron with velocity v and an ion with density n_i. If $f(v)$ is given by the Maxwellian distribution, the emissivity is

$$\epsilon_\nu \, d\nu = \frac{8}{3}\left(\frac{2\pi}{3}\right)^{1/2} \frac{Z_i^2 e^6}{m_e^2 c^3}\left(\frac{m_e}{kT_e}\right)^{1/2} n_i n_e \, g_{ff}(\nu, T_e) \, e^{-h\nu/kT_e} \, d\nu. \tag{2.3}$$

In this equation, Z_i is the atomic number of the ions, c is the speed of light in vacuum, e is the electron charge, m_e is the electron mass, h is the Planck constant, k is the Boltzmann constant, and g_{ff} is the Gaunt factor, a function that varies slowly with frequency. For $h\nu/kT_e \ll 1$, the Gaunt factor is given by

$$g_{ff}(\nu, T_e) = \frac{\sqrt{3}}{\pi}\left[\ln \frac{(2kT_e)^{3/2}}{\pi e^2 \nu \, m_e^{1/2}} - \frac{5\gamma}{2}\right], \tag{2.4}$$

where $\gamma = 0.577$ is the Euler constant. Numerically, we have

$$g_{ff}(\nu, T_e) = 9.77\left(1 + 0.130 \log \frac{T_e^{3/2}}{\nu}\right), \tag{2.5}$$

with ν in Hz. For $h\nu \ll kT_e$, $\epsilon_\nu \propto n_i n_e T_e^{-1/2}$, neglecting the variations in the Gaunt factor.

The total luminosity per unit volume of a plasma for free–free emission is given by

$$L = 4\pi \int \epsilon_\nu d\nu = \frac{32\pi e^6}{3 m_e c^3 h}\left(\frac{2\pi k T_e}{3 m_e}\right)^{1/2} n_i \, n_e \, Z_i^2 \, \bar{g}_{ff}, \tag{2.6}$$

with units of erg cm^{-3} s^{-1} and a mean value of the Gaunt factor $\bar{g}_{ff} \simeq 1.1 - 1.5$. The absorption coefficient for electron–ion bremsstrahlung is

$$k_\nu = \frac{\epsilon_\nu}{B_\nu(T_e)} = \frac{4}{3}\left(\frac{2\pi}{3}\right)^{1/2} \frac{Z_i^2 \, e^6 \, n_i \, n_e \, g_{ff}}{m_e^{3/2} c \, (kT_e)^{3/2} \, \nu^2}, \tag{2.7}$$

given in cm^{-1}, where we use (2.3) and the Rayleigh–Jeans approximation for the Planck function (Chap. 3). Applying (2.4), we have

$$k_\nu = 0.1731\left(1 + 0.130 \log \frac{T_e^{3/2}}{\nu}\right) \frac{Z_i^2 \, n_i \, n_e}{T_e^{3/2} \, \nu^2}. \tag{2.8}$$

Finally, if we neglect the g_{ff} dependency on ν and T_e, we obtain

$$k_\nu \propto \frac{n_\mathrm{i}\, n_\mathrm{e}}{T_\mathrm{e}^{3/2}\, \nu^2}. \tag{2.9}$$

We can introduce the optical depth of free–free radiation if we integrate along the whole extension l of the source,

$$\tau_\nu = \int k_\nu \,\mathrm{d}l \propto T_\mathrm{e}^{-3/2}\, \nu^{-2} \int n_\mathrm{e}^2 \,\mathrm{d}l, \tag{2.10}$$

where we consider $n_\mathrm{i} = n_\mathrm{e}$. The integral along the line of sight is named *emission measure*,

$$\mathrm{EM} = \int n_\mathrm{e}^2 \,\mathrm{d}l, \tag{2.11}$$

generally given in pc cm^{-6}. In a more quantitative way, the optical depth can be described by the expression

$$\tau_\nu = 8.235 \times 10^{-2}\, a(\nu, T_\mathrm{e})\, T_\mathrm{e}^{-1.35}\, \nu^{-2.1}\, \mathrm{EM}, \tag{2.12}$$

valid for $h\nu/kT_\mathrm{e} \ll 1$ and $T_\mathrm{e} < 9 \times 10^5$ K, where $a(\nu,T_\mathrm{e})$ is a function that varies slowly with frequency and temperature, T_e is in K, ν is in GHz, and the emission measure is in pc/cm^6.

2.2.2 Synchrotron Radiation (Magnetobremsstrahlung)

This is a type of nonthermal radiation, emitted by relativistic electrons being deflected by the magnetic field of the interstellar medium. It occurs in the diffuse interstellar medium and in supernova remnants.

A relativistic electron with energy E deflected by a magnetic field of intensity B moves along a spiral and emits radiation in a cone with angle $\phi = \mathrm{mc}^2/\mathrm{E}$ in the form of pulses that produce a continuum spectrum. The emissivity of a group of electrons is

$$\epsilon_\nu = \frac{1}{4\pi} \int P(\nu, E) N(E) \,\mathrm{d}E, \tag{2.13}$$

where P is the emitted power per unit frequency interval for an electron and $N(E)\mathrm{d}E$ is the number of electrons with energy between E and E+dE per unit volume along the light of sight. According to the observations, the distribution function of cosmic electrons can be approximately given by

$$N(E)\,\mathrm{d}E = KE^{-\gamma}\,\mathrm{d}E. \tag{2.14}$$

Typical values are $\gamma \simeq 2.6$ and $K \simeq 3.3 \times 10^{-17}$ erg$^{\gamma-1}$ cm^{-3} for a typical field of $B \simeq 3 \times 10^{-6}$ Gauss. If the electron distribution is homogeneous and isotropic and the magnetic field is also homogeneous, then we have

$$\epsilon_\nu = K\alpha(\gamma)\frac{\sqrt{3}}{8\pi}\frac{e^3}{m_e c^2}\left[\frac{3e}{4\pi m_e^3 c^5}\right]^{(\gamma-1)/2} B^{(\gamma+1)/2}\,\nu^{-(\gamma-1)/2}, \tag{2.15}$$

where $\alpha(\gamma)$ is a slow varying function of the order of one.

2.2.3 *Line Radiation*

Besides continuum radiation, several lines are also observed at radio wavelengths such as (1) the neutral H 21 cm line, produced in diffuse interstellar clouds and in regions along the Galaxy's spiral arms; (2) lines of various molecules, produced by thermal or maser emission, and (3) recombination lines, produced by H and He and formed by electron recombination in high-energy levels ($n > 50$). The study of interstellar emission lines will be addressed in Chaps. 3 and 4.

2.3 Cosmic Background Radiation

The cosmic background radiation was discovered in 1965 by Penzias and Wilson, researchers at Bell Telephone Laboratories, New Jersey, USA. This discovery, as frequently happens, was made quite by accident. Penzias and Wilson were doing 7 cm radio observations and after subtraction of all known sources they noticed a residual signal whose nature was unknown. At the same time, a team of Princeton University predicted that some radiation should be observed at that exact wavelength, as a result of theoretical models for the formation of the Universe. This radiation would essentially be the remnant of the Big Bang, the explosion that gave birth to the Universe. According to this prediction, the radiation field would be isotropic and would correspond to a blackbody microwave emission with temperature of around 2.7 K.

Atmospheric absorption at $\lambda < 6$ mm prevents observation from the ground of the points next to the maximum of the Planck curve. Until some years ago, just indirect evidence, given by CN molecule observations, pointed to the fact that this maximum existed at all. Since 1975, observations using air balloons and rockets allowed the determination of the curve for shorter wavelengths. Recently, results of the COBE satellite have led to the accurate determination of the complete curve, and nowadays, small temperature anisotropies, of the order of 10^{-5}, have been detected with important consequences for the Big Bang standard model and the formation of structures in the Universe. Apart from being very important to cosmological studies, the cosmic background radiation also bears a great significance for the present text because at any point of the interstellar medium, this radiation corresponds, at least, to the radiation of a blackbody with temperature 2.7 K.

2.4 Integrated Radiation Field

The interstellar radiation field in the optical and ultraviolet comes essentially from integrated stellar radiation and bears special significance for the study of the interstellar medium. Photons with wavelengths shorter than 2,000 Å play an important role in the ionization of elements, as well as in the heating of dust grains in interstellar clouds.

Near $\lambda = 912$ Å (Lyman limit), there is a cut in the number of available photons because H, the most abundant element, absorbs most of these photons. The column density of neutral H near the galactic plane can reach values of the order of $N_H \sim 10^{20}$–10^{21} cm^{-2}. Since the absorption cross section near the Lyman limit is of the order of $\sigma_H \sim 6.3 \times 10^{-18}$ cm^2, the optical depth is

$$\tau_H \simeq N_H \sigma_H \simeq 6.3 \times 10^2 - 6.3 \times 10^3 \gg 1, \tag{2.16}$$

so, all the radiation emitted at $\lambda \leq 912$ Å is absorbed near the source. Since the absorption cross section is proportional to λ^3, the quanta with wavelengths shorter than 100 Å are not completely absorbed anymore, and thus, an X-ray radiation field is observed. Above 912 Å photons can, for instance, ionize elements for which the ionization potential is less than 13.6 eV.

The stellar radiation field can be computed if we know its spatial density or the number of stars of each spectral type and the flux (or energy density) emitted by each stellar type, taking into account the modifications due to interstellar extinction. In the ultraviolet region, the field is essentially determined by hot stars, with spectral types earlier than K.

Calculations of the energy density at an average point of the interstellar medium, that is, a point which is not very close to any star, have been made since the 1930s, having greatly progressed since the 1970s when ultraviolet observations allowed their calibration and consequently the refinement of the theoretical models.

The radiation flux in the interstellar medium can be approximated by an ensemble of black bodies, affected by a dilution factor due to the dilution of the radiation over great distances. For instance, the model proposed by Werner and Salpeter (1969) can be written as

$$F_\lambda \simeq \sum_1^4 W_i B_\lambda(T_i). \tag{2.17}$$

Table 2.1 Coefficients of the Werner and Salpeter (1969) model for the interstellar radiation field

T_i (K)	14,500	7,500	4,000	2.7
W_i	4×10^{-16}	1.5×10^{-14}	1.5×10^{-13}	1

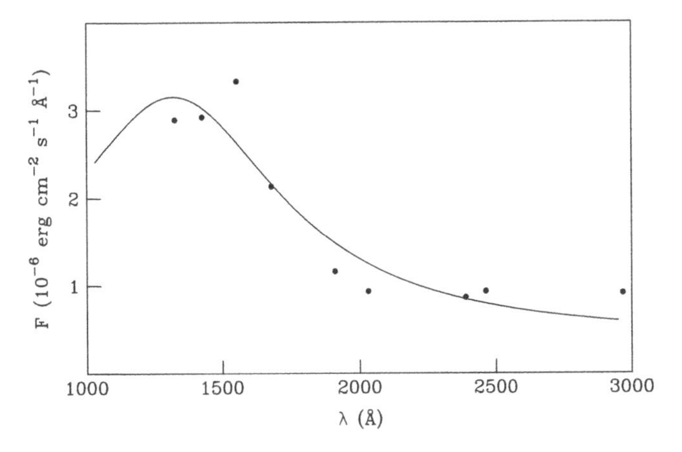

Fig. 2.2 The UV interstellar radiation field. *Dots* are satellite data and the *curve* shows a simple theoretical model

The values of the fluxes and the dilution factors of each of the four components are listed in Table 2.1.

More accurately, we must consider the real flux distributions, as well as a detailed analysis of the interstellar dust distribution, responsible for extinction. With the development of ultraviolet satellites (OAO, TD-1, Copernicus, IUE), the radiation field in this once unaccessible region has become well known. For instance, Witt and Johnson (1973) used OAO-2 observations to determine the field between 1,250 and 4,250 Å, and other works have complemented and extended these calculations until the 1990s.

Figure 2.2 shows an example of a recent model for the interstellar radiation field in the wavelength interval $1{,}000 < \lambda(\text{Å}) < 3{,}000$. The curve shows a recent theoretical model, and the points represent observational data obtained with ultraviolet satellites.

In a general way, the agreement between the models and the observational data is quite good, and the calculated fluxes are of the same order of magnitude of the observed fluxes, $F_\lambda \simeq 1\text{–}3 \times 10^{-6}$ erg cm^{-2} s^{-1} Å^{-1} for the considered wavelengths. Meanwhile, the flux distribution with galactic latitude is not entirely satisfactory, which is attributed to an unrealistic parameterization of the stellar distribution with galactic latitude and is a reflex of the non-homogeneity distribution of the interstellar matter.

2.5 Radiative Transfer

2.5.1 Radiative Transfer Equation

Let us consider the problem of radiative transfer in the interstellar field radiation and estimate the field energy density at a point in the galactic disk, influenced by stars, gas, and dust. If we assume the Galaxy to be a flat disk, the radiation-specific intensity I_ν (erg cm^{-2} s^{-1} Hz^{-1} sr^{-1}) at a given frequency is only a function of the height to the galactic plane, z. The one-dimensional steady-state radiative transfer equation without induced emissions and with coherent and isotropic scattering may be written as

$$\mu \frac{dI_\nu}{dz} = j_\nu - (k_\nu + \sigma_\nu)I_\nu + \sigma_\nu J_\nu, \tag{2.18}$$

where $\mu = \cos\theta$, θ being the angle between the direction of the beam propagation and the normal to the considered element; j_ν is the emission coefficient per volume (erg cm^{-3} s^{-1} Hz^{-1} sr^{-1}); k_ν is the absorption coefficient per volume (cm^{-1}); σ_ν is the scattering coefficient per volume (cm^{-1}); and J_ν is the radiation mean intensity. Let us define the optical depth by the expression

$$d\tau_\nu = (k_\nu + \sigma_\nu)dz = k_E\, dz, \tag{2.19}$$

where k_E is the total extinction coefficient. Let us take $\tau = 0$ at $z = 0$ and $\tau = \tau_H$ at $z = H$, the scale height of the disk. Replacing in (2.18), we have

$$\mu \frac{dI_\nu}{d\tau_\nu} = \frac{j_\nu}{k_\nu + \sigma_\nu} - I_\nu + \frac{\sigma_\nu}{k_\nu + \sigma_\nu} J_\nu, \tag{2.20}$$

which can be simplified to

$$\mu \frac{dI_\nu}{d\tau_\nu} = S_\nu - I_\nu + \gamma_\nu J_\nu, \tag{2.21}$$

where we define the dust grain albedo as

$$\gamma_\nu = \frac{\sigma_\nu}{k_\nu + \sigma_\nu} \tag{2.22}$$

and the source function as

$$S_\nu = \frac{j_\nu}{k_\nu + \sigma_\nu}. \tag{2.23}$$

Let us consider the part of the observed optical spectrum where emission essentially comes from the Galaxy's stars. Let n_*^s be the stellar density (cm^{-3} or pc^{-3}) of stars of spectral type s and L_ν^s the luminosity (erg s^{-1} Hz^{-1}) of these stars in the frequency interval between ν and ν+dν. We may write

$$j_\nu = \frac{1}{4\pi} \sum n_*^s \, L_\nu^s, \tag{2.24}$$

where the summation must be extended to all the stars that contribute to the considered frequency interval.

2.5.2 *Transfer Equation Solution*

Let us consider the solution of (2.21) using the two-beam Schuster approximation (1905, see also Mihalas 1978). In this case, we define

$$I_\nu^+ = \int_0^{\pi/2} I_\nu \sin\theta \, \mathrm{d}\theta = \int_0^1 I_\nu \, \mathrm{d}\mu \;\; (0 \le \mu \le 1) \tag{2.25a}$$

$$I_\nu^- = \int_{\pi/2}^{\pi} I_\nu \sin\theta \, \mathrm{d}\theta = \int_{-1}^0 I_\nu \, \mathrm{d}\mu \;\; (-1 \le \mu \le 0) \tag{2.25b}$$

Recalling that $\mu = 1$ for $\theta = 0$, $\mu = 0$ for $\theta = \pi/2$, and $\mu = -1$ for $\theta = \pi$. In this case, I_ν^+ and I_ν^- are independent of μ and

$$\begin{aligned} J_\nu &= \frac{1}{4\pi} \int I_\nu \, \mathrm{d}\omega = \frac{1}{4\pi} \int_0^{2\pi} \mathrm{d}\phi \int_0^{\pi} I_\nu \sin\theta \, \mathrm{d}\theta \\ &= \frac{1}{2} \int_0^{\pi} I_\nu \sin\theta \, \mathrm{d}\theta = \frac{1}{2} \int_{-1}^{1} I_\nu \, \mathrm{d}\mu \\ &= \frac{1}{2} \int_{-1}^{0} I_\nu \, \mathrm{d}\mu + \frac{1}{2} \int_0^1 I_\nu \, \mathrm{d}\mu \\ &= \frac{1}{2} \left(I_\nu^+ + I_\nu^- \right), \end{aligned} \tag{2.26}$$

where we use dω = sin θ dθ dϕ for the solid angle element around the considered direction. In the same way, the flux F_ν is given by

$$F_\nu = \frac{1}{\pi} \int I_\nu \cos\theta \, d\omega = 2 \int_{-1}^{1} I_\nu \, \mu \, d\mu = I_\nu^+ - I_\nu^-. \tag{2.27}$$

By analogy, this expression may be written as

$$F_\nu = F_\nu^+ - F_\nu^-, \tag{2.28}$$

where we define

$$F_\nu^+ = 2\int_0^1 I_\nu \,\mu \,\mathrm{d}\mu \tag{2.29a}$$

and

$$F_\nu^- = -2\int_{-1}^0 I_\nu \,\mu \,\mathrm{d}\mu. \tag{2.29b}$$

Multiplying (2.21) by $\mathrm{d}\mu$, integrating between 0 and 1, and applying (2.27) and (2.28), we obtain

$$\frac{\mathrm{d}}{\mathrm{d}\tau_\nu}\int_0^1 I_\nu \,\mu \,\mathrm{d}\mu = S_\nu \int_0^1 \mathrm{d}\mu - \int_0^1 I_\nu \,\mathrm{d}\mu + \gamma_\nu \,J_\nu \int_0^1 \mathrm{d}\mu$$

$$\frac{1}{2}\frac{\mathrm{d}I_\nu^+}{\mathrm{d}\tau_\nu} = S_\nu - I_\nu^+ + \gamma_\nu J_\nu, \tag{2.30}$$

where we assume the homogeneous case, in which j_ν, k_ν, σ_ν, and γ_ν are independent of μ. In the same way, multiplying (2.21) by $\mathrm{d}\mu$ and integrating between -1 and 0,

$$-\frac{1}{2}\frac{\mathrm{d}I_\nu^-}{\mathrm{d}\tau_\nu} = S_\nu - I_\nu^- + \gamma_\nu J_\nu. \tag{2.31}$$

Adding and subtracting (2.30) and (2.31) and considering (2.26) and (2.27), we obtain the relations

$$\frac{1}{2}\frac{\mathrm{d}}{\mathrm{d}\tau_\nu}\left(I_\nu^+ - I_\nu^-\right) = 2S_\nu - \left(I_\nu^+ + I_\nu^-\right) + 2\gamma_\nu J_\nu$$

$$\frac{\mathrm{d}F_\nu}{\mathrm{d}\tau_\nu} = 4S_\nu - 4J_\nu(1-\gamma_\nu) \tag{2.32}$$

$$\frac{1}{2}\frac{\mathrm{d}}{\mathrm{d}\tau_\nu}\left(I_\nu^+ + I_\nu^-\right) = -\left(I_\nu^+ - I_\nu^-\right)$$

$$\frac{\mathrm{d}J_\nu}{\mathrm{d}\tau_\nu} = -F_\nu. \tag{2.33}$$

Calculating the differential of (2.32) and using (2.33),

$$\frac{d^2F_\nu}{d\tau_\nu^2} = -4(1-\gamma_\nu)\,\frac{dJ_\nu}{d\tau_\nu} = 4(1-\gamma_\nu)F_\nu. \tag{2.34}$$

The expression (2.34) is an ordinary linear differential equation of second order, with constant coefficients. Its solution can be written as

$$F_\nu = A\cosh\left[2(1-\gamma_\nu)^{1/2}\tau_\nu\right] + B\sinh\left[2(1-\gamma_\nu)^{1/2}\tau_\nu\right]. \tag{2.35}$$

Constants A and B can be determined by boundary conditions. For $\tau_\nu = 0$, we have $F_\nu(\tau_\nu) = 0$ and thus $A = 0$. Expression (2.35) then becomes

$$F_\nu = B\sinh\left[2(1-\gamma_\nu)^{1/2}\tau_\nu\right]. \tag{2.36}$$

From (2.32) and (2.36),

$$J_\nu = \frac{S_\nu}{1-\gamma_\nu} - \frac{1}{4(1-\gamma_\nu)}\,\frac{dF_\nu}{d\tau_\nu}$$

$$\frac{dF_\nu}{d\tau_\nu} = 2(1-\gamma_\nu)^{1/2}B\cosh\left[2(1-\gamma_\nu)^{1/2}\tau_\nu\right]$$

in such a way that

$$J_\nu = \frac{S_\nu}{1-\gamma_\nu} - \frac{B}{2(1-\gamma_\nu)^{1/2}}\cosh\left[2(1-\gamma_\nu)^{1/2}\tau_\nu\right]. \tag{2.37}$$

Applying the second boundary condition at the edge of the disk, where there is no incident radiation, we have $\tau_\nu = \tau_{\nu H}$ for $z = H$. From (2.26) and (2.27)

$$J_\nu(\tau_{\nu H}) = \frac{1}{2}I_\nu^+(\tau_{\nu H})$$

$$F_\nu(\tau_{\nu H}) = I_\nu^+(\tau_{\nu H})$$

so that

$$J_\nu(\tau_{\nu H}) = \frac{1}{2}F_\nu(\tau_{\nu H}). \tag{2.38}$$

From (2.36), (2.37), and (2.38)

$$B = \frac{2S_\nu}{1-\gamma_\nu}\left\{\sinh\left[2(1-\gamma_\nu)^{1/2}\tau_{\nu H}\right] + \frac{\cosh\left[2(1-\gamma_\nu)^{1/2}\tau_{\nu H}\right]}{(1-\gamma_\nu)^{1/2}}\right\}^{-1}. \tag{2.39}$$

When we replace the value of B in (2.36), we may calculate the flux and the mean intensity. Finally, the energy density can be calculated by

$$U_\nu = \frac{1}{c}\int I_\nu \, \mathrm{d}\omega = \frac{4\pi}{c} J_\nu. \tag{2.40}$$

2.5.3 Numerical Example: Energy Density

Let us calculate the energy density at $\lambda = 5{,}500$ Å, in the middle of the visible spectrum, for a galactic disk model, assuming 2H $\simeq$ 200 pc. The energy density is determined by stars of spectral type earlier than M that tend to concentrate in the region near the disk. In these conditions, the emission coefficient j_{5500} is given by (2.24). The luminosity at 5,500 Å can be determined by

$$L_{5500} = f\, L_\odot\, 10^{0.4(4.83-M_\mathrm{v})}, \tag{2.41}$$

where M_V is the absolute visual magnitude of the considered star; $L_\odot = 3.83 \times 10^{33}$ erg s^{-1} is the Sun's luminosity, whose absolute bolometric magnitude is taken to be 4.83; and f is the fraction of stellar radiation in a small wavelength interval centered on $\lambda = 5{,}500$ Å. Since approximately 93 % of the solar luminosity is concentrated in a bandwidth of about 1,000 Å around 5,500 Å, $f \simeq 0.93 \times 10^{-3}$ Å^{-1}. Taking n_*^s in units of pc^{-3}, (2.24) becomes

$$j_{5500} = 9.61 \times 10^{-27} \sum n_*^s \; 10^{0.4(4.83-M_V)}. \tag{2.42}$$

Let us consider a typical luminosity function (Allen 1973, p. 247). In this case, the emission coefficient per unit wavelength j_{5500} (erg cm^{-3} s^{-1} Å^{-1} sr^{-1}) is given in Table 2.2 for the main stellar spectral types.

A more complete study of interstellar extinction will be presented in Chap. 9. For now, we will consider the total extinction coefficient adopting a mean extinction Δm/H $\simeq$ 1 mag kpc^{-1}. As

$$\tau_{5500H} = k_\mathrm{E}\, H = \frac{\Delta m}{1.086}, \tag{2.43}$$

Table 2.2 Average emission coefficients for stars with different spectral types

Spectral type	j_{5500}
O	5.96×10^{-31}
B	9.93×10^{-29}
A	1.01×10^{-28}
F	1.07×10^{-28}
G	8.09×10^{-29}
K	1.25×10^{-28}
M	3.29×10^{-29}
Total	5.47×10^{-28}

we obtain

$$k_{\rm E} = \frac{\Delta m/H}{1.086} \simeq 2.98 \times 10^{-22}\,{\rm cm}^{-1} \tag{2.44}$$

and thus $\tau_{5500{\rm H}} \simeq 9.21 \times 10^{-2}$. As we shall see in Chap. 9, the optical properties of the dust grains, such as the albedo, vary with wavelength in a complex way for different types of solid interstellar particles. Using a typical value of $\gamma_{5500} \simeq 0.2$, we have $\sigma_{5500} \simeq 5.96 \times 10^{-23}$ cm^{-1} and $k_{5500} \simeq 2.38 \times 10^{-22}$ cm^{-1}. The source function is then $S_{5500} \simeq 1.84 \times 10^{-6}$ erg cm^{-2} s^{-1} Å^{-1} sr^{-1} and the constant $B \simeq 3.55 \times 10^{-6}$ erg cm^{-2} s^{-1} Å^{-1} sr^{-1}. The mean intensity J_{5500} in the center of the disk where $\tau_{5500} = 0$ is $J_{5500} \simeq 3.15 \times 10^{-7}$ erg cm^{-2} s^{-1} Å^{-1} sr^{-1}. Finally, the energy density is $U_{5500} \simeq 1.32 \times 10^{-16}$ erg cm^{-3} Å^{-1}, which can be compared with the observational value, of the order of $U_{5500} \simeq 0.63 \times 10^{-16}$ erg cm^{-3} Å^{-1}. Note that this value corresponds to density U_λ. To obtain U_ν, we use the fact that $U_\lambda\, d\lambda = U_\nu\, d\nu$, so $U_\nu = (c/\nu^2)U_\lambda = (\lambda^2/c)U_\lambda$. In this case, we have $\nu U_\nu = \lambda U_\lambda \sim 7 \times 10^{-13}$ erg cm^{-3} (cf. Fig. 2.1).

2.6 High-Energy Radiation

The origin of high-energy radiation is not very well known. It can comprise an extragalactic component, as well as several galactic sources. The sources may be stable, variable, or even present violent explosions. As for the infrared part of the spectrum, the Galaxy's center presents considerable emission. High-energy radiation is usually divided in classes, according to the detection technique used, from soft X-rays, with energies up to approximately 10 keV, through hard X-rays, with energies up to 100 keV, to high-energy γ-rays, with energies higher than 10 MeV. High-energy detectors have been recently mounted on satellites such as COS-B, COMPTON, and CHANDRA.

Several galactic objects like pulsars, binary systems, novae, and supernovae remnants show high-energy emission. Besides this, there is an extragalactic contribution that includes active galactic nuclei and a non-identified component, attributed to a large number of non-resolved sources, such as supernovae, strong stellar winds, or a hot intergalactic medium. Finally, γ-rays can also be produced from the interaction between cosmic rays and interstellar gas, forming pions that decay into a pair of γ-rays.

Some of the probable mechanisms responsible for X-ray and γ-ray emission are bremsstrahlung, synchrotron emission, and inverse Compton scattering. The first two were considered in Sect. 2.2. In inverse Compton scattering, a photon (energy $h\nu$) collides with an electron or ion with velocity v and energy $E \gg h\nu$, resulting in energy transfer from the electron to the photon. The energy of the produced quantum is

$$E_\gamma \simeq \gamma^2 h\nu \quad (q \ll 1) \tag{2.45}$$

$$E_\gamma \simeq \gamma m_e c^2 \quad (q \gg 1), \tag{2.46}$$

where

$$q = \frac{\gamma h\nu}{m_e c^2} \tag{2.47}$$

and

$$\gamma = \frac{E}{m_e c^2} = \frac{1}{\sqrt{1 - \frac{v^2}{c^2}}}. \tag{2.48}$$

For instance, if we consider a photon from the stellar radiation field with $\lambda = 5{,}500$ Å, frequency $\nu = 5.45 \times 10^{14}$ Hz, and energy $h\nu = 3.62 \times 10^{-12}$ erg $= 2.26$ eV interacting with a cosmic electron with energy $E = 1$ GeV, we have $\gamma = E/m_e c^2 \simeq 1.95 \times 10^3$, $q \simeq 0.0086 \ll 1$, and $E_\gamma \simeq 1.38 \times 10^{-5}$ erg $= 8.59$ MeV, corresponding to $\nu_\gamma \simeq 2.08 \times 10^{21}$ Hz.

Exercises

2.1 Show that the cosmic microwave background radiation, remnant of the Big Bang, has a maximum given by the peak of curve B from Fig. 2.1. What is the wavelength corresponding to this maximum?

2.2 The emission measure in the direction of an H II region is 10^3 pc/cm^6 and the column density of hydrogen nuclei in the same direction is 10^{20} cm^{-2}. Estimate the electron density and the H II region dimensions.

2.3 Use the radiation field model with four components given in Table 2.1 and equation (2.17) and (a) calculate the energy density U_λ in the optical spectrum center, where $\lambda = 5{,}500$ Å. (b) Compare the result with the mean value obtained from the solution of the transfer equation.

2.4 Use the model obtained from the solution of the transfer equation and estimate the flux, the mean intensity, and the energy density in the rim of the galactic disk for $\lambda = 5{,}500$ Å.

2.5 (a) Assume that the radiation field at some point in the interstellar medium can be characterized by a blackbody with $T = 10^4$ K and dilution factor $W = 10^{-14}$. What would be the flux at $\lambda = 2{,}000$ Å in erg cm^{-2} s^{-1} Å^{-1}? (b) Compare the result with the value predicted by the four component model described in Sect. 2.4. (c) Which of the two above models better follows the observations, taking into account the observational data shown in Fig. 2.2?

Bibliography

Allen, C.W.: Astrophysical Quantities. Athlone, London (1973). Very useful table compilation of physical and astronomical data, constants and unit conversions. Table 2.2 is based on this reference. See also the recent revised version by Cox, A.N. (ed.). 2000. Allen's Astrophysical Quantities. New York, American Institute of Physics/Springer

Gondhalekar, P.M.: The ultraviolet starlight in the Galaxy. In: Bowyer, S., Leinert, C. (eds.) IAU Symposium 139, p. 49. Kluwer, Dordrecht (1990). Updated discussion on the interstellar radiation field. Figure 2.2 is based on this reference

Kaplan, S.A., Pikelner, S.B.: The Interstellar Medium. Harvard University Press, Cambridge (1970). Referred to in chapter 1. Includes a discussion on the interstellar radiation field

Kolb, E.W., Turner, M.S.: The Early Universe. Addison-Wesley, Reading (1990). A good discussion on the early Universe, the Big Bang and the cosmic microwave background radiation

Lang, K.R.: Astrophysical Formulae. Springer, Berlin (1999). Updated compilation of astrophysical data, including detailed formulae of the principal physical processes mentioned in section 2.2

Mezger, P.G.: In: Setti, G.G., Fazzio, G.G. (eds.) Infrared Astronomy, p. 1. Reidel, Dordrecht (1978). Review article about the interstellar medium. Figure 2.1 is based on this reference

Mihalas, D.: Stellar Atmospheres. Freeman, San Francisco (1978). Accurate discussion on radiative transfer, applied to stellar atmospheres

Pinkau, K. (ed.): The Interstellar Medium. Reidel, Dordrecht (1974). An interesting ensemble of articles by specialists about several aspects of the physics of the interstellar medium

Rybicki, G.B., Lightman, A.P.: Radiative Processes in Astrophysics. Wiley, New York (1979). Very complete discussion on the radiation field concepts and the main radiative processes mentioned in sections 2.2 and 2.6

Schuster, A.: Radiation through a foggy atmosphere. Astrophys J **21**, 1 (1905). Discussion of the classical method of the transfer equation solution

Verschuur, G.L., Kellermann, K.I. (eds.): Galactic and Extragalactic Radio Astronomy. Berlin, Springer (1988). Ensemble of basic articles on radio emission applied to the study of galaxies' structure

Werner, M.W., Salpeter, E.E.: Mon. Grain temperatures in interstellar dust clouds. Notices Roy. Astron. Soc. **145**, 249 (1969). Discussion on the interstellar radiation field. Table 2.1 is based on this reference

Witt, A.N., Johnson, M.W.: Astrophys. J. The interstellar radiation density between 1250 and 4250 Angstroms. **181**, 363 (1973). An example of the interstellar radiation field calculation in the ultraviolet

Wynn-Williams, G.: The Fullness of Space. Cambridge University Press, Cambridge (1992). Referred to in chapter 1. Includes a qualitative discussion on the interstellar radiation field and images at different wavelengths

Chapter 3
Spectral Line Formation

3.1 Introduction

In the last chapter, we saw how the mean energy density of the integrated interstellar radiation field corresponds to a radiation in thermodynamic equilibrium with a temperature of the order of 3 K or a bit higher. However, the mean energy of the photons in the radiation field is of the order of several eV, which corresponds to temperatures of the order of 10^3 K in stellar atmospheres. This temperature discrepancy points to the fact that there is no thermodynamic equilibrium (TE) in the interstellar space. Nevertheless, essentially due to the high abundance of H and He, the large number of elastic collisions between these elements allows equipartition of the gas kinetic energy, and consequently, a Maxwellian velocity distribution is established. Therefore, it is frequently possible to define a kinetic temperature for the different atoms, ions, and molecules in the interstellar gas, which means a considerable simplification. In this way, the relative populations of the several atomic and molecular energy levels have a tendency to reach the values obtained for TE, especially if excitation and de-excitation of these levels are achieved by collisional processes. For this reason, this chapter includes a general review of the principal processes and equations relevant to thermodynamic equilibrium.

It is convenient to define the equivalent thermodynamic equilibrium (ETE), that is, a TE state in which (1) the temperature is the kinetic temperature of the gas, defined by a Maxwellian velocity distribution; (2) the electron density is equal to the electron density of the gas n_e; and (3) the density of the atoms of element X in ionization state $r - 1$ is equal to the interstellar value, $n(X^{r-1})$. In the remainder of this chapter, we will use an asterisk (*) to indicate quantities in ETE. As we shall see later, we can in a general way use the ETE equations, defining the *deviation coefficients* relative to the thermodynamic equilibrium.

W.J. Maciel, *Astrophysics of the Interstellar Medium*,
DOI 10.1007/978-1-4614-3767-3_3,

3.2 Thermodynamic Equilibrium Equations

3.2.1 *Maxwellian Velocity Distribution Function*

Let $f(\mathbf{v})\mathrm{d}\mathbf{v}$ be the fraction of particles with mass m of a system whose velocity is in a three-dimensional interval spanning from $\mathbf{v}$ to $\mathbf{v}+\mathrm{d}\mathbf{v}$. In TE the function f is isotropic, and we can write $f(\mathbf{v}) = f(v)$, where $v = |\mathbf{v}|$. The *Maxwellian distribution function* is

$$f(v)\,\mathrm{d}v = \left(\frac{m}{2\pi kT}\right)^{3/2} \exp(-mv^2/2kT)\,\mathrm{d}v. \tag{3.1}$$

The velocity mean value (rms) is

$$\langle v^2 \rangle^{1/2} = \left(\frac{3kT}{m}\right)^{1/2}. \tag{3.2}$$

In terms of a velocity component, we have

$$f(v_x)\,\mathrm{d}v_x = \left(\frac{m}{2\pi kT}\right)^{1/2} \exp\left(-mv_x^2/2kT\right)\mathrm{d}v_x, \tag{3.3}$$

where $f(v_x)\mathrm{d}v_x$ is the fraction of particles whose velocity component in the x direction is in the interval v_x, $v_x+\mathrm{d}v_x$, independently of the other components. In this case, $\langle v_x \rangle = 0$ because the x component has the same probability of being positive or negative. However,

$$\langle v_x^2 \rangle^{1/2} = \left[\frac{\int f(v_x)\, v_x^2\, \mathrm{d}v_x}{\int f(v_x)\, \mathrm{d}v_x}\right]^{1/2} = \left(\frac{kT}{m}\right)^{1/2}. \tag{3.4}$$

Finally, if $f'(v)\mathrm{d}v$ is the fraction of particles with $v = |\mathbf{v}|$ between v and $v+\mathrm{d}v$, independently of the velocity vector direction, we have

$$\begin{aligned} f'(v)\,\mathrm{d}v &= 4\,\pi f(v)\,\mathrm{d}v\, v^2 \\ &= 4\pi\left(\frac{m}{2\pi kT}\right)^{3/2} v^2 \exp\left(-mv^2/2kT\right)\mathrm{d}v. \end{aligned} \tag{3.5}$$

In this case,

$$\langle v \rangle = \left(\frac{8kT}{\pi m}\right)^{1/2} \quad \text{(mean velocity)}, \tag{3.6a}$$

$$\langle v^2 \rangle^{1/2} = \left(\frac{3kT}{m}\right)^{1/2} \quad \text{(rms velocity)}, \tag{3.6b}$$

$$v_\mathrm{p} = \left(\frac{2kT}{m}\right)^{1/2} \quad \text{(most probable velocity)}. \tag{3.6c}$$

3.2.2 *Boltzmann Equation*

Let X be an element in ionization state r. In TE the relative populations of two energy levels j and k are related by the expression

$$\frac{n_j^*(\mathrm{X}^r)}{n_k^*(\mathrm{X}^r)} = \frac{g_{rj}}{g_{rk}} \exp\left[-(E_{rj} - E_{rk})/kT\right], \tag{3.7}$$

which is the *Boltzmann equation*, where g_{rj}, g_{rk} are the statistical weighs of levels j and k, and E_{rj}, E_{rk} are the levels' energies. In the more general case, outside TE, we define the deviation coefficients

$$b_j = \frac{n_j(\mathrm{X}^r)}{n_j^*(\mathrm{X}^r)}, \tag{3.8a}$$

$$b_k = \frac{n_k(\mathrm{X}^r)}{n_k^*(\mathrm{X}^r)}, \tag{3.8b}$$

and so (3.7) can be rewritten as

$$\frac{n_k(\mathrm{X}^r)}{n_j(\mathrm{X}^r)} = \frac{b_k}{b_j}\frac{g_{rk}}{g_{rj}} \exp\left(-h\nu_{jk}/kT\right), \tag{3.9}$$

where $\nu_{jk} = (E_{rk} - E_{rj})/h$ is the emitted or absorbed photon frequency in a radiative transition between levels j and k, assuming $E_{rj} < E_{rk}$ and T is the excitation temperature. From (3.9), we can see that the ratio between populations of levels j and k may be determined if we know the deviation coefficients b_j and b_k. The total density of particles in ionization state r is given by

$$\begin{aligned} n^*(\mathrm{X}^r) &= \sum_k n_k^*(\mathrm{X}^r) \\ &= \frac{n_j^*(\mathrm{X}^r)}{g_{rj}} \exp\left(E_{rj}/kT\right) \sum_k g_{rk} \exp\left(-E_{rk}/kT\right). \end{aligned} \tag{3.10}$$

Defining a partition function for atom X in ionization state r

$$f(\mathrm{X}^r) = f_r = \sum\nolimits_k g_{rk} \exp(-E_{rk}/kT), \tag{3.11}$$

we obtain

$$\frac{n_j^*(\mathrm{X}^r)}{n * (\mathrm{X}^r)} = \frac{g_{rj}}{f_r} \exp\left(-E_{rj}/kT\right). \tag{3.12}$$

3.2.3 Saha Equation

The Saha ionization equation gives the distribution of atoms of element X in the different ionization states in TE. It can be obtained from the Boltzmann excitation equation generalization. The *Saha equation* is

$$\frac{n^*(\mathrm{X}^{r+1})\, n_\mathrm{e}}{n^*(\mathrm{X}^r)} = \frac{f_{r+1} f_\mathrm{e}}{f_r}, \tag{3.13}$$

where the partition functions of atoms f_r and f_{r+1} are given by (3.11) and the partition function of free electrons per unit volume f_e is given by

$$f_\mathrm{e} = 2\left(\frac{2\pi m_\mathrm{e} kT}{h^2}\right)^{3/2}. \tag{3.14}$$

A usual approximation consists of only considering the first terms of the atomic partition functions

$$f_r \simeq g_{r,1} \exp\left(-E_{r,1}/kT\right), \tag{3.15a}$$

$$f_{r+1} \simeq g_{r+1,1} \exp\left(-E_{r+1,1}/kT\right). \tag{3.15b}$$

In this case, (3.13) becomes

$$\frac{n^*(\mathrm{X}^{r+1}) n_\mathrm{e}}{n^*(\mathrm{X}^r)} \simeq \left(\frac{2\pi m_\mathrm{e} kT}{h^2}\right)^{3/2} 2 \frac{g_{r+1,1}}{g_{r,1}} \mathrm{e}^{(-\Delta E_r/kT)}, \tag{3.16}$$

where $\Delta E_r = E_{r+1,1} - E_{r,1}$ is the energy needed to ionize element X^r from the ground state.

3.2.4 The Planck Function

Finally, in TE, the radiation specific intensity I_ν is given by the *Planck function*

$$I_\nu^* = B_\nu(T) = \frac{2h\nu^3}{c^2} \frac{1}{e^{h\nu/kT} - 1} \tag{3.17}$$

(units: erg cm^{-2} s^{-1} Hz^{-1} sr^{-1}). The Planck function has some approximations: Wien's law

$$B_\nu(T) \simeq \frac{2h\nu^3}{c^2} e^{-h\nu/kT} \quad (h\nu/kT \gg 1) \tag{3.18a}$$

and the Rayleigh–Jeans distribution

$$B_\nu(T) \simeq \frac{2\nu^2 kT}{c^2} \quad (h\nu/kT \ll 1). \tag{3.18b}$$

The energy density U_ν is given by

$$U_\nu^* = \frac{4\pi}{c} B_\nu(T) = \frac{8\pi h\nu^3}{c^3} \frac{1}{e^{h\nu/kT} - 1} \tag{3.19}$$

(units: erg cm^{-3} Hz^{-1}).

3.3 Radiative Transfer

Let us consider the one-dimensional steady-state radiative transfer equation without scattering. In this case, we may write

$$\frac{dI_\nu}{ds} = j_\nu - k_\nu I_\nu \tag{3.20}$$

for the radiation transfer equation along a direction characterized by element ds, where j_ν and k_ν are the volumetric emission and absorption coefficients, respectively (see Chap. 2). We define the optical depth by

$$d\tau_\nu = -k_\nu \, ds, \tag{3.21}$$

that is, we assume $\tau_\nu = 0$ in the region's edge closest to the observer (Fig. 3.1).

The total optical depth of the region is $\tau_{\nu r}$, corresponding to a possible incident intensity $I_\nu(s = 0)$. Taking into account (3.21), the transfer equation becomes

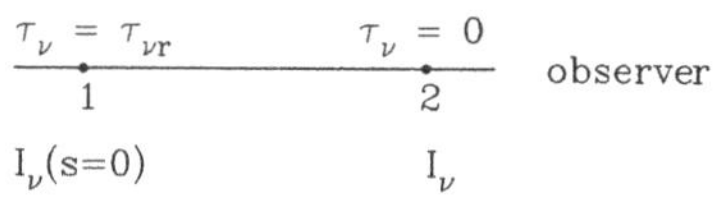

Fig. 3.1 Geometry used in the definition of the optical depth

$$\frac{dI_\nu}{d\tau_\nu} = I_\nu - \frac{j_\nu}{k_\nu}. \tag{3.22}$$

Multiplying both members of the equation by $e^{-\tau_\nu}$,

$$e^{-\tau_\nu}\frac{dI_\nu}{d\tau_\nu} = e^{-\tau_\nu}I_\nu - \frac{j_\nu}{k_\nu}e^{-\tau_\nu}.$$

But

$$\frac{d}{d\tau_\nu}(e^{-\tau_\nu}I_\nu) = e^{-\tau_\nu}\frac{dI_\nu}{d\tau_\nu} - e^{-\tau_\nu}I_\nu,$$

thus,

$$\frac{d}{d\tau_\nu}(e^{-\tau_\nu}I_\nu) = -\frac{j_\nu}{k_\nu}e^{-\tau_\nu}.$$

Integrating over the considered region, between points 1 and 2 of Fig. 3.1, we obtain

$$I_\nu = I_\nu(s=0)e^{-\tau_{\nu r}} + \int_0^{\tau_{\nu r}} \frac{j_\nu}{k_\nu}e^{-\tau_\nu}\,d\tau_\nu. \tag{3.23}$$

Therefore, the determination of the emergent intensity I_ν depends on the calculation of the source function j_ν/k_ν along the region. Finally, in TE, intensity $I_\nu{}^* = B_\nu(T)$, so that $dI_\nu/ds = 0$ and (3.20) become

$$\frac{j_\nu^*}{k_\nu^*} = B_\nu(T), \tag{3.24}$$

which is the Kirchhoff law.

3.4 Einstein Coefficients

3.4.1 *Emission Coefficient*

As seen in Chap. 2, the radiation emission and absorption processes can be described by macroscopic emission j_ν (erg cm^{-3} s^{-1} Hz^{-1} sr^{-1}) and absorption k_ν (cm^{-1}) coefficients. These coefficients can be related to the density of particles in the energy levels involved in the transition and with the Einstein coefficients for radiative transitions.

Fig. 3.2 Energy level diagram showing the emission and absorption transitions

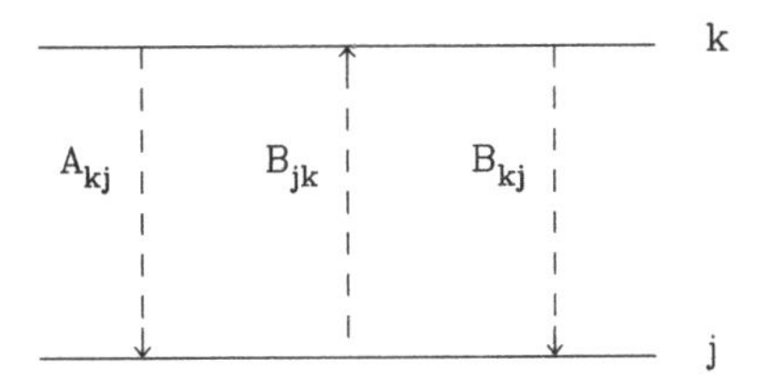

If we consider two levels j and k with energies E_j and E_k (Fig. 3.2), where $k > j$, the A_{kj} coefficient gives the probability per unit time that an atom X^r in excitation state k will decay spontaneously to state j emitting a photon. Therefore, A_{kj} units are s^{-1}, and the inverse of the emission coefficient ($A_{kj}{}^{-1}$) gives us essentially the lifetime of level k for level j emissions. The number of spontaneous emissions per unit volume and per unit time is

$$\frac{\text{number of spontaneous emissions } k \rightarrow j}{\text{cm}^3\text{ s}} = n_k(\text{X}^r)\,A_{kj}. \tag{3.25}$$

Spontaneous emission is generally isotropic, and so the total energy emitted spontaneously per unit volume, per unit time, and per unit solid angle is given by

$$\int j_\nu\,\text{d}\nu = \frac{1}{4\pi} h\,\nu_{jk}\,n_k(\text{X}^r)A_{kj} \tag{3.26}$$

(units: erg cm^{-3} s^{-1} sr^{-1}), where ν_{jk} is the central frequency of the line and the integral spans across the whole line. Sometimes it is useful to express the density $n_k(\text{X}^r)$ of (3.26) in terms of the density of particles in state $r + 1$. Using (3.8), (3.12), and (3.13), we may write

$$n_k(\text{X}^r) = \frac{b_k\,g_{rk}\,\text{e}^{-E_{rk}/kT} n * (X^{r+1})\,n_\text{e}}{f_{r+1} f_\text{e}}. \tag{3.27}$$

3.4.2 *Absorption Coefficient*

The absorption process can be characterized by the Einstein absorption coefficient B_{jk} modified by induced emissions, also called negative absorptions. These are characterized by coefficient B_{kj} and reflect the emissions due to photons absorption with energy $E_{jk} = E_k - E_j$.

In this case, the number of absorptions corrected from induced emissions per unit volume and per unit time is

$$\frac{\text{number of absorptions}}{\text{cm}^3\text{ s}} = \left[n_j(\text{X}^r)B_{jk} - n_k(\text{X}^r)B_{kj}\right]U_\nu, \tag{3.28}$$

where B_{jk} and B_{kj} have units erg^{-1} cm^3 s^{-1} Hz and energy density U_ν has units erg cm^{-3} Hz^{-1}. Note that the absorption coefficient can also be defined, for instance, using the radiation mean intensity.

By analogy to the emission case, the total energy absorbed in the line per unit volume, per unit time, and per unit solid angle is

$$\int k_\nu I_\nu \mathrm{d}\nu \simeq I_{\nu_{jk}} \int k_\nu \, \mathrm{d}\nu$$
$$= \frac{1}{c} h \, \nu_{jk} \left[n_j(\mathrm{X}^r) B_{jk} - n_k(\mathrm{X}^r) B_{kj} \right] I_{\nu_{jk}} \tag{3.29}$$

(units: erg cm^{-3} s^{-1} sr^{-1}), where we assume that I_ν is constant along the line intrinsic width and use the fact that $U_\nu = (4\pi/c) I_\nu$. The above approximation is generally valid, except for very strong absorption lines.

3.4.3 Relations Between the Einstein Coefficients

In thermodynamic equilibrium (TE), the energy emission and absorption rates are equal. Thus from (3.26) and (3.29),

$$\frac{n_k^*}{n_j^*} \frac{A_{kj}}{4\pi} = \left(B_{jk} - \frac{n_k^*}{n_j^*} B_{kj} \right) \frac{I_{\nu_{jk}}}{c},$$

$$\frac{n_k^*}{n_j^*} \left[\frac{A_{kj}}{4\pi} + \frac{B_{kj} I_{\nu_{jk}}}{c} \right] = B_{jk} \frac{I_{\nu_{jk}}}{c},$$

$$\frac{n_k^*}{n_j^*} = \left[B_{jk} \frac{I_{\nu jk}}{c} \right] \left[\frac{A_{kj}}{4\pi} + \frac{B_{kj} I_{\nu_{jk}}}{c} \right]^{-1} = \frac{g_k}{g_j} \mathrm{e}^{-h\nu_{jk}/kT},$$

where we use (3.7). From this equation, we can write

$$\frac{A_{kj}}{4\pi} = \frac{B_{jk} \frac{I_{\nu_{jk}}}{c} (g_j/g_k)}{\mathrm{e}^{-h\nu_{jk}/kT}} - \frac{B_{kj} I_{\nu_{jk}}}{c},$$

and using (3.17), we obtain

$$A_{kj} = \frac{8\pi h \nu_{jk}^3}{c^3} \frac{B_{jk} (g_j/g_k) \mathrm{e}^{h\nu_{jk}/kT} - B_{kj}}{\mathrm{e}^{h\nu_{jk}/kT} - 1}. \tag{3.30}$$

Since the Einstein coefficients only depend on atomic parameters and not on macroscopic quantities such as temperature, we should have $B_{jk}\, g_j/g_k = B_{kj}$ and

$$g_j\, B_{jk} = g_k\, B_{kj}, \tag{3.31}$$

$$A_{kj} = \frac{8\pi h\nu_{jk}^3}{c^3} B_{kj}, \tag{3.32}$$

$$A_{kj} = \frac{8\pi h\nu_{jk}^3}{c^3} \frac{g_j}{g_k} B_{jk}. \tag{3.33}$$

Although derived for TE, these relations are valid for the more general case.

3.5 Spectral Line Profile

3.5.1 Absorption Line Profile

Let us again invoke the process involving two energy levels j and k of an atom. The absorption coefficient may be written as

$$k_\nu = n_j(\mathrm{X}^r)\sigma_\nu, \tag{3.34}$$

where $\sigma_\nu(\mathrm{cm}^2)$ is the absorption cross section or the absorption coefficient per atom in level j, related to the integrated cross section σ by

$$\sigma = \int \sigma_\nu \mathrm{d}\nu. \tag{3.35}$$

If we define $\Delta\nu = \nu - \nu_{jk}$, we can introduce the *line profile function* $\phi(\Delta\nu)$ [units: Hz^{-1}] by

$$\sigma_\nu = \sigma\, \phi(\Delta\nu). \tag{3.36}$$

This function is *normalized*, that is,

$$\int \phi(\Delta\nu)\mathrm{d}\nu = 1, \tag{3.37}$$

where the integral applies to the whole line. Function $\phi(\Delta\nu)$ depends on the line intrinsic width, as well as on other broadening mechanisms that may occur. The *natural or intrinsic* broadening is a consequence of the Heisenberg uncertainty principle.

Energy levels like the ones in Fig. 3.2 are not infinitely thin, creating thus a small uncertainty in the energy value of each level and consequently in the lifetime of these levels. Therefore, the radiation absorption or emission frequencies are not limited to the central frequency, spreading somewhat into the "wings" of the spectral line. In the interstellar medium, the thermal motion of the atoms that produce the line induces an additional broadening due to the Doppler effect, that is, atoms with a velocity component in the direction of the observer produce a small shift toward shorter wavelengths, whereas atoms that move away from the observer produce a shift toward longer wavelengths.

In terms of optical depth, the profile can be written as

$$\tau_{\nu r} = \int k_\nu \, \mathrm{d}s = \int n_j \, \sigma_\nu \, \mathrm{d}s = \int n_j \, \sigma \, \phi(\Delta\nu) \mathrm{d}s,$$

that is,

$$\tau_{\nu r} = N_j \, \sigma \, \phi \, (\Delta\nu), \tag{3.38}$$

where we introduce the column density N_j measured in cm^{-2}:

$$N_j = \int n_j \, \mathrm{d}s. \tag{3.39}$$

The column density can be obtained integrating (3.38)

$$\int \tau_{\nu r} \, \mathrm{d}\nu = N_j \, \sigma \int \phi(\Delta\nu) \mathrm{d}\nu = N_j \sigma$$

$$N_j = \frac{1}{\sigma} \int \tau_{\nu r} \, \mathrm{d}\nu. \tag{3.40}$$

3.5.2 Doppler Profile

In the case of Doppler broadening due to thermal motions of the absorbing particles, we have

$$\phi(\Delta\nu) \, \mathrm{d}\nu = f(v) \mathrm{d}v, \tag{3.41}$$

where the velocity distribution function $f(v)$ is assumed to be Maxwellian and represents the fraction of atoms with radial velocity between v and $v+\mathrm{d}v$. For $v \ll c$, we have

$$\frac{\Delta\nu}{\nu_{jk}} = \frac{v}{c}, \tag{3.42}$$

$$\frac{dv}{c} = \frac{d\nu}{\nu_{jk}} = \frac{d\nu\,\lambda_{jk}}{c}. \tag{3.43}$$

Using (3.3),

$$\begin{aligned}
\phi(\Delta\nu) = f(v)\frac{dv}{d\nu} &= \left(\frac{m}{2\pi kT}\right)^{1/2} e^{-mv^2/2kT}\,\frac{c}{\nu_{jk}} \\
&= \frac{c}{b\sqrt{\pi}\,\nu_{jk}} e^{-(v/b)^2} \\
&= \frac{c}{b\sqrt{\pi}\,\nu_{jk}} e^{-(c\Delta\nu/b\nu_{jk})^2} \\
&= \frac{\lambda_{jk}}{b\sqrt{\pi}} e^{-(v/b)^2},
\end{aligned} \tag{3.44}$$

where we introduce parameter b given by

$$b = \left(\frac{2kT}{m}\right)^{1/2}, \tag{3.45}$$

with m being the mass of the absorbing atom. This parameter is related to the Doppler width $\Delta\nu_D$, that is, the value of $\Delta\nu$ for which the profile function decreases by a factor e. The $\phi(\Delta\nu)$ maximum is given by

$$\phi_M(\Delta\nu) = \frac{c}{b\sqrt{\pi}\,\nu_{jk}} = \frac{\lambda_{jk}}{b\sqrt{\pi}}, \tag{3.46}$$

so that

$$\begin{aligned}
\phi(\Delta\nu_D) &= \frac{\phi_M(\Delta\nu)}{e} \\
&= \frac{1}{e}\left(\frac{c}{b\sqrt{\pi}\,\nu_{jk}}\right) \\
&= \frac{c}{b\sqrt{\pi}\,\nu_{jk}} e^{-(c\Delta\nu_D/b\nu_{jk})^2},
\end{aligned}$$

that is,

$$\Delta\nu_D = b\frac{\nu_{jk}}{c} = \frac{b}{\lambda_{jk}}. \tag{3.47}$$

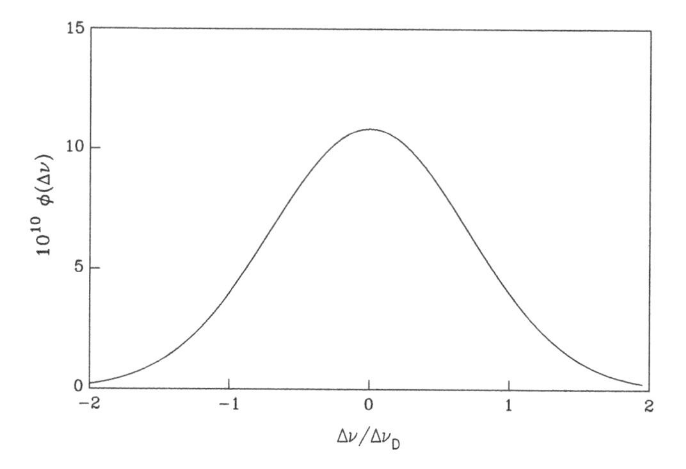

Fig. 3.3 An example of the Doppler profile for the K line of interstellar CaII

In terms of the Doppler width, profile $\Phi(\Delta\nu)$ can be written as

$$\phi(\Delta\nu) = \frac{1}{\sqrt{\pi}\,\Delta\nu_\mathrm{D}} \mathrm{e}^{-(\Delta\nu/\Delta\nu_\mathrm{D})^2}. \tag{3.48}$$

Note that the Doppler profile is normalized according to (3.37). The full width at half maximum (FWHM) represented by $\Delta\nu_\mathrm{h}$ may be calculated using

$$\phi(\Delta\nu_\mathrm{h}/2) = \frac{1}{2}\left(\frac{1}{\sqrt{\pi}\,\Delta\nu_\mathrm{D}}\right) = \frac{1}{\sqrt{\pi}\,\Delta\nu_\mathrm{D}} \mathrm{e}^{-(\Delta\nu_\mathrm{h}/2\Delta\nu_\mathrm{D})^2},$$

and the relation

$$\mathrm{e}^{(\Delta\nu_\mathrm{h}/2\Delta\nu_\mathrm{D})^2} = 2,$$

and the result is

$$\Delta\nu_\mathrm{h} = 2\Delta\nu_\mathrm{D}\sqrt{\ln 2} = \frac{2b\nu_{jk}\sqrt{\ln 2}}{c} = \frac{2b\sqrt{\ln 2}}{\lambda_{jk}}. \tag{3.49}$$

An example of the Doppler profile is shown in Fig. 3.3 for the interstellar Ca II K line, with $\lambda_{jk} = 3{,}933.66$ Å or $\nu_{jk} = 7.63 \times 10^{14}$ Hz. Assuming $T \simeq 100$ K for the cloud temperature, we obtain from (3.45) $b \simeq 2.0 \times 10^4$ cm s^{-1} with $m_\mathrm{Ca} \simeq 40\,\mathrm{m_H} = 6.68 \times 10^{-23}$ g. From (3.47), the Doppler width is $\Delta\nu_\mathrm{D} \simeq 5.2 \times 10^8$ Hz, so that the full width at half maximum (FWHM) is $\Delta\nu_\mathrm{h} \simeq 8.6 \times 10^8$ Hz. In terms of wavelength, the region characterized by a Doppler width is given by $\Delta\lambda \simeq \lambda_{jk}\,\Delta\nu_\mathrm{D}/\nu_{jk} \simeq 0.003$ Å, meaning that the line is quite narrow.

In the case of an absorption line, the profile of Fig. 3.3 gives us an idea of the reversed line profile with respect to the horizontal axis. The relation between the observed and original intensities is $I_\nu/I_\nu(0) = e^{-\tau_\nu}$, where the optical depth is $\tau_\nu \propto \phi(\Delta\nu)$, as seen in (3.38). For $|\Delta\nu| \gg \Delta\nu_D$, $\phi(\Delta\nu) \to 0$ and $I_\nu/I_\nu(0) \to 1$, that is, the radiation intensity is not affected by absorbing atoms at frequencies very far from the central frequency. In the center of the line where $\phi(\Delta\nu) = \phi(0) = 1/(\sqrt{\pi}\,\Delta\nu_D)$, we have a maximum optical depth, and values of $\tau_\nu \gg 1$ can be reached for completely dark lines, so that $I_\nu/I_\nu(0) \to 0$ in this region. For instance, for $\tau_\nu \simeq 1$ and $\tau_\nu \simeq 5$ in the center of the line, we would have $I_\nu/I_\nu(0) \simeq e^{-1} \simeq 0.37$ and $I_\nu/I_\nu(0) \simeq e^{-5} \simeq 0.01$, respectively. In intermediate regions, profile $\phi(\Delta\nu)$ decreases relatively to the central value. Since the number of absorbing atoms remains constant, ratio $I_\nu/I_\nu(0)$ increases up to the initial value $I_\nu/I_\nu(0) \simeq 1$ for frequencies far from the central frequency.

3.5.3 Lorentz Profile

From Heisenberg uncertainty principle, the finite mean life of energy levels implies a natural or intrinsic broadening of the spectral lines. This type of broadening can be characterized by a function called natural profile or *Lorentz profile*:

$$\phi(\Delta\nu) = \frac{\Gamma_k/4\pi^2}{(\nu - \nu_{jk})^2 + (\Gamma_k/4\pi)^2}, \tag{3.50}$$

where Γ_k is the quantum dissipation coefficient or damping constant,

$$\Gamma_k = \sum_j A_{kj}, \tag{3.51}$$

being essentially the inverse of the radiative lifetime of level k. Just like the Doppler profile, the Lorentz profile is normalized, according to (3.37). From (3.50), it is easy to demonstrate that in this case, the FWHM is $\Delta\nu_h = \Gamma_k/2\pi$.

3.5.4 Voigt Profile

In the more general case, besides a natural or radiative broadening and a thermal Doppler broadening (eventually turbulent), other mechanisms may contribute to the spectral line formation. For instance, magnetic fields can originate a type of Zeeman broadening, and high-density media can cause a collisional broadening, such as in stellar atmospheres. Lang (1999), Rybicki and Lightman (1979), and Jefferies (1968) discuss several types of broadening. In the interstellar medium, the most frequently types of broadening are Doppler (thermal or turbulent) and natural broadening. We can roughly say that expression (3.48) gives the profile in the central regions of the line, where the absorption coefficient is higher. Near the line wings, that is, in regions where $\Delta\nu \gg \Delta\nu_D$, natural broadening dominates and the profile is essentially given by (3.50). In the more general or intermediate case, we can combine these two expressions and thus obtain the *Voigt profile*

$$\phi(\Delta\nu) = \frac{1}{\sqrt{\pi}\Delta\nu_D} H(a,u), \tag{3.52}$$

where *H(a,u)* is the *Hjerting function*

$$H(a,u) = \frac{a}{\pi} \int_{-\infty}^{\infty} \frac{e^{-x^2} dx}{a^2 + (u-x)^2}. \tag{3.53}$$

In this relation, we use a dimensionless constant

$$a = \frac{\Gamma_k}{4\pi\,\Delta\nu_D}, \tag{3.54}$$

and we define

$$u = \frac{\nu - \nu_{jk}}{\Delta\nu_D} = \frac{\Delta\nu}{\Delta\nu_D}. \tag{3.55}$$

For most astrophysical cases, constant $a \ll 1$ and the Voigt profile becomes the Doppler profile in the center of the line and the Lorentz profile in the region of the radiative wings. For instance, for CaII K line, $\Gamma_k \sim 1.6 \times 10^8\ s^{-1}$, so that $a \sim 0.02$. We can estimate the transition region by making profiles (3.48) and (3.50) equal and thus obtaining

$$e^{-u^2}(u^2 + a^2) = \frac{a}{\sqrt{\pi}},$$

that can be written as

$$F(a,u) = (u^2 + a^2)e^{-u^2} - \frac{a}{\sqrt{\pi}} = 0. \tag{3.56}$$

For small values of a, Fig. 3.4 shows that $u \simeq 3$ or $\Delta\nu \simeq 3\ \Delta\nu_D$, that is, we have the Doppler profile for $\Delta\nu \lesssim 3\ \Delta\nu_D$ and the Lorentz profile for $\Delta\nu \gtrsim 3\ \Delta\nu_D$.

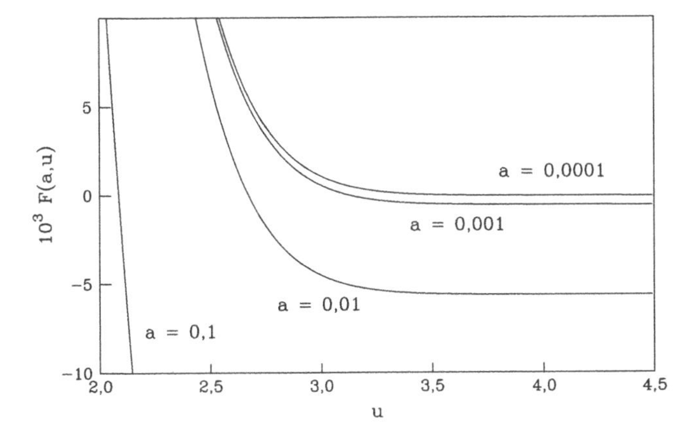

Fig. 3.4 Determination of the "u" parameter in the Voigt profile

3.5.5 *Integrated Absorption Cross Section*

We will now obtain an expression for the total cross section σ as a function of atomic constants. Replacing (3.34) in (3.29) and using (3.35), we obtain

$$I_{\nu jk} \int k_\nu \, \mathrm{d}\nu = I_{\nu jk} \int n_j \, \sigma_\nu \, \mathrm{d}\nu = I_{\nu jk} \, n_j \, \sigma. \tag{3.57}$$

Taking into account (3.29) and using (3.8), (3.31), and (3.7), we have

$$\frac{h\,\nu_j k(n_j B_{jk} - n_k B_{kj}) I_{\nu jk}}{c} = \frac{h\,\nu_{jk}\,n_j B_{jk}}{c}\left[1 - \frac{b_k}{b_j}\,\mathrm{e}^{-h\nu_{jk}/kT}\right] I_{\nu jk}$$

that can be written as follows, using (3.57):

$$\sigma = \sigma_u \left[1 - \frac{b_k}{b_j}\,\mathrm{e}^{-h\nu_{jk}/kT}\right], \tag{3.58}$$

where we introduce the total cross section without correction from induced emissions:

$$\sigma_u = \frac{h\nu_{jk} B_{jk}}{c}. \tag{3.59}$$

Cross section σ_u is related to the oscillator strength f_{jk} for transition $j \rightarrow k$ by

$$\sigma_u = \frac{\pi e^2}{m_e c} f_{jk}. \tag{3.60}$$

The oscillator strength for transition $k \rightarrow j$ is defined as

$$f_{kj} = -\frac{g_j}{g_k} f_{jk}, \tag{3.61}$$

and the total oscillator strength for absorptions from level j is

$$f_{\text{total}} = \sum_k f_{jk}.$$

The oscillator strengths must be computed from quantum mechanics or otherwise measured in laboratory.

3.5.6 *Relations Between the Einstein Coefficients and the Oscillator Strength*

If we consider, for instance, (3.59), (3.60), (3.31), and (3.32), we have the relations

$$B_{jk} = \frac{\pi e^2}{m_e h\nu_{jk}} f_{jk}, \tag{3.62}$$

$$B_{kj} = \frac{\pi e^2}{m_e h\nu_{jk}} \frac{g_j}{g_k} f_{jk}, \tag{3.63}$$

$$A_{kj} = \frac{8\pi^2 e^2 \nu_{jk}^2}{m_e c^3} \frac{g_j}{g_k} f_{jk}. \tag{3.64}$$

3.5.7 *Induced Emissions*

The integrated absorption cross section (3.58) includes two terms, the second one taking into account induced emissions. We are, generally, interested in two approximations for this equation. In the first case, we consider $h\nu_{jk}/kT \gg 1$. In this case, which generally occurs in the optical part of the spectrum, induced emissions are negligible and

$$\sigma \simeq \sigma_u \ (h\nu_{jk}/kT \gg 1), \tag{3.65}$$

where σ_u is given by (3.59). In the second case, we consider $h\nu_{jk}/kT \ll 1$. In this case, from (3.58), we have

$$\sigma \simeq \sigma_u \left[1 - \frac{b_k}{b_j} \left(1 - \frac{h\nu_{jk}}{kT} \right) \right]$$
$$= \sigma_u \frac{h\nu_{jk}}{kT} \left[\frac{b_k}{b_j} + \frac{kT}{h\nu_{jk}} \left(1 - \frac{b_k}{b_j} \right) \right]. \tag{3.66}$$

Defining

$$\chi = \frac{b_k}{b_j} + \frac{kT}{h\,\nu_{jk}} \left(1 - \frac{b_k}{b_j} \right), \tag{3.67}$$

we have

$$\sigma \simeq \sigma_u \frac{h\nu_{jk}}{kT} \chi \quad (h\nu_{jk}/kT \ll 1), \tag{3.68}$$

where the following terms are explicit: $h\nu_{jk}/kT$, a correction for induced emissions in TE, and χ, a function for deviations relative to TE.

3.6 The Source Function

We have seen that in TE the source function is given by Kirchhoff's law (3.24):

$$S_\nu^* = \frac{j_\nu^*}{k_\nu^*} = B_\nu(T). \tag{3.69}$$

In the more general case, the source function depends on the TE deviation coefficients. Assuming that the emission profile is equal to the absorption profile, (3.26), (3.34), and (3.36) allow us to write

$$S_\nu = \frac{j_\nu}{k_\nu} = \frac{h\,\nu_{jk}\, n_k\, A_{kj}\, \phi(\Delta\nu)}{4\pi\, n_j\, \sigma\, \phi(\Delta\nu)}. \tag{3.70}$$

Using (3.33),

$$S_\nu = \frac{h\nu_{jk}}{4\pi\sigma} \frac{n_k}{n_j} \frac{8\pi h\nu_{jk}^3}{c^3} \frac{g_j}{g_k} B_{jk}.$$

Using (3.9),

$$S_\nu = 2 \frac{h\nu_{jk} B_{jk}}{c} \frac{h\nu_{jk}^3}{\sigma c^2} \frac{b_k}{b_j} \mathrm{e}^{-h\nu_{jk}/kT}.$$

Using (3.58) and (3.59) in this expression, it is easy to demonstrate that

$$S_\nu = \frac{2h\,\nu_{jk}^3}{c^2}\left[\frac{b_j}{b_k}\,\mathrm{e}^{h\nu_{jk}/kT} - 1\right]^{-1}. \tag{3.71}$$

For the case $h\nu_{jk}/kT \ll 1$, we can use the Rayleigh–Jeans distribution (3.18b), and it can be shown that

$$S_\nu = \frac{2\nu_{jk}^2\,kT}{c^2}\,\frac{b_k}{b_j}\,\frac{1}{\chi}, \tag{3.72}$$

where χ is defined in (3.67).

Exercises

3.1 Show that the free electrons partition function per unit volume is given by relation (3.14).

3.2 Show that the Doppler profile (3.48) is normalized.

3.3 Show that the Lorentz profile (3.50) is normalized and that in this case the FWHM is $\Delta\nu_{\mathrm{h}} = \Gamma_k/2\pi$.

3.4 Collisional processes, radiation absorption, recaptures from the continuum, etc., maintain the population of 10^8 atoms at a certain energy level k. The Einstein emission coefficient relative to a lower level j is $A_{kj} \simeq 10^8\ \mathrm{s}^{-1}$. (a) What is the number of spontaneous emissions per second for level j? (b) What is the radiative lifetime of level k relatively to emissions to level j?

3.5 Calculate the oscillator strengths f_{jk} and f_{kj} for the neutral H 21 cm line, which has an emission coefficient $A_{kj} \simeq 2.9 \times 10^{-15}\ \mathrm{s}^{-1}$.

Bibliography

Drake, G.W.F.: Atomic, Molecular and Optical Physics Handbook. American Institute of Physics, New York (1996). Includes detailed tables of atomic and molecular data

Finn, G.D., Mugglestone, D.: Mon. Notices Roy. Tables of the line broadening function H(a,v). Astron. Soc. **129**, 221 (1965). Tables of the line broadening function H(a,v). Includes tables for the Hjerting function under several conditions. Other tabulations may be found in Harris, D.L. 1948. Astrophys. J. 108:112; and Hjerting, F. 1938. Astrophys. J. 88:508

Jefferies, J.T.: Spectral Line Formation. Blaisdell, Waltham (1968). Complete treatment of spectral line formation in astrophysical conditions and broadening processes

Lang, K.R.: Astrophysical Formulae. Springer, Berlin (1999). Referred to in Chapter 2. Includes basic equations of TE, Einstein coefficients definitions, and references to original works

Maciel, W.J.: Introdução à Estrutura e Evolução Estelar. Edusp, São Paulo (1999). Introduction to stellar structure and evolution, including a discussion on the main radiation field concepts, such as intensity and flux

Mihalas, D.: Stellar Atmospheres. W.H. Freeman, San Francisco (1978). Discussion on radiative transfer in stellar atmospheres and spectral line profiles. Includes problems related to line formation in atmospheres and expanding envelopes

Morton, D.C., Dinerstein, H.L.: Astrophys. J. **204**, 1 (1976). Tables with oscillator strength values for lines of astrophysical interest. See also Astrophys. J. Suppl. vol. 26, p.333, 1973

Reif, F.: Fundamentals of Statistical and Thermal Physics. McGraw-Hill, New York (1965). Basic text of thermodynamics and statistical physics, including a good discussion of TE equations

Rybicki, G.B., Lightman, A.P.: Radiative Processes in Astrophysics. Wiley, New York (1979). Referred to in Chapter 2. Good discussion of TE equations, absorption and emission coefficients, and radiative transfer

Shu, F.H.: The Physics of Astrophysics, vol. 1. University Science Books, Mill Valley (1991). Advanced treatment of radiative transfer with astrophysical applications

Unsöld, A.: Physik der Sternatmosphären. Springer, Berlin (1955). Classical text on stellar atmospheres, with a good discussion of spectral line formation in stars and of the Voigt profile

Chapter 4
Interstellar Emission and Absorption Lines

4.1 Introduction

The formation of interstellar emission and absorption lines depends, in general, on the excitation conditions of the atoms. However, some lines can be treated without a deep knowledge of these conditions. The neutral H 21 cm line, for instance, is produced by transitions between collisionally occupied states, which translates into applying the Boltzmann excitation equation, taking the excitation temperature to be equal to the kinetic temperature of the gas. Other interstellar lines come from transitions between the ground state and the first states of excitation (resonance lines), and observation of the relative intensities of these lines allows us to determine the physical conditions of the interstellar region without a detailed knowledge of the excitation conditions. As a means of applying the theory sketched in Chap. 3, we will consider some examples of emission and absorption lines observed in the interstellar medium.

4.2 Optical Recombination Lines

Optical recombination lines are lines produced in the recombination process between electrons and ions, in particular the Balmer series lines, which come from electron–proton recombinations. A more detailed analysis of the ionization equilibrium will be given in Chap. 6. These lines can be observed in H II regions related to hot stars or in the general interstellar medium. In this latter case, the emission apparently coincides with the 21 cm emission, tracing the Galaxy's spiral arms.

W.J. Maciel, *Astrophysics of the Interstellar Medium*,
DOI 10.1007/978-1-4614-3767-3_4,

4.2.1 Recombination Coefficient

In the process of radiation emission by recombination, the line emissivity, from (3.36), is

$$\int j_\nu \, d\nu = \frac{1}{4\pi} h\nu_{mn} \, n_m(\mathrm{H}^0) \, A_{mn} \tag{4.1}$$

(units: erg cm^{-3} s^{-1} sr^{-1}), where we use total quantum numbers m and n instead of k and j and consider H lines. According to (3.27), $n_m(\mathrm{H}^0)$ may be expressed in terms of $n(\mathrm{H}^+)n_e$ or $n_p n_e$:

$$n_m(\mathrm{H}^0) = \frac{b_m \, g_m \, e^{-E_1/m^2kT} n_p \, n_e}{f_p f_e}, \tag{4.2}$$

where E_1/m^2 is the energy of level m, being zero in the infinity. Let us define the *effective recombination coefficient*

$$\begin{aligned} \alpha_m &= \frac{b_m \, g_m \, A_{mn} \, e^{-E_1/m^2kT}}{f_e} \\ &= 4.14 \times 10^{-16} \, b_m m^2 \, A_{mn} \, e^{158{,}000/m^2T} T^{-3/2} \end{aligned} \tag{4.3}$$

(units: cm^3 s^{-1}). Here, we use relation (3.14), and we have $E_1 = -13.6$ eV and $g_m = 2\,m^2$. A typical value for this coefficient is $\alpha \simeq 3 \times 10^{-14}$ cm^3 s^{-1} for the Hβ line with temperature $T \simeq 10^4$ K. Since $A_{mn} \simeq 8.4 \times 10^6$ s^{-1}, we have approximately $b \simeq 0.20$. From (4.1),

$$4\pi \int j_\nu \, d\nu = h\,\nu_{mn} \, \alpha_{mn} \, n_p \, n_e \tag{4.4}$$

because $f_p = 1$. The quantity $\alpha_{mn} n_p n_e$ corresponds to the number of recombinations per cubic centimeter per second, so that the total emissivity can be calculated by multiplying this number by the transition energy. From (4.4), the optical recombination line intensity is proportional to $n_p n_e$ and is also a function of temperature. The intensity ratio between two lines of the Balmer series is only a function of temperature as the terms in $n_p n_e$ cancel out. Thus, once the line intensities are known, we can estimate T in the emissive region. However, since the interstellar extinction affects each line differently, observations can be used to determine the extinction.

4.2.2 Recombination Line Analysis

We will now obtain an expression for the radiation specific intensity I_ν, considering the absorption due to emission of optical lines to be negligible. In this case, the transfer equation solution can be simplified, and intensity I_ν, corrected from interstellar extinction, may be written as [see for instance (3.23)]

$$I_\nu \propto \int \frac{j_\nu}{k_\nu} e^{-\tau_\nu} d\tau_\nu \propto \int \frac{j_\nu}{k_\nu} \mathrm{e}^{-\tau_\nu}\, k_\nu\, \mathrm{d}s \propto \mathrm{e}^{-\tau_\nu} \int j_\nu\, \mathrm{d}s,$$

that is,

$$\int I_\nu\, \mathrm{d}\nu = \int_{\text{line}} \mathrm{d}\nu \int_0^L j_\nu\, \mathrm{d}s, \tag{4.5}$$

where the last integration is done along the line of sight, of length L. From (4.4) and (4.5), assuming α_{mn} and $n_\mathrm{p}/n_\mathrm{e}$ constants are along the line of sight,

$$\begin{aligned} \int I_\nu\, \mathrm{d}\nu &= \frac{h\nu_{mn}}{4\pi} \int_0^L \alpha_{mn}\, n_\mathrm{p}\, n_\mathrm{e}\, \mathrm{d}s \\ &= \frac{h\nu_{mn}\, \alpha_{mn}}{4\pi} \frac{n_\mathrm{p}}{n_\mathrm{e}} \int_0^L n_\mathrm{e}^2\, \mathrm{d}s. \end{aligned} \tag{4.6}$$

Using the emission measure [see Chap. 2, (2.11)],

$$\mathrm{EM} = \int_0^L n_\mathrm{e}^2\, \mathrm{d}s, \tag{4.7}$$

we have

$$\int I_\nu\, \mathrm{d}\nu = 2.46 \times 10^{17} h\, \nu_{mn}\, \alpha_{mn}\, \frac{n_\mathrm{p}}{n_\mathrm{e}} \mathrm{EM}, \tag{4.8}$$

where the following units are used: erg cm^{-2} s^{-1} sr^{-1} and the emission measure is expressed in pc cm^{-6}. Therefore, the observation of optical recombination line intensities (e.g., Hα and Hβ) allows the determination of the emission measure, which gives us an idea of the electron density in the emitter medium. Roughly, $n_\mathrm{p}/n_\mathrm{e} \simeq 1$ and the other quantities in the second term of (4.8) can be determined by knowing the atomic structure and setting the temperature.

4.2.3 Results

For normal H II regions, formed around O type stars, the emission measure has typical values EM ~ 10^3–10^4 pc cm^{-6}. Lower values, of the order of 5 pc cm^{-6} or higher, are estimated for diffuse galactic emission, and certain ionized nebulae, such as the Orion Nebula, have higher emission measures, EM ~ 10^7 pc cm^{-6}.

Finally, once the emission measure is determined, we can obtain the rms mean value $\langle n_\mathrm{e}^2 \rangle^{1/2}$, if we know the mean size of the emitter region. For typical H II regions, we have

$\langle n_e^2 \rangle^{1/2} \sim 10 - 100\,\text{cm}^{-3}$. In the case of the Orion Nebula, $\langle n_e^2 \rangle^{1/2} \sim 10^3\,\text{cm}^{-3}$, which corresponds to an emitter region of about 10 pc.

Diffuse galactic emission at Hα with emission measure EM ~ 3–15 pc cm^{-6} involves dimensions of the order of 1 kpc, which gives us $\langle n_e^2 \rangle \sim 0.005 - 0.015\,\text{c}$ m^{-6} and $\langle n_e^2 \rangle^{1/2} \sim 0.07 - 0.12\,\text{cm}^{-3}$. This emission is probably produced by a large number of H II regions and planetary nebulae, associated with OB stars.

4.3 The H 21 cm Emission Line

4.3.1 Introduction

The H 21 cm line is produced by a radiative transition between two hyperfine levels of the ground state $n = 1$. These levels correspond to the two possible electron *spin* orientations, parallel (higher level) or antiparallel (lower level), relative to the nucleus' *spin*. The two levels have an energy difference of $h\nu_{jk} = 5.9 \times 10^{-6}$ eV, and the emitted radiation frequency is $\nu_{jk} = 1.420 \times 10^9$ Hz = 1,420 MHz, corresponding to a wavelength $\lambda_{jk} = 21.11$ cm. The spontaneous transition probability is $A_{kj} = 2.9 \times 10^{-15}\,\text{s}^{-1}$, which is very low, much lower than the electron collisions rate, of the order of $10^{-9}\,n_e$. This low transition probability corresponds to a very long lifetime for the high level, $t_k \sim 1/A_{kj} \sim 10^7$ years. Given these conditions, the levels are collisionally occupied, with typical TE populations. The relative populations are given by the Boltzmann Equation (3.7):

$$\frac{n_k}{n_j} = \frac{g_k}{g_j}\,\text{e}^{-h\nu_{jk}/kT}. \tag{4.9}$$

In this case,

$$\frac{h\nu_{jk}}{kT} = \frac{1.44}{\lambda_{jk}(\text{cm})\,T(K)} = \frac{0.07}{T} \ll 1, \tag{4.10}$$

and for typical interstellar temperatures,

$$\frac{n_k}{n_j} \simeq \frac{g_k}{g_j} = \frac{3}{1} = 3, \tag{4.11}$$

that is, about 3/4 of the H atoms are in high-level k and 1/4 are in low-level j and the level populations do not depend on temperature.

4.3.2 Brightness Temperature

The radiation intensity can be obtained from the transfer equation solution (3.23), by replacing j_ν/k_ν with $B_\nu(T)$ and assuming T to be independent of τ_ν (homogeneous case):

$$I_\nu = I_\nu(0)\,\mathrm{e}^{-\tau_{\nu r}} + B_\nu(T)\int_0^{\tau_{\nu r}} \mathrm{e}^{-\tau_\nu}\,\mathrm{d}\tau_\nu. \tag{4.12}$$

Thus,

$$I_\nu = B_\nu(T)(1-\mathrm{e}^{-\tau_{\nu r}}), \tag{4.13}$$

where we neglect the radiation intensity falling on the region opposite to the observer and ν_τ is again the total optical depth of the emitting region. Intensity $B_\nu(T)$ is given by the Rayleigh-Jeans approximation (3.18b), that is, $B_\nu(T)$ is proportional to the gas kinetic temperature T. It is convenient to define the *brightness temperature* T_b as the temperature for which $B_\nu(T_\mathrm{b}) = I_\nu$. Using (4.13) and (3.18b),

$$T_\mathrm{b} = T(1-\mathrm{e}^{-\tau_{\nu r}}). \tag{4.14}$$

4.3.3 H Column Density Calculation

The total optical depth $\tau_{\nu r}$ is given by (3.38):

$$\tau_{\nu r} = N_j\,\sigma\,\phi(\Delta\nu), \tag{4.15}$$

where the $\phi(\Delta\nu)$ profile is essentially determined by the motion of the atoms responsible for the emission. Using (3.58) or (3.68), with $\chi \simeq 1$,

$$\tau_{\nu r} \simeq N_j\,\sigma_u\,\frac{h\nu_{jk}}{kT}\,\phi(\Delta\nu) \tag{4.16}$$

because $h\nu_{jk}/kT \ll 1$. Using (3.59), (3.33), (3.41), and (3.43),

$$\begin{aligned}\tau_{\nu r} &= \frac{hc^3\,A_{kj}\,g_k}{32\,\pi\,k\,g_j\,\nu_{jk}^2}\,\frac{N(\mathrm{HI})f(v)}{T}\\ &= 5.5\times 10^{-14}\frac{N(\mathrm{HI})f(v)}{T},\end{aligned} \tag{4.17}$$

where $N(\mathrm{HI}) = 4N_j$ is the total number of atoms of neutral H in a column with section 1 cm^2 along the line of sight. In a similar way to what was done to obtain (3.40), the column density of neutral H may be obtained by integrating the optical depth for all atomic velocities:

$$\int T\,\tau_{\nu r}\,\mathrm{d}v = 5.5\times 10^{-14}\,N(\mathrm{HI})\int f(v)\,\mathrm{d}v. \tag{4.18}$$

Using (4.14),

$$N(\mathrm{HI}) = 1.8\times 10^{18}\int T_{\mathrm{b}}(v)\Big[\frac{\tau_{\nu r}}{1-\mathrm{e}^{-\tau_{\nu r}}}\Big]\mathrm{d}v, \tag{4.19}$$

where we clearly see T_{b} dependency on v and measure v in km s^{-1}. Note that (4.19) is valid for an homogeneous medium where T does not depend on the position along the line of sight. For $\tau_{\nu r} \ll 1$, that is, for the optically thin or transparent case,

$$\frac{\tau_{\nu r}}{1-\mathrm{e}^{-\tau_{\nu r}}} \simeq 1,$$

and we obtain from (4.19)

$$N(\mathrm{HI}) = 1.8\times 10^{18}\int T_{\mathrm{b}}(v)\,\mathrm{d}v. \tag{4.20}$$

where $N(\mathrm{HI})$ is in cm^{-2}, T_{b} is in K, and v is in km s^{-1}. As we can see, for this case, it is only possible to derive information about column density and not temperature.

Measurements of T_{b} as a function of v for different positions in the Galaxy, characterized by galactic coordinates l and b, allow us to determine the neutral H column density for the optically thin case. These measurements are particularly important for mapping the hydrogen along the Galaxy's spiral arms. Figure 4.1

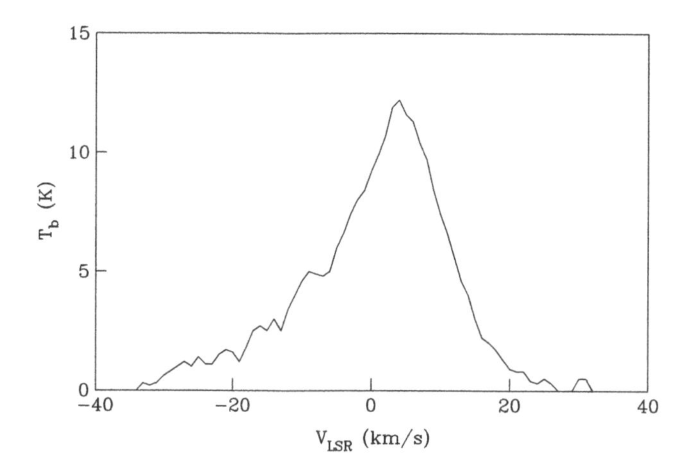

Fig. 4.1 Brightness profile in the direction of the planetary nebula NGC 2371

shows a typical profile, in the direction of planetary nebula NGC 2371, with coordinates $l \simeq 189°$ and $b \simeq 19°$, obtained from Berkeley survey data. The figure shows brightness temperature variation for the considered region as a function of velocity relative to the Local Standard of Rest (LSR). The LSR is a reference system suitable for the solar neighborhood, situated on the galactic plane at a distance R_0 between 6 and 8 kpc of the center, with mean value $R_0 \simeq 7.6$ kpc. At least three features can be identified in the figure, centered on velocities $v_{LSR} \simeq 4$ km s^{-1}, -10 km s^{-1}, and -15 km s^{-1}, with a possible feature at -25 km s^{-1}. These features are identified with different interstellar clouds, situated in the considered direction. Applying (4.20) to this profile, we obtain a column density $N(\text{HI}) \simeq 4.4 \times 10^{20}$ cm^{-2}.

To determine the particle volumetric density, it is necessary to know the position of the H atoms and thus of the distances involved. These distances can be geometrically determined for internal regions of the solar circle, or assuming a galactic differential rotation model, that is, using a rotation curve. In this case, results are limited by the presence of random components and other deviations of the rotation circular velocity.

4.3.4 *Results*

The distribution of neutral H obtained from the 21 cm line is quite similar to the distribution of the stellar optical emission in galaxies. In a general way, gas emission extends beyond optical emission and presents a higher concentration in the spiral arms region. These results further allow to determine, from the Doppler effect, the regions where rotation radial velocities are approaching or moving away relative to the Sun.

The galactic emission at the 21 cm line has a distribution comparable to the one of the optical emission, with a concentration in the galactic plane similar to the interstellar dust infrared emission, showing a link between atomic gas and interstellar dust grains.

For volumetric density, we obtain $n(\text{HI}) \simeq 0.7$ cm^{-3} for the solar neighborhood, decreasing for larger distances to the Galactic center. In the spiral arms, the density is higher, $n(\text{HI}) \simeq 1$–2 cm^{-3}. The neutral H layer thickness varies with R, being of the order of 250 pc for regions near the solar neighborhood. The neutral H total mass is estimated to be $5 \times 10^9\ M_\odot$, corresponding to about 5 % of the Galaxy's total mass.

Detailed studies of high-resolution emission of 21 cm for some sky regions are consistent with the existence of a background diffuse emitter region, where density contrasts (clouds) occur. The mean characteristics of these components are $n(\text{HI}) \sim 0.2$ cm^{-3} for the background and $n(\text{HI}) \sim 2$–4 cm^{-3} for the cloud region, where $N(\text{HI}) \sim 10^{19}$–$10^{21}$ cm^{-2}, supposing dimensions of the order of 3–15 pc. Meanwhile, more recent works show an extremely filamentary structure, which is corroborated in Hubble Space Telescope (HST) images, for instance. So these "typical" dimensions should be viewed with caution.

Only low-resolution data are available for most parts of the sky, and thus, statistical analysis is required. In this case, observational data are, in general, consistent with a two component medium: *clouds* and *intercloud medium*. If this is really the case, then the parameters' mean values of these regions are given by

Clouds: $T \simeq 60$ K, $n(\mathrm{HI}) \simeq 0.1-10\ \mathrm{cm}^{-3}$ and

Intercloud medium: $T \gg 60$ K, $n(\mathrm{HI}) < 0.1\ \mathrm{cm}^{-3}$.
The dimensions of the clouds are variable, of the order of 10 pc, typically. Naturally, the above exceptions also apply to these results, particularly taking into account that much denser clouds exist in the interstellar medium, with an H nuclei density of the order of, or superior to, $10^4\ \mathrm{cm}^{-3}$, where, besides dust, relatively complex molecules are observed, as we shall see in Chap. 10.

4.4 The H 21 cm Absorption Line

4.4.1 Observations

The H 21 cm line can be seen in absorption when the radio telescope is pointing toward galactic and extragalactic radio sources. Generally, these sources have very high brightness temperatures in comparison with H I clouds diffuse emission, so that observation of the line profile allows us to determine the optical depth $\tau_{\nu r}$ of the absorbing region. Generally, the observed profiles give us the antenna temperature T_A, which depends on the instrument properties and is proportional to the brightness temperature T_b, as a function of the radial velocity relative to the LSR, as seen in Fig. 4.1.

To estimate the kinetic temperature T and the neutral H column density N(HI) of the H I region, we need to know the emission profile of this region, which is not accurately possible, since we obviously have the radio source in the considered direction. We can, however, observe emission coming from the source surrounding regions and obtain mean values for the $T_b \times v$ distribution of the H I region.

4.4.2 Results

The H I region kinetic temperature can be determined from (4.14), using the brightness temperature of the surrounding regions by an interpolation process. The neutral H column density can be calculated from (4.19), using the computed optical depth from 21 cm absorption line observations. The results for clouds with different optical depths show a mean value of $T \simeq 80$ K. Besides diffuse H I clouds, we also observe a background emission at 21 cm with a low optical depth, which temperature is much higher, $T \gtrsim 1{,}000$ K. This emission probably comes from the intercloud medium, where densities are much lower than the ones for interstellar clouds.

The H I clouds column densities (regions where the central optical depth of the line is $\tau_0 \gtrsim 0.2$) are similar to the ones obtained for the emission line. Observations of clouds in external galaxies indicate $N(\mathrm{HI}) \simeq 3 \times 10^{20}\ \mathrm{cm}^{-2}$. For the observed sources in our Galaxy, we have the mean value

$$\frac{\langle N(\mathrm{HI})\rangle}{\langle L\rangle} \simeq 1.2 \times 10^{21}\ \mathrm{cm}^{-2}\ \mathrm{kpc}^{-1}, \tag{4.21}$$

which corresponds to $T \simeq 80$ K and $n(\mathrm{HI}) \simeq 0.4\ \mathrm{cm}^{-3}$. This value should be compared with the mean $n(\mathrm{HI}) \simeq 0.3\ \mathrm{cm}^{-3}$, obtained from emission line measurements. Just as in the previous case, we observe great variation in the volumetric density, which can reach up to about $100\ \mathrm{cm}^{-3}$. The mean distance of the observed galactic sources is $\langle L\rangle \simeq 2.6$ kpc. Supposing that the column density in each cloud is $3 \times 10^{20}\ \mathrm{cm}^{-3}$, this value indicates the existence of about four clouds per kpc in our Galaxy.

Observations of the 21 cm absorption line in the direction of strong radio sources can be used to estimate the magnetic field in the clouds, from the components separation produced by the Zeeman effect. In this case, separation $\Delta\nu_B$ is given by

$$\Delta\nu_B = \frac{eB}{2\pi\, m_e\, c} = 2.8 \times 10^6\ \mathrm{B\ Hz}, \tag{4.22}$$

where B is the interstellar magnetic field component parallel to the line of sight. Results show that $B \lesssim 10\ \mu\mathrm{G}$, in agreement with determinations of the magnetic field from interstellar polarization, as we shall see in Chap. 9.

4.5 Broad Absorption Lines

4.5.1 Broad Absorption Lines Intensity: H and H_2

An absorption line profile depends on the absorbing particle density, which is clearly indicated by curves of growth, as we shall see in the next section. In a general way, the Doppler profile (3.48) dominates the central regions of the line, where the absorption is higher. However, if the density of the absorbing particles is high enough, the profile of the line wings will essentially be given by the natural profile or the Lorentz profile (3.50). In these conditions, the intensity I_ν in the line wings where $\Delta\nu \gg \Gamma/4\pi$ is

$$I_\nu = I_\nu(0)\, \mathrm{e}^{-\tau_{\nu r}}, \tag{4.23}$$

where

$$\tau_{\nu r} = N_j\, \sigma\, \phi(\Delta\nu) = N_j \frac{\pi e^2}{m_e c} f_{jk} \frac{\Gamma_k/4\pi^2}{(\Delta\nu)^2}, \tag{4.24}$$

neglecting induced emissions (see 3.38, 3.50, and 3.60). The optical depth can be expressed in terms of $\Delta\lambda$, the wavelength separation relative to the center of the line. If we consider that

$$\Delta\nu = \nu - \nu_{jk} = c\left(\frac{1}{\lambda} - \frac{1}{\lambda_{jk}}\right) = c\,\frac{\lambda_{jk} - \lambda}{\lambda\,\lambda_{jk}} \simeq \frac{c\,\Delta\lambda}{\lambda^2}, \tag{4.25}$$

we have

$$\tau_{\nu r} = \frac{e^2\,\lambda^4\,f_{jk}\,N_j\,\Gamma_k}{4\,\pi\,m_e\,c^3(\Delta\lambda)^2}, \tag{4.26}$$

or even

$$\frac{I_\nu}{I_\nu(0)} = \exp\left[-\frac{e^2\,\lambda^4\,f_{jk}\,N_j\,\Gamma_k}{4\,\pi\,m_e\,c^3\,(\Delta\lambda)^2}\right]. \tag{4.27}$$

Therefore, the observation of broad absorption line intensities in the direction of a few stars allows the determination of column density N_j. This is possible for H, using the Lyman-α line ($\lambda = 1{,}215.67$ Å), and for the H_2 molecule, using the Lyman band lines ($\lambda < 1{,}100$ Å) observed in the direction of very hot stars with high ultraviolet emission.

4.5.2 Example: Interstellar Lyman-α Line

We will initially define the line equivalent width W_λ according to Fig. 4.2:

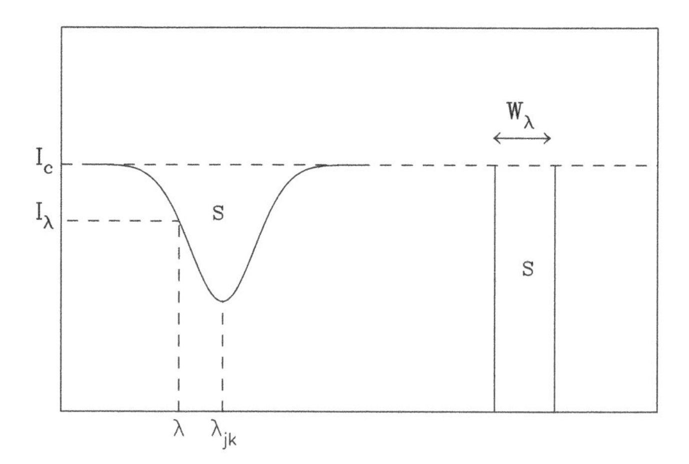

Fig. 4.2 Definition of the equivalent width

$$W_\lambda = \int \frac{I_c - I_\lambda}{I_c} \mathrm{d}\lambda = \int \left(1 - \frac{I_\lambda}{I_c}\right) \mathrm{d}\lambda, \tag{4.28}$$

where I_c is the continuum intensity and the integral is calculated along the line. Defining the line depth as

$$A_\lambda = 1 - \frac{I_\lambda}{I_c}, \tag{4.29}$$

we have in the center of the line $I_\lambda \ll I_c$ and $A_\lambda \to 1$, whereas in the line wings, $I_\lambda \simeq I_c$ and $A_\lambda \to 0$. In terms of line depth, the equivalent width can be written as

$$W_\lambda = \int A_\lambda \, \mathrm{d}\lambda. \tag{4.30}$$

Considering the transfer equation solution (3.23) with $I_\nu(0) = 0$ and letting τ_ν be the optical depth,

$$W_\lambda = \int (1 - \mathrm{e}^{-\tau_\nu}) \, \mathrm{d}\lambda = \frac{\lambda_{jk}^2}{c} \int (1 - \mathrm{e}^{-\tau_\nu}) \, \mathrm{d}\nu. \tag{4.31}$$

The equivalent width gives a direct measurement of the total energy absorbed in the spectral line and depends on the particle density at the levels responsible for the absorption.

We will now rewrite (4.24) in the form

$$\tau_{\nu r} = \frac{1}{u^2}, \tag{4.32}$$

where

$$u^2 = \frac{4\pi m_e c (\Delta\nu)^2}{N_j e^2 f_{jk} \Gamma_k}, \tag{4.33}$$

$$u = \left(\frac{4\pi m_e c}{N_j e^2 f_{jk} \Gamma_k}\right)^{1/2} \Delta\nu, \tag{4.34}$$

$$\mathrm{d}u = \left(\frac{4\pi m_e c}{N_j e^2 f_{jk} \Gamma_k}\right)^{1/2} \mathrm{d}\nu. \tag{4.35}$$

Considering (4.31) and (4.35),

$$W_\lambda = \frac{\lambda_{jk}^2}{c} \left(\frac{N_j e^2 f_{jk} \Gamma_k}{4\pi m_e c}\right)^{1/2} \int_{-\infty}^{\infty} (1 - e^{-1/u^2}) \, \mathrm{d}u. \tag{4.36}$$

The integral in (4.36) is equal to $2\sqrt{\pi}$, so that

$$W_\lambda = \frac{\lambda_{jk}^2}{c}\left(\frac{N_j e^2 f_{jk}\Gamma_k}{4\pi m_e c}\right)^{1/2} 2\sqrt{\pi}. \tag{4.37}$$

Considering a two-level atom, we have

$$\Gamma_k = A_{kj}, \tag{4.38}$$

and using (4.38) and (3.64) in (4.37), we obtain

$$W_\lambda = \frac{2\pi e^2}{m_e c^2} f_{jk}\,\lambda_{jk}\left(2\frac{g_j}{g_k}N_j\right)^{1/2}. \tag{4.39}$$

Applying (4.39) to the Lyman-α line and remembering that excited state level $n = 2$ corresponds to a doublet with levels $2^2P_{1/2}$ and $2^2P_{3/2}$, we have $g_j = 2$, $g_k = 6$, and $g_j/g_k = 1/3$. Using also $f_{jk} = 0.4162$ and $\lambda_{jk} = 1{,}215.67$ Å, we have

$$W_\lambda = 7.3 \times 10^{-10}\sqrt{N_j}, \tag{4.40}$$

$$N_j = 1.9 \times 10^{18}\, W_\lambda^2, \tag{4.41}$$

where the equivalent width W_λ is in Å and the column density $N_j \simeq N(\mathrm{HI})$ is in cm^{-2}.

An example of the Lyman-α absorption line of interstellar origin is shown in Fig. 4.3, where we present an ultraviolet spectrum segment in the direction of

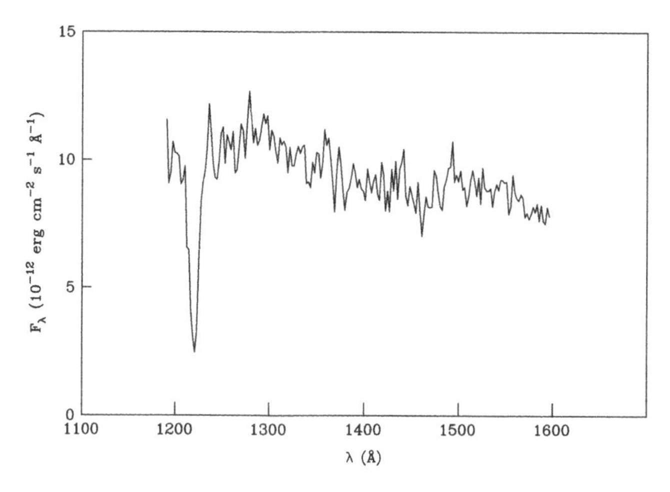

Fig. 4.3 IUE UV spectra in the direction of the planetary nebula NGC 2392

the central star of planetary nebula NGC 2392, obtained with the *International Ultraviolet Explorer* (IUE). The star is very hot, with a temperature above 70,000 K, and thus presents considerable emission in this part of the spectrum.

4.5.3 Results

Analysis of H broad interstellar lines allows us to have estimates of the column density N(HI) in the interval 10^{19}–10^{21} cm^{-2} and of the mean volumetric density along the line of sight n(HI) ~ 0.01–2 cm^{-3}. The equivalent width can be measured in Fig. 4.3, resulting in $W_\lambda \sim 10$ Å, which corresponds to $N(\mathrm{HI}) \simeq 1.9 \times 10^{20}$ cm^{-2} using (4.41). Assuming that the distance to the nebula is $d \simeq 1.0$ kpc and that the gas is uniformly distributed, we obtain a mean volumetric density $N(\mathrm{HI})/d \simeq 0.1$ cm^{-3}. Considering a "typical" cloud with dimensions 10 pc, the resulting density is 6 cm^{-3}.

One of the most important results from the H column density analysis is the observed correlation between this density and the star reddening, measured by the color excess in magnitudes

$$E(B-V) = (B-V) - (B-V)_0, \tag{4.42}$$

where $(B–V)$ is the observed color index and $(B–V)_0$ is the intrinsic color index, corresponding to a star without reddening of the same spectral type. This correlation can be approximated by

$$N_\mathrm{H} \simeq K\,E(B-V), \tag{4.43}$$

where constant K, obtained for different directions, has a mean value

$$K \simeq 6 \times 10^{21}\ \mathrm{cm^{-2}\,mag^{-1}}, \tag{4.44}$$

valid for $0.0 \lesssim E(B–V) \lesssim 0.6$. In relation (4.43), we consider the sum of N(HI) given by the Lyman-α line and 2N(H_2) given by H_2 molecular lines. This corresponds to the number of H nuclei that we simply represent by N_H (or n_H). This correlation points to the connection between gas and dust in interstellar clouds and, as we shall see in Chap. 9, can be used to determine the ratio between the masses of gas and dust in the interstellar medium.

Determinations of H_2 column density indicate that the abundance of this molecule lies between 10 and 70 % of the total H in the line of sight, being higher in the direction of stars with more reddening. In H_2 Lyman bands, the populations of the two lower rotational levels $J = 0$ (parahydrogen) and $J = 1$ (orthohydrogen) can be roughly given by the Boltzmann equation if the lines are strong enough. However, the observation of line intensity allows to estimate the absorbing region temperature, whose mean value is $T \simeq 80$ K, in agreement with the mean temperature determined from the neutral H 21 cm line. The mean column density is $N_H \simeq 5 \times 10^{20}$ cm^{-2}, which is similar to the value obtained from the 21 cm line.

4.5.4 *Lines of Highly Ionized Elements*

An interesting result related to broad interstellar lines concerns the discovery of lines of highly ionized elements, such as O VI (1,031.9, 1,037.6 Å), Si IV (1,400 Å), and C IV (1,550 Å) in the spectrum of hot stars. These lines suggest the existence of a *coronal gas* with very high temperatures, $T \sim 10^6$ K, surrounding the diffuse interstellar gas. The coronal gas estimated density is at least a thousand times lower than the one of the other interstellar regions, and the ionization processes are of collisional origin, in opposition to photoionization of the H II gas. Even though there has been evidence for the existence of this gas for 20 years, its properties are not well understood, because it is not detectable in optical, radio, or infrared wavelengths.

4.6 Curve of Growth

The determination of stellar and interstellar abundances frequently uses a relation between the spectral line equivalent width and the number of atoms in the energy level where the absorption took place. This relation between W_λ (or W_λ/λ_{jk}) and the effective number of absorbing atoms $N_j f_{jk}$ defines the so-called *curve of growth*, whose general expression depends on the solution of the radiative transfer equation in the spectral line. At the same time, a simple expression can be obtained for weak lines, where the optical depth is $\tau_\nu \ll 1$ along the whole line, both in the Doppler center and in the wings. In this case, from (4.31),

$$\frac{W_\lambda}{\lambda_{jk}} = \frac{\lambda_{jk}}{c} \int (1 - e^{-\tau_\nu})\, d\nu \simeq \frac{\lambda_{jk}}{c} \int \tau_\nu\, d\nu. \tag{4.45}$$

Using (4.15),

$$\frac{W_\lambda}{\lambda_{jk}} = \frac{\lambda_{jk}}{c} N_j \sigma \; (\tau_\nu \ll 1). \tag{4.46}$$

Neglecting induced emissions, $\sigma = \sigma_u$, and using (3.60),

$$\frac{W_\lambda}{\lambda_{jk}} = \frac{\pi e^2}{m_e c^2} N_j f_{jk}\, \lambda_{jk}, \tag{4.47}$$

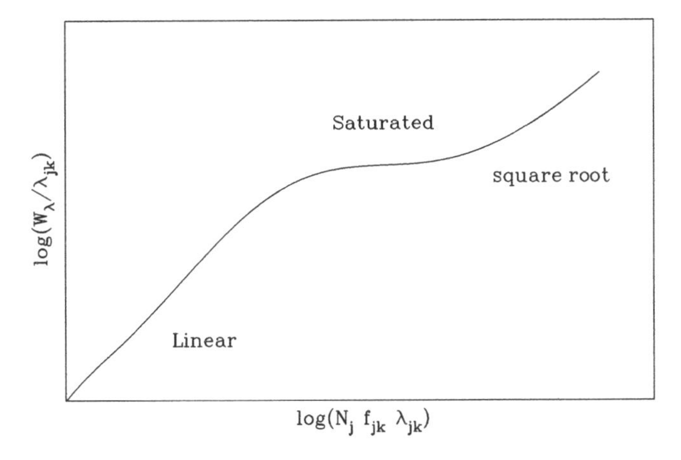

Fig. 4.4 A schematic view of the curve of growth

which is therefore valid for *weak lines* with $\tau_\nu \ll 1$. This means that W_λ/λ_{jk} is proportional to $N_j f_{jk}$, corresponding to the *linear part* of the curve of growth, as can be seen in the schematic curve of Fig. 4.4.

In the more general case, the integral (4.31) involving τ_ν is more complex and depends on the $\phi(\Delta\nu)$ profile assumed. In the case of the Doppler profile (3.44), (4.31) becomes

$$\frac{W_\lambda}{\lambda_{jk}} = \frac{\lambda_{jk}}{c} \int \left\{ 1 - \exp\left[-N_j \frac{\pi e^2}{m_e c} f_{jk} \frac{\lambda_{jk}}{b\sqrt{\pi}} \mathrm{e}^{-(v/b)^2} \right] \right\} \mathrm{d}v. \tag{4.48}$$

If τ_0 is the optical depth in the line center,

$$\begin{aligned} \tau_0 = \tau(\Delta v = 0) &= N_j \frac{\pi e^2}{m_e c} f_{jk} \frac{\lambda_{jk}}{b\sqrt{\pi}} \\ &= 1.50 \times 10^{-2}\, b^{-1}\, N_j\, \lambda_{jk} f_{jk}, \end{aligned} \tag{4.49}$$

where b is in cm s^{-1}, N is in cm^{-2}, and λ is in cm, and so

$$\frac{W_\lambda}{\lambda_{jk}} = \frac{\lambda_{jk}}{c} \int \left\{ 1 - \exp\left[-\tau_0\, \mathrm{e}^{-(v/b)^2} \right] \right\} \mathrm{d}v. \tag{4.50}$$

Defining

$$F(\tau_0) = \int_0^\infty \left[1 - \exp\left(-\tau_0\, \mathrm{e}^{-x^2} \right) \right] \mathrm{d}x, \tag{4.51}$$

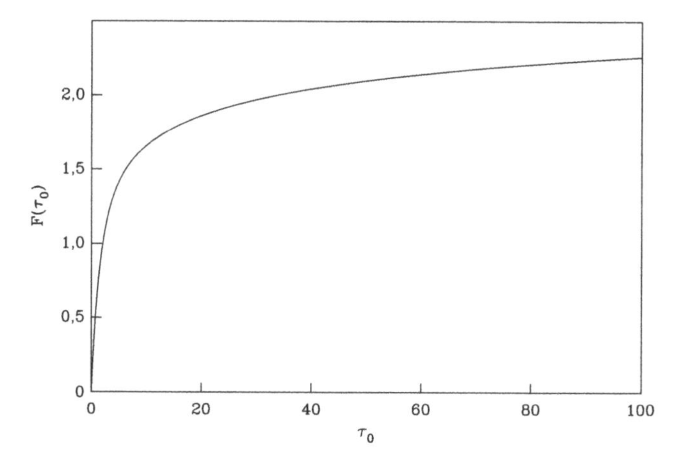

Fig. 4.5 The function F as defined by (4.51)

we obtain

$$\frac{W_\lambda}{\lambda_{jk}} = \frac{2bF(\tau_0)}{c} \text{ (Doppler profile).} \tag{4.52}$$

Figure 4.5 shows the behavior of function $F(\tau_0)$ for $\tau_0 < 100$. When $\tau_\nu \ll 1$, it can be shown that (4.52) becomes (4.47).

We will now consider relatively strong lines, where the number of absorbing particles along the line of sight is high enough so that $\tau_0 \gg 1$, but $\tau_\nu \lesssim 1$ in the line wings.

Let $\nu_{jk} \pm \delta\nu$ be the transition frequency where optical depth $\tau(\nu_{jk} \pm \delta\nu) \simeq 1$. From (4.31), we may write

$$\frac{W_\lambda}{\lambda_{jk}} \simeq \frac{\lambda_{jk}}{c} \int_{v_{jk}-\delta v}^{v_{jk}+\delta v} (1 - \mathrm{e}^{-\tau_0})\, \mathrm{d}v \simeq \frac{\lambda_{jk}}{c} 2\delta v, \tag{4.53}$$

where we neglect the contribution of the integral second term. If we again consider a Doppler profile in the center region of the line, (3.38) and (3.60) allow us to write the condition $\tau(\nu_{jk} \pm \delta\nu) \simeq 1$ in the form

$$N_j \frac{\pi e^2}{m_e c} f_{jk}\, \phi_D(\delta v) \simeq 1. \tag{4.54}$$

Using (3.47), (3.48), and (4.49), we obtain for width $\delta\nu$

$$\delta v \simeq \frac{b}{\lambda_{jk}} \sqrt{\ln \tau_0}. \tag{4.55}$$

Replacing (4.55) in (4.53), we finally have

$$\frac{W_\lambda}{\lambda_{jk}} = \frac{2b}{c}\sqrt{\ln \tau_0} \simeq \frac{2b}{c}\left[\ln\left(\frac{\pi e^2}{m_e c b\sqrt{\pi}} N_j f_{jk}\lambda_{jk}\right)\right]^{1/2}, \tag{4.56}$$

valid for relatively strong lines, with $\tau_0 \gg 1$, $\tau_\nu \lesssim 1$ and a Doppler profile. We notice that W_λ/λ_{jk} slowly increases with $N_j f_{jk}$, because all available energy in the line central regions has been absorbed and the increase in number of absorbing atoms does not have a considerable effect. We say the line is *saturated* and (4.56) corresponds to the *saturation part or flat* of the curve of growth (see Fig. 4.4). Note that in this region, W_λ/λ_{jk} is proportional to parameter b, defined in (3.45).

Finally, for very strong lines, $\tau_0 \gg 1$ and $\tau_\nu \gg 1$, even for the line radiative wings, and $\tau_\nu \sim 1$ far from the center ν_{jk}. In this case, we can use variable u so that [see (3.60) and (4.33)]

$$\tau_\nu = \frac{N_j \sigma \Gamma_k}{4\pi^2 \Delta\nu^2} = \frac{1}{u^2}, \tag{4.57}$$

$$u = \left[\frac{4\pi^2}{N_j \sigma \Gamma_k}\right]^{1/2} \Delta\nu, \tag{4.58}$$

so that (4.31) allows us to write

$$\frac{W_\lambda}{\lambda_{jk}} = \frac{\lambda_{jk}}{c}\left[\frac{N_j \sigma \Gamma_k}{4\pi^2}\right]^{1/2} \int \left(1 - e^{-1/u^2}\right) du, \tag{4.59}$$

that is,

$$\frac{W_\lambda}{\lambda_{jk}} = \frac{\lambda_{jk}}{c}\left(\frac{N_j \sigma \Gamma_k}{\pi}\right)^{1/2}, \tag{4.60}$$

which is a valid expression for *very strong lines, where* $\tau_\nu \gg 1$. In terms of W_ν,

$$W_\nu = \frac{c}{\lambda_{jk}^2} W_\lambda = \left(\frac{N_j \sigma \Gamma_k}{\pi}\right)^{1/2}. \tag{4.61}$$

Relations (4.60) and (4.61) correspond to the *square root* part of the curve of growth (see Fig. 4.4). For an atom with two levels j and k, we obtain the equivalent expression (4.39), which can be written as

$$\frac{W_\lambda}{\lambda_{jk}} = \frac{2\pi e^2}{m_e c^2} f_{jk}\left[2\frac{g_j}{g_k} N_j\right]^{1/2}. \tag{4.62}$$

For a multilevel atom, we can write

$$\frac{W_\lambda}{\lambda_{jk}} = \frac{2\pi e^2}{m_e c^2}\left[2N_j \frac{f_{jk}}{g_k}\sum_j g_j f_{jk}\right]^{1/2}. \quad (4.63)$$

In this section, we have considered a Doppler profile for the line central regions and a Lorentz natural profile for the wings region. In the more general case, other broadening mechanisms can be important, having an effect on the profile $\phi(\Delta\nu)$ and on the equivalent width W_λ. Applying the above profiles together, we obtain the Voigt profile (3.52), and we may write the equivalent width in terms of this profile.

4.7 Narrow Absorption Lines

4.7.1 Introduction

As we have seen in Chap. 1, interstellar absorption lines are observed since Hartmann's pioneering work in 1904. In interstellar medium conditions, most atoms are in their lower energy state levels, particularly in the ground state. Absorption lines initially observed in the optical were essentially produced by elements or ions with low abundances, whose transitions involving ground states occur in that region of the spectrum. The most important elements, such as atomic and molecular H, have their resonance lines in the ultraviolet ($\lambda \lesssim 3{,}000$ Å), which cannot be observed from the ground, so the study of these presumably more important lines only began in a systematic and detailed way with the development of ultraviolet astronomy from the end of the 1960s onward. Basically, a big step forward occurred with the launch of ultraviolet instruments and satellites, such as TD1-A, ANS, OAO, Copernicus, IUE, IMAPS, and more recently HST and FUSE.

Table 4.1 shows some observed optical absorption lines, the ones that produced the major part of the data about interstellar clouds up to the 1970s.

Table 4.1 Examples of interstellar optical absorption lines

Ion/molecule	λ (Å)	Transition
Na I	(D2) 5,889.95	$3^2S_{1/2} - 3^2P^0_{3/2}$
Na I	(D1) 5,895.92	$3^2S_{1/2} - 3^2P^0_{1/2}$
Ca II	(K) 3,933.66	$4^2S_{1/2} - 4^2P^0_{3/2}$
Ca II	(H) 3,968.47	$4^2S_{1/2} - 4^2P^0_{1/2}$
CN	3,874.61	$B^2\Sigma^+ \leftarrow X^2\Sigma^+(0,0)\,R(0)$
CH	4,300.31	$A^2\Delta \leftarrow X^2\Pi(0,0)\,R_2(1)$
CH^+	4,232.58	$A^1\Pi \leftarrow X^1\Sigma^+(0,0)\,R(0)$
CH^+	3,957.74	$A^1\Pi \leftarrow X^1\Sigma^+(1,0)\,R(0)$

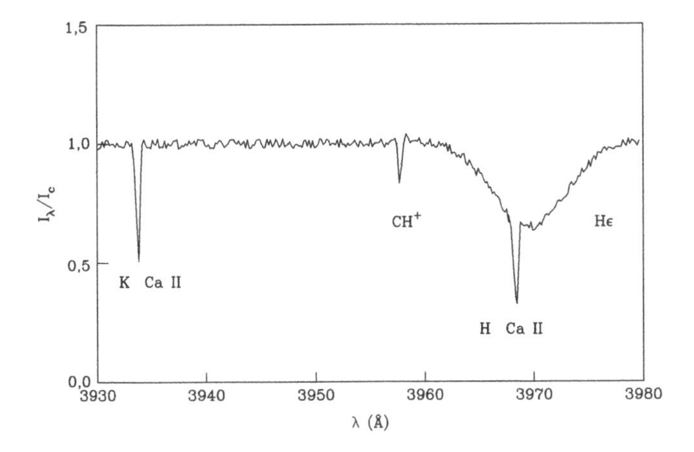

Fig. 4.6 Examples of interstellar optical absorption lines in the spectrum of ζ Oph

Table 4.2 Examples of interstellar ultraviolet lines

Ion	λ(Å)	Ion	λ(Å)
H I	1,215.67	Mg I	2,852.13
H I	1,025.72	Mg II	2,795.53
C I	1,328.83	Si I	2,514.32
C I	1,277.24	Si II	1,193.29
C II	1,334.53	Si II	1,260.42
C IV	1,548.20	Si III	1,206.51
N I	1,200.71	S II	1,259.52
N II	1,134.17	S III	1,190.21
O I	1,302.17	Fe II	2,599.40
O I	1,039.23	Fe II	2,382.03

Figure 4.6 shows an example of optical absorption lines of interstellar origin superimposed on the spectrum of Oph, a O9.5 V type star with $E(B\text{–}V) = 0.32$. The Ca II K (3,933.66 Å) and H (3,968.47 Å) lines as well as the 3,957.74 Å CH$^+$ line can be observed. Note that the Ca II H line occurs inside the Hϵ hydrogen line, which is much broader and of stellar origin.

The analysis of Ca II and Na I lines leads to the determination of the density of these ions and also allows to estimate the effective width of the galactic disk, found to be $2H \simeq 240$ pc, similar to the value obtained from 21 cm line observations. The column densities can also be determined by the doublet ratio method. In lines of the same element but with different f-values, the observed ratio between intensities allows us to directly determine the optical depth in the line center and the column density of the considered atom.

In a similar way, Table 4.2 shows some of the ultraviolet lines that have been revealed to be quite important in the study of the physical properties of interstellar clouds. Figure 4.7 shows an example of a high-resolution ultraviolet

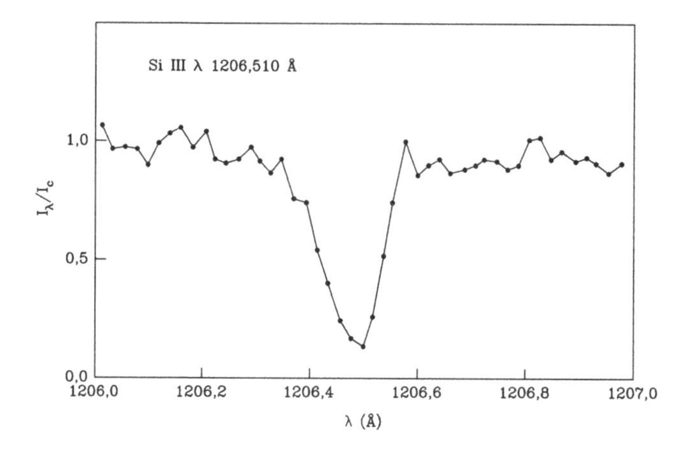

Fig. 4.7 An example of the SiIII UV absorption line in the direction of ζ Oph

absorption line profile for Si III in the direction of the star ζ Oph. This is an example of a non-saturated line, whereas lines of more abundant elements, such as the H Lyman-α line, are saturated in this star's direction.

4.7.2 Problems in the Analysis of the Curves of Growth

The analysis of absorption lines from ground observations—mainly Na I and Ca II lines—indicates the presence of several components (clouds) with radial velocities up to 100 km s^{-1}, though denser clouds have velocities of the order of 10 km s^{-1} and are correlated with components observed at 21 cm. The number of components along the line of sight is of the order of 4–8 kpc^{-1}.

Besides macroscopic velocities, the interstellar medium generally presents a (micro)turbulent component, that tends to complicate the absorption line profile. If we represent the turbulence by a mean velocity (rms), $\langle v_t \rangle^{1/2}$, and the velocity has a Maxwellian distribution, the thermal-turbulent Doppler profile will still be the profile given by (3.48), but the Doppler width is now

$$\Delta \nu_D = \frac{\nu_{jk}}{c} \left[\frac{2kT}{m} + \langle v_t^2 \rangle \right]^{1/2}. \tag{4.64}$$

Therefore, the higher the turbulent velocity, the broader the line Doppler width, that is, the difference relative to the central frequency ν_{jk} so that the intensity decreases by a factor e. In this case, we need a higher column density of the absorbing atoms for the line to become saturated. The same effect occurs if we have two or more identical clouds in the line of sight instead of one, and the cloud velocity separation is higher than the internal velocity dispersion:

$$\sigma_v = \left(\frac{kT}{m}\right)^{1/2} = \frac{b}{\sqrt{2}} = \left(\frac{c}{\nu_{jk}\sqrt{2}}\right)\Delta\nu_D. \tag{4.65}$$

In this case, the effective width of the line will be broader, and more atoms per square centimeter will be necessary for line saturation to occur.

4.7.3 Empirical Curve of growth

The theoretical curve of growth discussed in Sect. 4.6 assumes a Doppler profile, which is not necessarily realistic. In practice, we use the *empirical curve of growth* method, putting together atoms that probably have the same velocity distribution and calculating a curve of growth (or parts of a curve of growth) for these atoms. For instance in H I regions, ions in the dominating ionization state must be N I, O I, Mg II, Si II, Al I, Ar I, S II, Fe II, C II, and Na II; ions in nondominating ionization states are Mg I, Si I, C I, Na I, S I, K I, and Fe I. These two groups must in principle define two curve of growth if their relative abundances in the clouds are equal.

For each group, we have plotted the equivalent width observed values, log (W_λ/λ_{jk}) as a function of $\log(f_{jk}\ \lambda_{jk})$ for the available lines. For each ion, we have a part of the curve of growth, and N_j can be determined by shifting the different sections of the curve along the x-axis, until a match between different ions is achieved. Once we have the curve, we can, for instance, adjust a theoretical profile and determine the cloud temperature. This method was initially applied to the interstellar medium by Strömgren in 1948. More recently, it has been applied to satellite ultraviolet observations.

Figure 4.8 shows an empirical curve of growth along the line of sight of ζ Oph star. In this figure, full circles represent atoms in the dominating ionization state, and crosses are the corresponding neutral atoms. The solid line is a theoretical curve characterized by a Maxwellian velocity distribution, with parameter $b = 6.5$ km s^{-1} [see (3.45)].

4.7.4 Interstellar Abundances

The chemical composition of interstellar clouds determined from curve of growth analysis presents deviations relative to the so-called cosmic or universal chemical composition, essentially based on observations of the Sun, stars, and meteorites.

These differences are illustrated in Table 4.3, where we present cosmic (ε_c) and interstellar (ϵ_i) abundances per number atoms, taken in the ζ Oph direction, which is considered a typical reddened star. Also included is the *depletion factor* f_d. Parameters ϵ_c, ϵ_i, and f_d are defined by

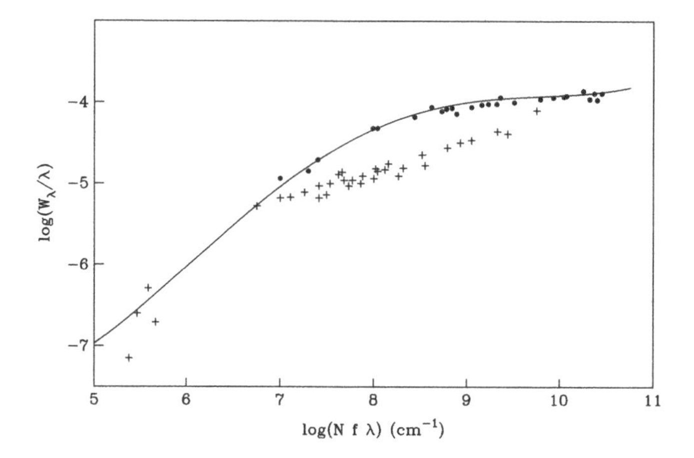

Fig. 4.8 The empirical curve of growth in the direction of Zeta Oph

Table 4.3 Cosmic abundances, interstellar abundances, and depletion factors of some chemical elements

Element	ϵ_c	ϵ_i	f_d
H	12.0	–	–
He	11.0	–	–
C	8.6	7.9	−0.7
N	8.0	7.3	−0.7
O	8.9	8.0	−0.9
Na	6.3	5.4	−0.9
Mg	7.6	6.0	−1.6
Al	6.5	3.1	−3.4
Si	7.5	5.9	−1.6
P	5.4	4.3	−1.1
S	7.3	6.9	−0.4
Ca	6.4	2.7	−3.7
Fe	7.5	5.4	−2.1

$$\epsilon_c = \log \left[\frac{N_X}{N_H}\right]_c + 12, \tag{4.66}$$

$$\epsilon_i = \log \left[\frac{N_X}{N_H}\right]_i + 12, \tag{4.67}$$

$$f_d = \epsilon_i - \epsilon_c. \tag{4.68}$$

We see that elements such as C, N, and O are less abundant relative to the cosmic values by a factor of the order of $10^{0.7} \simeq 5$ and Fe, Ca, and Al by a factor of the order of 100–5,000. These elements are probably condensed in the form of interstellar dust grains and do not contribute in a significant way to the observed absorption. In Chap. 9, we will consider the chemical composition of solid dust grains in the interstellar medium in a more detailed way.

Exercises

4.1 Make a Gaussian fit to the brightness temperature profile of Fig. 4.1, assuming four interstellar clouds in the considered direction. What would be the column density for each of these clouds?

4.2 The H Lyman-α line involves a transition between two levels j and k, whose parameters are $\lambda_{jk} = 1{,}215.67$ Å, $g_j = 2$, $g_k = 6$, and $A_{kj} = 6.265 \times 10^8\ \mathrm{s}^{-1}$. (a) Calculate the oscillator strength f_{jk} for this line. (b) Calculate the dissipation constant Γ_k. (c) Calculate the Doppler width $\Delta\nu_\mathrm{D}$, assuming a kinetic temperature $T = 80$ K for the H cloud. (d) Consider a region in the line radiative wings where $\Delta\nu \simeq 10\ \Delta\nu_\mathrm{D}$. Show that in this case $(\Delta\nu)^2 \gg (\Gamma_k/4\pi)^2$. (e) Estimate the optical depth in the line wings, if the H column density is $N_\mathrm{H} = 3 \times 10^{20}\ \mathrm{cm}^{-2}$. Which fraction of the original intensity is absorbed in this region? Suppose that all H is in the ground state. (f) Calculate the optical depth in the line center. (g) Calculate the FWHM. (h) In which region the line becomes so weak that $\tau \simeq 1$?

4.3 Show that (4.52) becomes (4.47) when $\tau_\nu \ll 1$.

4.4 A spectral line with central wavelength λ is formed in a region characterized by kinetic temperature T and microturbulence velocity v_t. (a) Assuming that the broadening of the line is due to the Doppler process, how could the Doppler width of this line be written? (b) Considering the SI III line with $\lambda = 1{,}206$ Å in a cloud with $T = 80$ K, what must be the value of the turbulent velocity so that the Doppler width increases by a factor of two?

4.5 Measurements of the equivalent width of the absorption NaI D lines at $\lambda = 5{,}890$ Å in the direction of star HD 190066 (type B1I) give the result $W \sim 400$ mÅ. (a) Assume this is a weak line and calculate the column density of neutral Na atoms in the direction of the star. Show that in this case, the following relation is valid:

$$N \simeq \frac{11.3\,W}{\lambda^2 f},$$

where N is in cm^{-2}, W is in mÅ, and λ is in cm. Use $f = 0.65$. (b) Analysis of the line saturation suggests a correction factor of the order of 6 for the column density. Apply this factor to result (a) and estimate the Na total column density, assuming that 99 % of the sodium atoms are ionized.

Bibliography

Burton, W.B.: In: Pfenniger, D., Bartholdi, P. (eds.) The Galactic Interstellar Medium, p. 1. Berlin, Springer (1992). Discussion on the H 21 cm line and its application to the study of the interstellar medium and the Galaxy's structure. See also the article by W.B. Burton in Verschuur, G.L. & Kellermann, K.I. (eds.). Galactic and Extragalactic Radio Astronomy, referred to in Chapter 2, page 61

Fuhr, J.R., Wiese, W.L.: In: Lide, D. (ed.) Atomic Transition Probabilities. Handbook of Chemistry and Physics, p. 10. CRC, Boca Raton (1991). Basic reference for numerical values of oscillator strengths and other parameters. See also the previous work by W. L. Wiese et al. Atomic Transition Probabilities. Washington, NBS, 1966

Grevesse, N., Noels, A., Sauval, A.J.: In: Holt, S.S., Sonne-Born, G.G. (eds.) Cosmic Abundances. ASP Conference Series, p. 117. American Society of Pacific, San Francisco (1996). Contains an updated discussion on "cosmic" abundance

Jenkins, E.B.: In: Houziaux, L., Butler, H.E. (eds.) IAU Symposium 36, p. 281. Reidel, Dordrecht (1970). Includes an updated discussion on the interstellar Lyman-α line

Mihalas, D., Binney, J.: Galactic Astronomy. Freeman, San Francisco (1981). Excellent discussion on kinematic aspects of the Galaxy's structure and the LSR. It also includes a summary of the main properties of stars and interstellar gas

Morton, D.C.: Atomic data for resonance absorption lines. I – Wavelengths longward of the Lyman limit. Astrophys J Suppl **77**, 119 (1991). Interstellar line f-values compilation. For some examples of ultraviolet interstellar line analysis see also Astrophys. J. vol. 147, p.1017, 1967 and Astrophys. J. vol. 197, p.85, 1975. Figure 4.7 is based on data taken from this reference

Osterbrock, D.: Astrophysics of Gaseous Nebulae and Active Galactic Nuclei. University Science Books, Mill Valley (1989). Referred to in Chapter 1. Excellent discussion on emission lines observed in interstellar photoionized nebulae

Scheffler, H., Elsässer, H.: Physics of the Galaxy and Interstellar Matter. Springer, Berlin (1988). Referred to in Chapter 1. Includes a good discussion on the principal emission and absorption interstellar lines and examples. Figure 4.6 is based on data taken from this reference

Spitzer, L.: Physical Processes in the Interstellar Medium. Wiley, New York (1978). Referred to in Chapter 1. Excellent discussion on the main characteristics of interstellar emission and absorption lines and of curves of growth. Figure 4.8 is based on data taken from this reference

Weaver, H., Williams, D.R.W.: The Berkeley low-latitude survey of neutral hydrogen Part I. Profiles. Astron. Astrophys. **8**, 1 (1973). Example of a large survey (Berkeley) of neutral H emission in the interstellar gas. See also Astron. Astrophys. vol. 17, p.1, 1974. Figure 4.1 is based on data taken from this reference

Wilson, T.L., Matteucci, F.: Abundances in the interstellar medium. Astron. Astrophys. Rev. **4**, 1 (1992). Review article on interstellar abundances. See also B.D. Savage e K.R. Sembach. Annual Rev. Astron. Astrophys. vol. 34, p.279, 1996

Chapter 5
Excitation in the Interstellar Medium

5.1 Introduction

We have seen in Chap. 3 that in thermodynamic equilibrium, densities relative to two energy levels j and k can be calculated from the Boltzmann equation, if the atom parameters and temperature T are known. In this case, n_k/n_j is determined from the global or macroscopic properties of the medium. In the interstellar medium, as we have already mentioned, there are great deviations relative to the TE. As a consequence, we need to introduce deviation coefficients b_j and b_k, which depend on the detailed processes that trigger excitation and de-excitation of the different atoms and ions. Therefore, to obtain relative densities

$$\frac{n_k}{n_j} = \frac{b_k}{b_j}\frac{g_k}{g_j}\exp\left(-h\nu_{jk}/kT\right) \tag{5.1}$$

[see (3.9)], we need to consider in detail each process of population and depopulation of the atom energy levels. In this chapter, we will consider the excitation and de-excitation processes in statistical equilibrium, with a simple application to atoms with two or three levels.

5.2 Statistical Equilibrium

In the more general case, density n_j of level j is a function of time. Meanwhile, we are frequently interested in steady-state processes where $n_j(t) = n_j$, meaning that the density has a constant value of equilibrium. In this case, we should have

$$\frac{\mathrm{d}n_j}{\mathrm{d}t} = 0, \tag{5.2}$$

W.J. Maciel, *Astrophysics of the Interstellar Medium*,
DOI 10.1007/978-1-4614-3767-3_5,

that is, the total number of transitions that populate the level, taking into account all possible processes per unit volume and per unit time, must balance the transitions that depopulate the level. If $(R_{jk})_y$ is the probability per unit time that a particle in level j will move to level k triggered by process y, this condition can be written as

$$\frac{dn_j}{dt} = -n_j \sum_y \sum_k (R_{jk})_y + \sum_y \sum_k n_k (R_{kj})_y = 0. \tag{5.3}$$

In this equation, level k can be higher or lower than j, both being bound states of the atom. Equation (5.3) is the *statistical equilibrium equation*, a less restrictive condition than TE. In a general way, if we have m levels, we will have $m - 1$ linearly independent equations, whose solution depends on knowing rates R_{jk} and allows to obtain relative populations for all levels. For instance, in the case of a two-level atom, with levels j and k, where only one process is important, with rates R_{jk} and R_{kj}, we have

$$-n_j R_{jk} + n_k R_{kj} = 0,$$

that is,

$$\frac{n_k}{n_j} = \frac{R_{jk}}{R_{kj}}. \tag{5.4}$$

5.3 Collisional Excitation

When only collisional processes are important, that is, when we can neglect radiative excitations and de-excitations, the statistical equilibrium equation solution is trivial, and the relative populations of the levels are given by the Boltzmann equation [see for instance (3.7)]. In Chap. 4, we use this fact when considering the H 21 cm emission line. For a nontrivial solution of (5.3), excitation and de-excitation radiative processes must be included.

5.4 Collisional/Radiative Excitation

5.4.1 The Statistical Equilibrium Equation

Let us write the statistical equilibrium equation including collisional excitations and de-excitations with rates $(R_{jk})_c$, spontaneous transitions with rates A_{kj}, and induced transitions with coefficients B_{jk}. From (5.3)

$$n_j\left[\sum_k (R_{jk})_c + \sum_k B_{jk}U_\nu + \sum_{k<j} A_{jk}\right] = \sum_k n_k(R_{kj})_c + \sum_k n_k B_{kj}U_\nu + \sum_{k>j} n_k A_{kj}$$

or

$$n_j\left\{\sum_k \left[(R_{jk})_c + B_{jk}U_\nu\right] + \sum_{k<j} A_{jk}\right\} = \sum_k n_k\left[(R_{kj})_c + B_{kj}U_\nu\right] + \sum_{k>j} n_k A_{kj}. \quad (5.5)$$

In the interstellar medium, we generally want to consider collisional processes involving electrons or H atoms. It is thus usual to write the probability $(R_{jk})_c$ in terms of the *collisional excitation rate* or *collisional excitation coefficient* γ_{jk}, which is given by

$$\gamma_{jk} = \frac{(R_{jk})_c}{n_c}, \quad (5.6)$$

where n_c is the field particle density. The usual units of γ_{jk} are cm^3 s^{-1}. For collisions with electrons, we have

$$\gamma_{jk} = \frac{(R_{jk})_c}{n_e}, \quad (5.7)$$

and for collisions with H atoms

$$\gamma_{jk} = \frac{(R_{jk})_c}{n_H}. \quad (5.8)$$

For collisions with electrons, the statistical equilibrium equation (5.5) may be written as

$$n_j\left[\sum_k (n_e\gamma_{jk} + B_{jk}U_\nu) + \sum_{k<j} A_{jk}\right] = \sum_k n_k(n_e\gamma_{kj} + B_{kj}U_\nu) + \sum_{k>j} n_k A_{kj}. \quad (5.9)$$

5.4.2 *Relation Between Rates* γ_{jk} *and* γ_{kj}

We will now deduce a relation between excitation and de-excitation collisional rates γ_{jk} and γ_{kj} using the fact that in TE, the collisional processes that tend to populate and depopulate a given level j must balance each other. Writing (5.9) for an atom with two levels j and k, where $k > j$, we have

$$n_j(n_e\gamma_{jk} + B_{jk}U_\nu) = n_k(n_e\gamma_{kj} + B_{kj}U_\nu) + n_kA_{kj}. \tag{5.10}$$

From the *detailed balance* condition,

$$n_j^* n_e \gamma_{jk} = n_k^* n_e \gamma_{kj}, \tag{5.11}$$

$$\gamma_{jk} = \frac{n_k^*}{n_j^*}\gamma_{kj}. \tag{5.12}$$

Using the Boltzmann equation (3.7), we obtain

$$g_j\gamma_{jk} = g_k\gamma_{kj}\,\mathrm{e}^{-E_{jk}/kT}, \tag{5.13}$$

where $E_{jk} = E_k - E_j$. Note that if we apply the detailed balance condition to radiative transitions, we obtain once again the Boltzmann equation.

5.4.3 Collisional Excitation Rate

Let us consider a test particle whose velocity $\mathbf{v}_t$ and excitation conditions are modified by collisions with particles of the medium with velocity $\mathbf{v}_c$, in such a way that the relative velocity of these particles is $\mathbf{u} = \mathbf{v}_t - \mathbf{v}_c$. If n_c is the field particle density, the probability that a test particle will collide in the time interval dt may be written as

$$P(u)\mathrm{d}t = n_c u\sigma(u)\mathrm{d}t, \tag{5.14}$$

where $\sigma(u)$ is the collisional cross section given in cm^2. Therefore, the probability that a test particle in state j moves to state k triggered by a collision in the time interval dt can be written as

$$P_{jk}(u)\mathrm{d}t = n_c u\,\sigma_{jk}(u)\mathrm{d}t. \tag{5.15}$$

From (5.6), we see that coefficient γ_{jk} gives the collision probability per unit time per field particle so that $n_c\,\gamma_{jk}$ gives the number of excitations per cubic centimeter per second, and the product $n_t\,n_c\,\gamma_{jk}$ gives the number of excitations, where n_t is the test particle density. Comparing (5.6) and (5.15), we note that γ_{jk} can be written as

$$\gamma_{jk} = \langle u\,\sigma_{jk}\rangle, \tag{5.16}$$

where we consider the mean of probability $P(u)$ for all relative velocities. For a Maxwellian velocity distribution $f(u)$, if σ_{jk} is independent of the **u** direction, we have

$$\gamma_{jk} = \frac{u\,\sigma_{jk}(u)f(u)\mathrm{d}u}{f(u)\mathrm{d}u} = \frac{4\ell^3}{\sqrt{\pi}}\int_0^{\infty} u^3\sigma_{jk}(u)\,\mathrm{e}^{-\ell^2u^2}\mathrm{d}u, \quad (5.17)$$

where

$$\ell^2 = \frac{m_\mathrm{r}}{2kT} \quad (5.18)$$

and the reduced mass m_r is

$$m_\mathrm{r} = \frac{m_\mathrm{c}m_\mathrm{t}}{m_\mathrm{c}+m_\mathrm{t}}, \quad (5.19)$$

being m_t the mass of the test particles and m_c the mass of the field particles.

5.4.4 Relations Between Cross Sections

We will now obtain a relation between the collisional excitation cross section $\sigma_{jk}(u_j)$ for particles in state j and the collisional de-excitation cross section $\sigma_{kj}(u_k)$ for particles in state k, where u_j and u_k are the relative velocities between these particles and the field particles before and after excitation. In TE, the number of collisions per cubic centimeter per second for particles in the interval between u_j and $u_j + \mathrm{d}u_j$ must balance the corresponding number in the interval between u_k and $u_k + \mathrm{d}u_k$:

$$n_j^* n_c u_j \sigma_{jk}(u_j) f(u_j)\mathrm{d}\mathbf{u}_j = n_k^* n_c u_k \sigma_{kj}(u_k) f(u_k)\mathrm{d}\mathbf{u}_k. \quad (5.20)$$

Considering

$$E_{jk} = E_k - E_j, \quad (5.21)$$

we have

$$\frac{1}{2}m_\mathrm{r}u_k^2 = \frac{1}{2}m_\mathrm{r}u_j^2 - E_{jk}, \quad (5.22)$$

that is, to have excitation, the initial kinetic energy must be higher than the threshold value E_{jk}. Using (5.20) with the Maxwellian distribution, the Boltzmann equation, and (5.22), we obtain

$$g_j \sigma_{jk}(u_j) u_j^2 = g_k \sigma_{kj}(u_k) u_k^2. \tag{5.23}$$

5.4.5 *Cross Section and Collision Strength*

In the excitation of neutral atoms by electrons, the de-excitation cross section $\sigma_{kj}(u_k)$ has a threshold finite value, that is, when

$$\frac{1}{2} m_r u_j^2 = E_{jk}, \tag{5.24}$$

and therefore,

$$\frac{1}{2} m_r u_k^2 = 0 \quad \text{or} \quad u_k = 0 \tag{5.25}$$

[see (5.22)]. In this case, from (5.23), $\sigma_{jk}(u_j) = 0$ in the threshold, that is, when ${u^2}_j = 2E_{jk}/m_r \neq 0$. For electron–ion collisions, there is electrostatic attraction and $\sigma_{jk} \neq 0$, which means that it is finite. In this case, from (5.23),

$$\sigma_{kj}(u_k) \propto \frac{1}{u_k^2}. \tag{5.26}$$

For energies near the threshold, the excitation cross section is essentially constant. In the case of electron–ion collisions, we can define the *collision intensity* or *strength* $\Omega(j,k)$ by

$$\sigma_{jk}(v_j) = \frac{\pi}{g_j} \left(\frac{h}{2\pi m_e v_j} \right)^2 \Omega(j,k), \tag{5.27}$$

where we use v_j for u_j, because the electron velocity is generally much higher than the ion velocity and $\Omega(j,k)$ is a dimensionless number. Considering (5.23),

$$g_j \frac{\pi}{g_j} \left(\frac{h}{2\pi m_e v_j} \right)^2 \Omega(j,k) v_j^2 = g_k \frac{\pi}{g_k} \left(\frac{h}{2\pi m_e v_k} \right)^2 \Omega(k,j) v_k^2,$$

and we finally obtain

$$\Omega(j,k) = \Omega(k,j), \tag{5.28}$$

that is, the excitation collision strength is equal to the de-excitation collision strength.

5.4.6 Relation Between Coefficients γ_{kj} and Ω(j,k)

We will now obtain a relation between the de-excitation collision coefficient γ_{kj} and the collision strength $\Omega(j,k)$. From (5.16) and (5.17), we may write

$$\begin{aligned}\gamma_{kj} &= \langle u_k \sigma_{kj}(u_k) \rangle \\ &= \frac{4\ell^3}{\sqrt{\pi}} \int_0^\infty u_k^3 \sigma_{kj}(u_k)\, \mathrm{e}^{-\ell^2 u_k^2} \mathrm{d}u_k. \end{aligned} \tag{5.29}$$

Using (5.27) and (5.28),

$$\begin{aligned}\gamma_{kj} &= \frac{4\ell^3}{\sqrt{\pi}} \int_0^\infty u_k^3 \frac{\pi}{g_k} \left(\frac{h}{2\pi m_e v_k}\right)^2 \Omega(j,k)\, \mathrm{e}^{-\ell^2 u_k^2} \mathrm{d}u_k \\ &= \frac{4\ell^3 \pi h^2 \Omega(j,k)}{\sqrt{\pi} g_k (2\pi m_e)^2} \int_0^\infty u_k\, \mathrm{e}^{-\ell^2 u_k^2} \mathrm{d}u_k, \end{aligned} \tag{5.30}$$

because $u_k \simeq v_k$. Given that

$$\int_0^\infty \mathrm{e}^{-ax^2} x\, \mathrm{d}x = \frac{1}{2a}, \tag{5.31}$$

we obtain

$$\gamma_{kj} = \frac{h^2 \Omega(j,k)}{g_k (2\pi m_e)^{3/2} (kT)^{1/2}}. \tag{5.32}$$

The collision strength $\Omega(j,k)$ can vary with electron velocity, but in general, we consider an essentially constant mean value so that from (5.32), the variation of the excitation collision coefficient with temperature is $\gamma_{kj} \propto T^{-1/2}$.

A detailed discussion on collisional excitation processes, classical and recent references, and numerical values may be found in Osterbrock (1989).

5.4.7 Example: O II and O III in Photoionized Nebulae

In photoionized nebulae, such as planetary nebulae, O II and O III ions are particularly important, not just because of the high relative abundance of oxygen, which is of the order of $10^{-4}\, n_\mathrm{H}$, but also because of the energy level arrangement of these ions. Figure 5.1 shows a schematic diagram of the first five levels of these ions. N II belongs to the same isoelectronic series of O III, presenting a similar configuration of energy levels, and thus, some of its lines are also indicated between

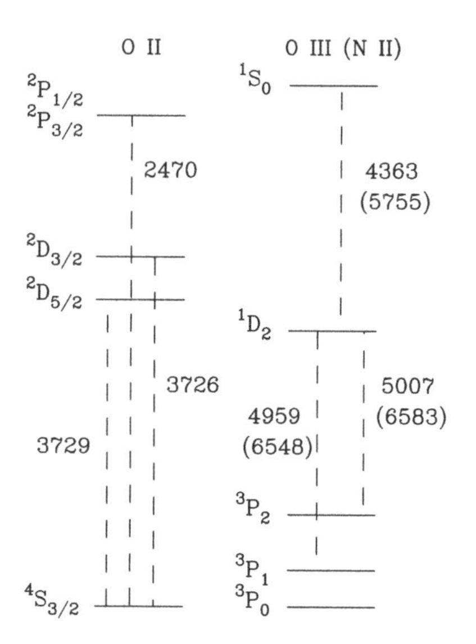

Fig. 5.1 Energy level diagram for O II and O III (NII)

brackets in Fig. 5.1. Excitation energies of these levels are $E \lesssim 5$ eV, of the same order of magnitude as the mean kinetic energy of the abundant electrons in H ionized regions, where the electron temperature is $T_e \simeq 10^4$ K.

Values of the collision strength, or of cross section, are calculated quantum mechanically and in the general case depend on the particle velocity. In astrophysical applications, the mean values of Ω are often used, considering a Maxwellian velocity distribution at a certain temperature T. Table 5.1 relates some O II, O III, and N II ion lines with mean collision strengths typical of photoionized nebulae, where $T_e \simeq 10^4$ K. The usual notation indicates multiplicity and orbital and total angular momentum. For instance, for level $^3P_2$, we have:

- Multiplicity $= 2S + 1$, where $S = 1$ is the *total spin*.
- Orbital angular momentum P or $L = 1$.
- Total angular momentum $J = 2$ or $g_J = 2J + 1 = 5$.

Table 5.1 shows the relation between wavelengths, energies, and spontaneous emission probabilities A_{kj}. These probabilities are very low, because electric dipole transitions are forbidden for these levels. In other words, once excited, a given level will be occupied for a long time. In laboratory conditions, there will be collisions with particles from the medium quite rapidly, and thus, the atom will be collisionally de-excitated. However, in H II regions and planetary nebulae, the low density implies radiative de-excitation instead. For example, transitions between O III levels $^1D_2$–$^3P_2$ and $^1D_2$–$^3P_1$ correspond to photons with wavelengths $\lambda(^1D_2–^3P_2)$ = 5,006.9 Å and $\lambda(^1D_2–^3P_1)$ = 4,958.9 Å. These are N_1 and N_2 *nebular lines*, originally attributed to element *nebulium* (see Chap. 1), for which the notation "forbidden lines," [OIII] λ4959/5007 Å, is used.

Table 5.2 shows the electron configuration for the first three oxygen ions.

Table 5.1 Examples of OII, OIII and NII lines with some atomic data

Level	λ_{jk} (Å)	E_{jk} (eV)	$\Omega(j, k)$	A_{kj} (s^{-1})
O II				
$^4S_{3/2}$–$^2D_{5/2}$	3,728.8	3.325	0.80	3.6×10^{-5}
$^4S_{3/2}$–$^2D_{3/2}$	3,726.0	3.328	0.55	1.8×10^{-4}
$^2D_{3/2}$–$^2D_{5/2}$	–	0.0025	1.17	1.3×10^{-7}
$^4S_{3/2}$–$^2P_{1/2}$	2,470.2	5.019	0.40	2.4×10^{-2}
$^4S_{3/2}$–$^2P_{3/2}$	2,470.3	5.019	Total	5.8×10^{-2}
$^2D_{3/2}$–$^2P_{1/2}$	7,329.6	1.692	0.28	9.4×10^{-2}
$^2D_{5/2}$–$^2P_{1/2}$	7,318.8	1.694	0.30	5.6×10^{-2}
$^2D_{3/2}$–$^2P_{3/2}$	7,330.7	1.691	0.41	5.8×10^{-2}
$^2D_{5/2}$–$^2P_{3/2}$	7,319.9	1.694	0.73	1.1×10^{-1}
O III				
$^3P_0$–$^3P_1$	88.4 μm	0.014	0.54	2.6×10^{-5}
$^3P_0$–$^3P_2$	32.7 μm	0.038	0.27	3.0×10^{-11}
$^3P_1$–$^3P_2$	51.8 μm	0.024	1.29	9.8×10^{-5}
$^3P_0$–$^1D_2$	4,931.0	2.515	2.17	2.7×10^{-6}
$^3P_1$–$^1D_2$	4,958.9	2.500	Total	6.7×10^{-3}
$^3P_2$–$^1D_2$	5,006.9	2.476	Total	2.0×10^{-2}
$^3P_1$–$^1S_0$	2,321.0	5.342	0.28	2.2×10^{-1}
$^3P_2$–$^1S_0$	2,331.4	5.318	Total	7.8×10^{-4}
$^1D_2$–$^1S_0$	4,363.2	2.842	0.62	1.8
N II				
$^3P_0$–$^3P_1$	204 μm	0.006	0.40	2.1×10^{-6}
$^3P_0$–$^3P_2$	76 μm	0.016	0.28	1.2×10^{-12}
$^3P_1$–$^3P_2$	122 μm	0.010	1.13	7.5×10^{-6}
$^3P_0$–$^1D_2$	6,527.1	1.900	2.68	5.4×10^{-7}
$^3P_1$–$^1D_2$	6,548.1	1.894	Total	1.0×10^{-3}
$^3P_2$–$^1D_2$	6,583.4	1.883	Total	3.0×10^{-3}
$^3P_1$–$^1S_0$	3,062.8	4.048	0.35	3.4×10^{-2}
$^3P_2$–$^1S_0$	3,070.8	4.038	Total	1.5×10^{-4}
$^1D_2$–$^1S_0$	5,754.6	2.155	0.41	1.1

Table 5.2 Electronic configuration for the OI, OII and OIII ions

Ion	K: 1s	L: 2s	2p
O I	2	2	4
O II	2	2	3
O III	2	2	2

We will use (5.32) to obtain the collisional de-excitation coefficient γ_{kj} for electron–ion collisions for O III transition $^3P - {}^1D_2$, which gives rise to the λ4959/5007 Å lines. Considering $T \simeq 10^4$ K with $\Omega(j,k) = 2.17$ and $g_k = 5$, we obtain $\gamma_{kj} \simeq 3.7 \times 10^{-8}$ cm^3 s^{-1}. In an H ionized region with electron density $n_e \sim 10^4$ cm^{-3}, typical for planetary nebulae, the probability per unit time of an O^{++} ion to experience a collisional de-excitation is $n_e\, \gamma_{kj} \simeq 3.7 \times 10^{-4}$ s^{-1}. From Table 5.1, we see that the sum of spontaneous radiative transition probabilities from level $^1D_2$ to lower levels is of the order of 0.027 s^{-1}, much higher than the

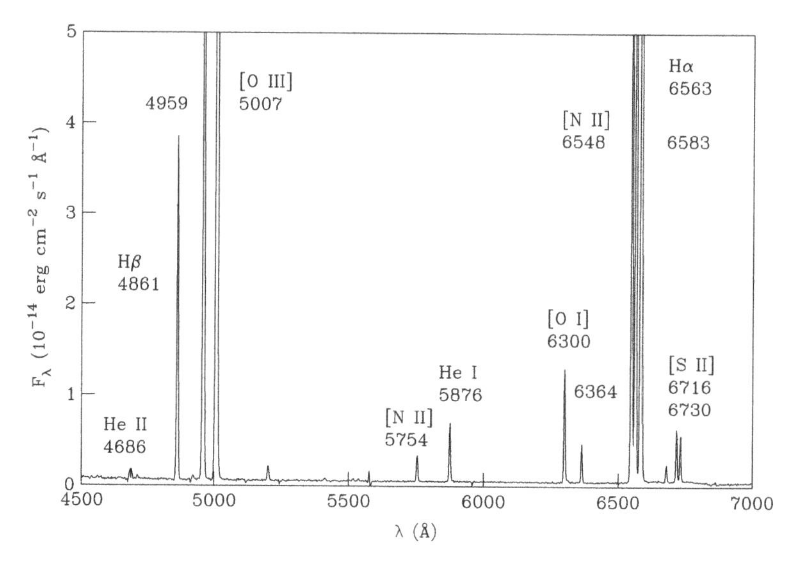

Fig. 5.2 The spectrum of NGC 2346 (Roberto Costa, IAG/USP)

collisional de-excitation probability. Therefore, de-excitation will be radiative and the spectral line will be observed in emission. In fact, a large part of photoionized nebulae radiation is concentrated in forbidden lines, and the analysis of these line intensities in H II regions and planetary nebulae spectra is essential to determine electron density, temperature, and element abundances in these regions (Chap. 8).

Figure 5.2 shows a spectrum of NGC 2346, a planetary nebula where a faint continuum is observed superimposed on several emission lines. Some of these lines are identified as Hβ, Hα and He, N, O, and S lines, and some of which are forbidden. Note that [OIII] λ4959/5007 Å and [NII] λ6548/6583 Å forbidden lines are very strong, being saturated in the shown spectrum. The Hα line appears between the [N II] doublet lines.

5.4.8 Example: H Collisional Excitation in H I Regions

In H I regions, the high abundance of H causes collisions with this atom to become more important than collisions with electrons, because in these clouds the n_e/n_H ratio is relatively small. One of the processes involving collisions with neutral H is the H hyperfine level excitation, a process where the electron spin is reversed by collisions with other H atoms. Table 5.3 relates some values of the de-excitation coefficients as a function of temperature for typical temperatures of interstellar regions.

Table 5.3 Hydrogen deexcitation coefficients for some typical interstellar temperatures

T (K)	γ_{kj} (cm^3 s^{-1})
10	2.3×10^{-12}
30	3.0×10^{-11}
100	9.5×10^{-11}
300	1.6×10^{-10}
1,000	2.5×10^{-10}

Using relation (5.13) between coefficients γ_{kj} and γ_{jk}, we can calculate the excitation coefficient γ_{jk} with $g_j = 1$ and $g_k = 3$ for a temperature interval like in Table 5.3. From (5.16), the de-excitation cross section may be written

$$\gamma_{kj} \simeq \langle u \rangle \sigma_{kj}, \tag{5.33}$$

because σ_{kj} is approximately constant for low temperatures in collisions involving neutral particles. Considering (5.13),

$$\begin{aligned} \gamma_{jk} &= \frac{g_k}{g_j} \gamma_{kj}\, \mathrm{e}^{-E_{jk}/kT} \\ &= \frac{g_k \sigma_{kj}}{g_j} \langle u \rangle\, \mathrm{e}^{-E_{jk}/kT} \\ &= \frac{g_k \sigma_{kj}}{g_j} \langle u \rangle 10^{-5040 E_{jk}/T}. \end{aligned} \tag{5.34}$$

In the last relation, E_{jk} is in eV. The mean relative velocity is given by

$$\langle u \rangle = \left[\frac{8kT}{\pi m_{\mathrm{r}}}\right]^{1/2} = \left[\frac{8kT}{\pi (m_{\mathrm{H}}/2)}\right]^{1/2} \simeq 2.1 \times 10^4 T^{1/2}, \tag{5.35}$$

where T is in K and $<u>$ is in cm s^{-1}. From (5.34) and (5.35), if σ_{kj} is approximately constant for low temperatures, γ_{kj} is essentially proportional to $T^{1/2}$.

As an example, assuming $T \simeq 80$ K, the data of Table 5.3 provide $\gamma_{kj} \simeq 7.5 \times 10^{-11}$ cm^3 s^{-1}. From (5.13) with $g_j = 1$, $g_k = 3$, and $E_{jk} = 5.9 \times 10^{-6}$ eV $= 9.4 \times 10^{-18}$ erg, we obtain $\gamma_{jk} \simeq 2.2 \times 10^{-10}$ cm^3 s^{-1}. From (5.35), the mean relative velocity between H atoms is $<u> \simeq 1.9 \times 10^5$ cm s^{-1} so that the mean de-excitation cross section is $\sigma_{kj} \simeq 4 \times 10^{-16}$ cm^{-2}. In an H I cloud with density $n_{\mathrm{H}} \sim$ 1–100 cm^{-3}, the collisional de-excitation probability will be $n_{\mathrm{H}}\, \gamma_{kj} \sim 10^{-10}$–$10^{-8}$ s^{-1}, thus much higher than the spontaneous radiative transition probability $A_{kj} \sim 10^{-15}$ s^{-1} (see Chap. 4). If the collisional process is the only one to populate/depopulate the given levels, the relative population of these levels will be given by the Boltzmann equation. Of course, the high H abundance in interstellar clouds and the large dimensions of these clouds contribute to the existence of an observable radiation in the 21 cm line.

5.4.9 Other Processes

Other processes involving H atoms or molecules may be important in H I regions. We can cite among them:

- C II fine structure level $^2P_{3/2}$ de-excitation by means of collisions with H atoms, with rates of the order of $\gamma_{kj} \simeq 8 \times 10^{-10}$ cm^3 s^{-1} for $T \simeq 100$ K.
- H_2 molecule rotational level $J = 2$ de-excitation by means of collisions with H atoms, with $\gamma_{kj} \simeq 3 \times 10^{-12}$ cm^3 s^{-1} for $T \simeq 100$ K.
- CO rotational level $J = 1$ de-excitation by means of collisions with H_2 molecules, with $\gamma_{kj} \simeq 4 \times 10^{-12}$ cm^3 s^{-1} for $T \simeq 100$ K.

Besides the collisional and radiative processes considered, other processes may populate a certain excited level k, such as:

- *Population from radiative recombination*: in the electron–ion recombination process, with emission of a photon, an atom of a certain excited level is formed, which then decays to the ground level (Chap. 6). For the steady-state case, several energy levels of the atom will be populated.
- *Population from dielectronic recombination*: this process occurs in two phases—firstly, the electron and the ion form an atom in an excited state, and the emitted energy is used to excite another internal electron of the atom, resulting in two excited electrons with the same energy as free particles and without emission of radiation. Secondly, the atom may return to the original configuration or it may cascade to a lower level while emitting energy. In the interstellar medium, this process is important for temperatures of the order of 10^4 K, because in order to excite an internal electron of an atom, the energy must be high. Ionization through internal layers can also lead to the production of excited ions, whose de-excitation produces an additional ion, in the so-called Auger effect.
- *Population from two photons emission*: in this process, there is emission of two photons instead of one, due to the existence of forbidden transitions. For instance, a proton can capture an electron in an excited level and then decay to state 2^2S. The transition to state 1^2S being forbidden, the de-excitation can occur with emission of two photons, whose added energies are equal to the energy of a Lyman-α photon. This process is important for continuum emission in planetary nebulae.
- *Excitation by photon pumping*: after photon absorption, an atom may become excited and then cascades to a low-energy excited state, unreachable by direct radiative excitation from the ground state. In this way, there will be an excited level occupation triggered by the absorbed photon. This process occurs for instance in the excitation of H hyperfine levels of the ground state by radiation in the H Lyman-α line.
- *Excitation by charge exchange reactions*: In charge exchange reactions of the type

$$A + B^+ \rightarrow A^+ + B^* \rightarrow A^+ + B + h\nu,$$

excited state ions may form, denoted above by an asterisk.

5.5 Two-Level Atoms

5.5.1 Deviation Coefficients

Let us consider the statistical equilibrium equation (5.9), taking into account collisions between electrons in a two-level atom, with levels denoted by 1 (lower) and 2 (upper). We have

$$n_1(n_e\gamma_{12} + B_{12}U_\nu) = n_2(n_e\gamma_{21} + B_{21}U_\nu) + n_2A_{21}, \tag{5.36}$$

that is,

$$\frac{n_2}{n_1} = \frac{n_e\gamma_{12} + B_{12}U_\nu}{n_e\gamma_{21} + B_{21}U_\nu + A_{21}}. \tag{5.37}$$

We may rewrite (5.1) in the form

$$\frac{n_2}{n_1} = \frac{b_2}{b_1}\frac{g_2}{g_1}e^{-h\nu/kT}, \tag{5.38}$$

where we have left out frequency index (1,2) to simplify. From (5.37) and (5.38), we obtain

$$\frac{b_2}{b_1} = \frac{g_1}{g_2}e^{h\nu/kT}\frac{n_e\gamma_{12} + B_{12}U_\nu}{n_e\gamma_{21} + B_{21}U_\nu + A_{21}}. \tag{5.39}$$

Using the relations between the Einstein coefficients (3.32) and (3.33),

$$\frac{b_2}{b_1} = \frac{g_1e^{h\nu/kT}}{g_2}\frac{n_e\gamma_{12} + \frac{g_2c^3A_{21}U_\nu}{g_1 8\pi h\nu^3}}{n_e\gamma_{21} + \frac{c^3A_{21}U_\nu}{8\pi h\nu^3} + A_{21}}. \tag{5.40}$$

Using (5.13), simplifying and dividing by A_{21}, we get

$$\frac{b_2}{b_1} = \frac{n_e\gamma_{21}}{A_{21}} + \frac{e^{h\nu/kT}c^3U_\nu}{8\pi\, h\nu^3}\Bigg/ 1 + \frac{n_e\gamma_{21}}{A_{21}} + \frac{c^3U_\nu}{8\pi\, h\nu^3}. \tag{5.41}$$

Therefore, knowing the radiation field U_ν and the probability of spontaneous emission A_{21}, besides the de-excitation coefficient γ_{21}, we may determine the ratio between deviation coefficients b_2/b_1. Replacing in (5.1), the relative populations of the two levels may be obtained for several values of n_e. Equation (5.41) only includes collisions with electrons. If other particles are considered, we should include a summation in the other possible collisional de-excitation coefficients.

5.5.2 Application to a Planck Radiation Field

We saw in Chap. 2 that a reasonable approximation for the interstellar radiation field is a four-component Planck model diluted by a certain factor W. In a more simplified manner, we can consider a one-component model with a certain temperature T_r, known as the *radiation temperature*, with a dilution factor W. This model is particularly useful when we consider the cosmic background radiation, for which $T_r = 2.7$ K and $W = 1$. The energy density may be written

$$\begin{aligned} U_\nu &= \frac{W}{c} \int I_\nu \, \mathrm{d}\omega \\ &= \frac{4\pi W}{c} \frac{2h\nu^3}{c^2} \frac{1}{\mathrm{e}^{h\nu/kT_r} - 1} \\ &= \frac{8\pi h\nu^3}{c^3} \frac{W}{\mathrm{e}^{h\nu/kT_r} - 1}, \end{aligned} \tag{5.42}$$

and thus,

$$\frac{c^3 U_\nu}{8\pi h\nu^3} = \frac{W}{\mathrm{e}^{h\nu/kT_r} - 1}. \tag{5.43}$$

Replacing in (5.41) yields

$$\frac{b_2}{b_1} = \frac{n_e \gamma_{21}}{A_{21}} + \frac{W \,\mathrm{e}^{h\nu/kT}}{\mathrm{e}^{h\nu/kT_r} - 1} \Bigg/ 1 + \frac{n_e \gamma_{21}}{A_{21}} + \frac{W}{\mathrm{e}^{h\nu/kT_r} - 1}. \tag{5.44}$$

As we have seen in Chap. 2, the interstellar radiation field corresponds to $T_r \sim 10^4$ K with a dilution factor $W \sim 10^{-14}$, for optical transitions. In this case, the terms of (5.41) involving U_ν are negligible, and we may write

$$\frac{b_2}{b_1} = \frac{1}{1 + \frac{A_{21}}{n_e \gamma_{21}}}. \tag{5.45}$$

Let us consider two cases:

1. If $\gamma_{21} n_e \gg A_{21}$, that is, collisional processes are comparatively more important and $b_2/b_1 \to 1$, that is, the relative populations are determined by the Boltzmann equation.
2. If $\gamma_{21} n_e \ll A_{21}$, from (5.45), we have

$$\frac{b_2}{b_1} = \frac{n_e \gamma_{21}}{n_e \gamma_{21} + A_{21}} \simeq \frac{n_e \gamma_{21}}{A_{21}} \propto n_e, \tag{5.46}$$

that is, b_2/b_1 is small (because $\gamma_{21} n_e / A_{21} \ll 1$) and proportional to the electron density n_e. Collisional de-excitations are negligible, and ratio n_2/n_1 is determined by the equilibrium between collisional excitations, which are proportional to n_e, and radiative de-excitations. Thus, each collisional excitation ends up producing a photon.

5.5.3 *H Excitation in H I Regions*

Let us consider H hyperfine level excitation triggered not only by collisions with other atoms but also by radiative transitions caused by the cosmic background radiation. In this case, we replace n_e for n_H in (5.41), we expand the exponential terms of (5.41) and (5.42), and we obtain

$$\frac{b_2}{b_1} \simeq 1 \tag{5.47}$$

for

$$\chi \simeq 1 + \frac{1}{23 n_H}, \tag{5.48}$$

where χ is the correction term for the absorption cross section σ, for case $h\nu \ll kT$ (Chap. 3),

$$\sigma = \sigma_u \frac{h\nu}{kT} \chi, \tag{5.49}$$

and we use $T \simeq 80$ K and $\gamma_{21} \simeq 8 \times 10^{-11}$ cm^3 s^{-1} (see Table 5.3). Since χ is just a little higher than 1, the total cross section is a bit higher than it would be in TE, when $\chi = 1$, and the radiative transition effect will generally be small.

5.6 Three-Level Atoms

Finally, let us consider an atom with three levels, 1, 2, and 3, in the particular case when (i) transitions between levels 2 and 3 are negligible, (ii) induced radiative transitions are negligible, and (iii) only collisions with electrons are considered. In this case, using (5.45) for b_2/b_1 and b_3/b_1, we can show that the emitted intensity ratio I_{31}/I_{21} in lines 3–1 and 2–1 is given by

$$\begin{aligned} \frac{I_{31}}{I_{21}} &= \frac{n_3 A_{31} h\nu_{31}}{n_2 A_{21} h\nu_{21}} \\ &= \frac{g_3 A_{31} \nu_{31}}{g_2 A_{21} \nu_{21}} \left[\frac{1 + \frac{A_{21}}{n_e \gamma_{21}}}{1 + \frac{A_{31}}{n_e \gamma_{31}}} \right] e^{-E_{23}/kT}, \end{aligned} \tag{5.50}$$

where intensity is generally given in units: erg $\mathrm{cm^{-2}\ s^{-1}\ cm^{-1}}$.

Let us consider two cases:

(i) If n_e is very high, the term between brackets approaches 1, and the relative populations are given by the Boltzmann equation:

$$\frac{I_{31}}{I_{21}} = \frac{n_3^*}{n_2^*}\frac{A_{31}\nu_{31}}{A_{21}\nu_{21}}. \tag{5.51}$$

(ii) If n_e is low, we may neglect terms $n_e\gamma_{21}$ and $n_e\gamma_{31}$ relatively to A_{21} and A_{31}, respectively, in (5.50), thus obtaining

$$\frac{I_{31}}{I_{21}} = \frac{g_3 A_{31}\nu_{31}A_{21}n_e\gamma_{31}}{g_2 A_{21}\nu_{21}A_{31}n_e\gamma_{21}}\,\mathrm{e}^{-E_{23}/kT}. \tag{5.52}$$

Using (5.13), we have

$$\frac{I_{31}}{I_{21}} = \frac{g_3\gamma_{31}\,\mathrm{e}^{-E_{13}/kT}\nu_{31}}{g_2\gamma_{21}\,\mathrm{e}^{-E_{12}/kT}\nu_{21}} = \frac{\gamma_{13}\nu_{31}}{\gamma_{12}\nu_{21}}. \tag{5.53}$$

Therefore, the emission rate per $\mathrm{cm^{-3}}$ and per second is proportional to γ_{13}/γ_{12}, and each excitation produces a photon. Equation (5.50) can be used to determine n_e and electron temperature T_e in H II regions and planetary nebulae, by means of the analysis of ion forbidden lines such as O II, O III, N II, S I, and Ne III, particularly [O II] λ3726/3729 Å and [S II] λ6716/6730 Å lines for n_e and [O III] λ4363/5007 Å and [N II] λ5754/6583 Å lines for T_e.

5.6.1 Example: Electron Density in Planetary Nebulae

We can apply (5.50) to estimate the electron density in planetary nebulae from intensity measurements of S II lines. S^+ has an energy level diagram similar to the one of O^+ (see Fig. 5.1), and in this case, we may write

$$\begin{aligned} &Level\ 3 \rightarrow {}^2D_{5/2}\\ &Level\ 2 \rightarrow {}^2D_{3/2}\\ &Level\ 1 \rightarrow {}^4S_{3/2}. \end{aligned}$$

The transitions correspond to $\lambda(2,1) = 6{,}730.8$ Å and $\lambda(3,1) = 6{,}716.4$ Å lines, with $A_{21} = 8.8 \times 10^{-4}\ \mathrm{s^{-1}}$ and $A_{31} = 2.6 \times 10^{-4}\ \mathrm{s^{-1}}$. We can obtain γ_{21} and γ_{31} using (5.32) with $g_2 = 4$, $g_3 = 6$, $\Omega \simeq 7$, and $T \simeq 10^4$ K. Applying (5.50), we see that $E_{23} \simeq 6.3 \times 10^{-15}$ erg $\ll$ kT $\simeq 1.38 \times 10^{12}$ erg so that the exponential term is essentially equal to 1. For instance, Costa et al. (1996) have estimated a ratio

$I_{6716}/I_{6731} \simeq 0.89$ for planetary nebula NGC 3132. Replacing this value in (5.50) and using the above parameters, we obtain $n_e \simeq 620$ cm^{-3}. This result may be compared with $n_e = 710$ cm^{-3}, a value obtained by a more accurate method. In Chap. 8, we will return to the discussion of the determination of density and electron temperature in photoionized nebulae.

Exercises

5.1 Consider the emission lines [N II] λ6548/6583 Å in a nebula with $T_e = 10^4$ K. (a) Calculate the collisional de-excitation coefficient γ_{kj} from level $^1D_2$ to levels ^3P. Use data from Table 5.1. (b) What is the total probability per unit time of spontaneous emissions from level $^1D_2$ to lower levels? (c) How will de-excitation of level $^1D_2$ be accomplished in a region with $n_e \simeq 10^3$ cm^{-3}?

5.2 In a two-level atom, with levels j and k, de-excitation will be radiative if $A_{kj} \gg n_e\, \gamma_{kj}$ and collisional if $A_{kj} \ll n_e\, \gamma_{kj}$. We may define the critical density n_c that separates the two regimes by

$$n_c = \frac{A_{kj}}{\gamma_{kj}}.$$

This expression can be easily generalized for atoms with many levels. Determine the critical density for [O III] λ4959/5007 Å and [N II] λ6548/6583 Å transitions in a nebula with $T_e = 10^4$ K.

5.3 The hydrogen hyperfine level de-excitation coefficient for collisions with other H atoms can be approximated by the expression

$$\gamma_{kj} \simeq 0.005\, T + 0.400$$

in the approximated interval $30 \lesssim T(\mathrm{K}) \lesssim 300$, typical of diffuse interstellar clouds, where T is in K and γ_{kj} is given in units of 10^{-10} cm^3 s^{-1}. What is the error introduced by applying this expression to an interstellar cloud with $T = 100$ K?

5.4 Demonstrate relation (5.50).

5.5 Consider the three-level model mentioned in Sect. 5.6 for ion S^+. (a) What are the energies E_{21} and E_{31} that correspond to lines 6,730.8 Å and 6,716.4 Å? (b) Estimate the collisional de-excitation coefficients γ_{21} and γ_{31} for levels $2D_{3/2}$ and $2D_{5/2}$, respectively. (c) Consider the probability of spontaneous emission in level ^2D given by $A_{31} \simeq 2.6 \times 10^{-4}$ s^{-1}. How will de-excitation be accomplished in NGC 3132, assuming $T_e \simeq 10^4$ K and $n_e \sim 620$ cm^{-3}?

Bibliography

Burke, P.G., Eissner, W.B., Hummer, D.G., Percival, I.C. (eds.): Atoms in Astrophysics. Plenum, New York (1983). Includes several articles about excitation conditions and numerical values of collision coefficients and collision strengths

Costa, R.D.D., Chiappini, C., Maciel, W.J., Freitas-Pacheco, J.A.: New abundances of southern planetary nebulae. Astron. Astrophys. Suppl. **116**, 249 (1996). Determination of the chemical composition of a sample of galactic planetary nebulae, their densities and electron temperatures. See also Astron. Astrophys. vol. 276, p.184, 1993

Czyzak, S.J.: In: Middlehurst, B.M., Aller, L.H. (eds.) Nebulae and Interstellar Matter, p. 403. University of Chicago Press, Chicago (1968). Good discussion on excitation processes, determination of collision coefficients and strengths in a classical collection of astrophysical texts

Osterbrock, D.: Astrophysics of Gaseous Nebulae and Active Galactic Nuclei. University Science Books, Mill Valley (1989). Referred to in chapter 1. Includes an excellent discussion on the excitation conditions in interstellar photoionized nebulae, numerical values and references to classical and recent works about collision strengths determination. Table 5.1 is based on this reference

Rybicki, G.B., Lightman, A.P.: Radiative Processes in Astrophysics. Wiley, New York (1979). Referred to in chapter 2. Basic text about radiative processes, with a discussion of TE equations

Scheffler, H., Elsässer, H.: Physics of the Galaxy and Interstellar Matter. Springer, Berlin (1988). Referred to in chapter 1. Examples and applications of interstellar line excitation conditions

Spitzer, L.: Physical Processes in the Interstellar Medium. Wiley, New York (1978). Referred to in chapter 1. Includes a good discussion on statistical equilibrium, excitation conditions and applications to the interstellar medium, with some numerical values and references. Table 5.3 is based on this reference

Chapter 6
Ionization in the Interstellar Medium

6.1 Ionization Equilibrium

There are many clues that point to the existence of ionized gas in the interstellar medium, such as H II regions, planetary nebulae, and supernova remnants. Atom ionization and recombination processes, as well as formation and dissociation of molecules, may be described in steady state by an equation similar to the statistical equilibrium equation (5.3),

$$\frac{dn_j}{dt} = -n_j \sum_Y \sum_k (R_{jk})_Y + \sum_Y \sum_k n_k (R_{kj})_Y = 0, \tag{6.1}$$

as long as bound-free transitions in Y are considered. We will now discuss the principal physical processes related to the ionization of H and heavy elements for interstellar medium conditions.

Let us consider a relatively simple situation where ionizations originate from photon absorption, that is, *photoionization*, and recombinations are radiative, that is, an electron is captured by an ion to form an atom in a certain state j, with emission of a photon. In this case, the probability rates R are defined in terms of coefficients:

$$R_{j\mathrm{ph}}(\text{photoionization}) \rightarrow \beta_{j\mathrm{ph}}$$

$$R_{\mathrm{ph}j}(\text{radiative recombination}) \rightarrow n_e \alpha_j,$$

that is, $\beta_{j\mathrm{ph}}$ (s^{-1}) is the probability per unit time of a radiative transition from state j to the continuum and α_j is the recombination coefficient (cm^3 s^{-1}) defined in such a way that the probability per unit time of forming an atom in state j by electron recombination is $n_e \alpha_j$. Given these conditions, the *ionization equilibrium equation* may be written

W.J. Maciel, *Astrophysics of the Interstellar Medium*,
DOI 10.1007/978-1-4614-3767-3_6, © Springer Science+Business Media New York 2013

$$\sum_j n_j(\mathrm{X}^r)\beta_{j\mathrm{ph}} = \sum_j n\left(\mathrm{X}^{r+1}\right) n_\mathrm{e}\alpha_j. \tag{6.2}$$

Therefore, the product $n(\mathrm{X}^{r+1})n_\mathrm{e}\alpha_j$ gives the number of ion X^{r+1} recombinations with electrons per unit volume per unit time, originating the formation of atoms at level j. Applying the summation we obtain the total number of recombinations per cm^3 per second, which must balance the total number of ionizations per cm^3 per second.

In general only the ground state is populated in the interstellar medium and photoionizations are made essentially from that level. However, recombinations can be done for every level j of the atom, followed by decay until the ground level is reached. In these conditions (6.2) may be written

$$n(\mathrm{X}^r)\beta_{1\,\mathrm{ph}} = n\left(\mathrm{X}^{r+1}\right)n_\mathrm{e}\sum_j \alpha_j. \tag{6.3}$$

In the general case, atom X exists in various ionization states X^0, X^1, etc., and

$$n(\mathrm{X}) = \sum_r n(\mathrm{X}^r). \tag{6.4}$$

Frequently, only two consecutive ionization states contribute in a significant way to the total abundance of an ion, the others being negligible. Denoting these two states by r and $r + 1$, we have

$$n(\mathrm{X}) \simeq n(\mathrm{X}^r) + n\left(\mathrm{X}^{r+1}\right). \tag{6.5}$$

Defining the *degree of ionization* of element X by $x(\mathrm{X})$ or simply x,

$$x = \frac{n\left(\mathrm{X}^{r+1}\right)}{n(\mathrm{X})} = \frac{n\left(\mathrm{X}^{r+1}\right)}{n(\mathrm{X}^r) + n\left(\mathrm{X}^{r+1}\right)}. \tag{6.6}$$

In this case, if only X^{r+1} is abundant, $x \to 1$, and if only X^r is abundant, $x \to 0$. We can also write

$$\frac{x}{1-x} = \frac{n\left(\mathrm{X}^{r+1}\right)}{n(\mathrm{X}^r)}. \tag{6.7}$$

Replacing (6.7) in (6.3) and defining the total recombination coefficient for all levels

$$\alpha = \sum_j \alpha_j, \tag{6.8}$$

we obtain

$$(1-x)\beta_{1\,\mathrm{ph}} = x n_{\mathrm{e}} \sum_j \alpha_j = x\, n_{\mathrm{e}}\, \alpha, \tag{6.9}$$

which is the ionization equilibrium equation generally used for the interstellar medium. In terms of densities, (6.7) and (6.9) allow us to write

$$\frac{n(\mathrm{X}^{r+1})n_{\mathrm{e}}}{n(\mathrm{X}^{r})} = \frac{\beta_{1\,\mathrm{ph}}}{\alpha}, \tag{6.10}$$

which must be compared with the Saha equation (3.13) or (3.16), valid for thermodynamic equilibrium (TE):

$$\frac{n^{*}(\mathrm{X}^{r+1})n_{\mathrm{e}}}{n^{*}(\mathrm{X}^{r})} = \frac{f_{r+1}f_{\mathrm{e}}}{f_r}. \tag{6.11}$$

From (6.10) and (6.11) it remains clear that in absence of TE, detailed ionization and recombination processes must be taken into account, by means of rate $\beta_{1\mathrm{ph}}$ and coefficient α, respectively.

6.2 Photoionization Rate

The photoionization rate or photoionization probability rate or photoionization probability per unit time $\beta_{j\mathrm{ph}}$ can be related to photoionization cross section $\sigma_{\nu\mathrm{ph}}$ for radiative transitions producing a free electron or continuum. To obtain this relation we will initially consider the probability per unit time $\beta_{j\mathrm{ph}}$ of a bound–bound transition between states j and k. As we have seen in Chap. 3, the total energy absorbed in a line per cubic centimeter per second per unit solid angle is $\int I_\nu k_\nu \mathrm{d}\nu$, generally given in erg cm^{-3} s^{-1} sr^{-1}. Considering the relation between energy density U_ν and intensity I_ν,

$$U_\nu = \frac{1}{c}\int I_\nu\, \mathrm{d}\omega, \tag{6.12}$$

we have that the total energy absorbed per cm^3 and per second is $\int c U_\nu k_\nu \mathrm{d}\nu$ (erg cm^{-3} s^{-1}), that is, the number of transitions $j \rightarrow$ k per cm^3 per second is $\int (cU_\nu k_\nu / h\nu)\mathrm{d}\nu$ (cm^{-3} s^{-1}). The probability per unit time β_{jk} for transitions departing from level j is

$$\beta_{jk} = \frac{1}{n_j}\int \frac{cU_\nu k_\nu}{h\nu}\,\mathrm{d}\nu, \tag{6.13}$$

where the integral is done along the whole line. As $k_\nu = n_j\sigma_\nu$,

$$\beta_{jk} = \int \frac{cU_\nu\sigma_\nu}{h\nu}\mathrm{d}\nu \quad \text{(bound – bound)}. \tag{6.14}$$

By analogy with (6.14), the photoionization probability per unit time $\beta_{j\mathrm{ph}}$ may be written

$$\beta_{j\mathrm{ph}} = \int_{\nu_0}^{\infty} \frac{cU_\nu\sigma_{\nu f}}{h\nu}\mathrm{d}\nu \quad \text{(bound – free)}. \tag{6.15}$$

The integral extends along the whole frequency interval of the continuum absorption spectrum, that is, from a critical frequency $\nu = \nu_0$, below which the photons do not have enough energy for the transition, to $\nu \to \infty$.

6.3 Radiative Recombination Coefficient

The radiative recombination coefficient α_j introduced in (6.2) can be related to the recombination cross section or recapture cross section σ_{cj} of an electron by an ion to form an atom at excitation level j. We may write

$$\alpha_j = \langle\sigma_{cj}\upsilon\rangle = \int_0^\infty \sigma_{cj}\upsilon f(\upsilon)\mathrm{d}\upsilon, \tag{6.16}$$

where $f(\upsilon)$ is again the fraction of particles (electrons) in velocity interval υ to $\upsilon + \mathrm{d}\upsilon$. For a Maxwellian distribution,

$$\begin{aligned}\alpha_j &= \int_0^\infty \sigma_{cj}\left[4\pi\left(\frac{m_\mathrm{e}}{2\pi kT}\right)^{3/2}\upsilon^3\mathrm{e}^{-m_\mathrm{e}\upsilon^2/2kT}\right]\mathrm{d}\upsilon \\ &= \frac{4\ell^3}{\sqrt{\pi}}\int_0^\infty \sigma_{cj}\upsilon^3\mathrm{e}^{-\ell^2\upsilon^2}\mathrm{d}\upsilon,\end{aligned} \tag{6.17}$$

where

$$\ell^2 = \frac{m_\mathrm{e}}{2kT}. \tag{6.18}$$

Equations (6.17) and (6.18) must be compared to (5.17) and (5.18) from the previous chapter. In the same way, the total recombination factor becomes

$$\alpha = \sum_j \alpha_j = \frac{4\ell^3}{\sqrt{\pi}}\sum_j \int_0^\infty \sigma_{cj}\upsilon^3\mathrm{e}^{-\ell^2\upsilon^2}\mathrm{d}\upsilon. \tag{6.19}$$

6.3.1 Milne Relation

We will now obtain a relation between the recombination cross section σ_{cj} and the photoionization cross section $\sigma_{\nu ph}$. In TE we may apply the detailed balance condition, where the number of ionizations per cm^3 per second from level j, triggered by photons in the energy interval $h\nu$ to $h\nu$ + d($h\nu$), must balance the number of recombinations to level j per cm^3 per second, producing electrons with velocities between υ and υ + dυ:

$$4\pi n_j^*(\mathrm{X}^r)\sigma_{\nu\,\mathrm{ph}}\left(1-\mathrm{e}^{-h\nu/kT}\right)\frac{I_\nu^*\mathrm{d}\nu}{h\nu}=n^*\left(\mathrm{X}^{r+1}\right)n_\mathrm{e}\sigma_{cj}\upsilon f(\upsilon)\mathrm{d}\upsilon. \tag{6.20}$$

Term $(1-\mathrm{e}^{-h\nu/kT})$ refers to induced emissions (see Chap. 3). From (6.20)

$$\sigma_{cj}=\frac{n_j^*(\mathrm{X}^r)}{n^*\left(\mathrm{X}^{r+1}\right)n_\mathrm{e}}\frac{4\pi}{f(\upsilon)}\frac{I_\nu^*(\mathrm{d}\nu/\mathrm{d}\upsilon)}{h\,\nu\,\upsilon}\sigma_{\nu\mathrm{ph}}(1-\mathrm{e}^{-h\nu/kT}). \tag{6.21}$$

Using relation (3.12) between n_j* and n*,

$$\sigma_{cj}=\frac{n^*(\mathrm{X}^r)}{n^*\left(\mathrm{X}^{r+1}\right)n_\mathrm{e}}\frac{g_{rj}}{f_r}\frac{4\pi}{f(\upsilon)}\frac{I_\nu^*(\mathrm{d}\nu/\mathrm{d}\upsilon)}{h\,\nu\,\upsilon}\frac{\sigma_{\nu f}(1-\mathrm{e}^{-h\nu/kT})}{\mathrm{e}^{E_{rj}/kT}}. \tag{6.22}$$

Using the Saha equation, with $f_r\simeq g_{r,1}\,\mathrm{e}^{-E_{r,1}/kT}$,

$$\sigma_{cj}=\frac{g_{r,1}}{2g_{r+1,1}}\left(\frac{h^2}{2\pi m_\mathrm{e}kT}\right)^{3/2}\frac{g_{rj}}{g_{r,1}}\frac{4\pi}{f(\upsilon)}\frac{I_\nu^*(\mathrm{d}\nu/\mathrm{d}\upsilon)}{h\,\nu\,\upsilon}\frac{\sigma_{\nu f}(1-\mathrm{e}^{-h\nu/kT})}{\mathrm{e}^{-\phi_r/kT}\mathrm{e}^{(E_{rj}-E_{ri})/kT}}, \tag{6.23}$$

where ϕ_r is the ion-bound energy in level r. Replacing the Maxwell and Planck functions and after some rearranging, we obtain

$$\sigma_{cj}=\frac{g_{rj}}{g_{r+1,1}}\frac{h^3\nu^2}{m_\mathrm{e}^3c^2\upsilon^3}\frac{\mathrm{d}\nu}{\mathrm{d}\upsilon}\frac{\sigma_{\nu f}\mathrm{e}^{m_\mathrm{e}\nu^2/2kT}}{\mathrm{e}^{h\nu/kT}\mathrm{e}^{-\phi_r/kT}\,\mathrm{e}^{(E_{rj}-E_{ri})/kT}}. \tag{6.24}$$

The exponential terms cancel out,

$$\frac{\mathrm{e}^{m_\mathrm{e}\upsilon^2/2kT}}{\mathrm{e}^{h\nu/kT}\,\mathrm{e}^{-\phi_r/kT}\,\mathrm{e}^{(E_{rj}-E_{ri})/kT}}=1, \tag{6.25}$$

because, by energy conservation, the electron kinetic energy is given by

$$\frac{1}{2}m_\mathrm{e}\upsilon^2=h\nu-\phi_j, \tag{6.26}$$

where ϕ_j, the bound energy of level j, is

$$\phi_j = \phi_r - (E_{rj} - E_{r,1}) \tag{6.27}$$

and

$$\frac{1}{2} m_e v^2 = h\nu - \phi_r + (E_{rj} - E_{r,1}). \tag{6.28}$$

From (6.26) we have

$$m_e \, v \, dv = h \, d\nu, \tag{6.29}$$

thus (6.24) becomes

$$\sigma_{cj} = \frac{g_{rj}}{g_{r+1,1}} \frac{h^2 \nu^2}{m_e^2 c^2 v^2} \sigma_{\nu f}, \tag{6.30}$$

which is the *Milne relation* between the recombination cross section and the photoionization cross section.

6.3.2 *Oscillator Strength for Bound-Free Transitions*

The bound-free absorption cross section $\sigma_{\nu \mathrm{ph}}$ may be given in terms of oscillator strength, as for the case of bound–bound transitions, with a proper definition of the f-values. For the bound–bound case, we have seen in Chap. 3 that

$$\sigma_u = \frac{\pi e^2}{m_e c} f_{jk}, \tag{6.31}$$

$$\sigma = \sigma_u \left[1 - \frac{b_k}{b_j} \mathrm{e}^{-h\nu/kT} \right]. \tag{6.32}$$

In bound-free transitions, neglecting induced emissions, we define the continuum oscillator strength in such a way that

$$\sigma_{\nu \mathrm{ph}} = \frac{\pi e^2}{m_e c} \frac{df_\nu}{d\nu}. \tag{6.33}$$

The $\sigma_{\nu \mathrm{ph}}$ variation with frequency is determined by atomic structure, and so this variation is introduced in the definition of the continuum oscillator strength f_ν.

6.4 Photoionization of Hydrogen

Due to its high abundance, hydrogen is the most important element of the interstellar medium. Let us consider the ionization equilibrium of this element or, in a more general way, of an hydrogen-like atom with nuclear charge Z (H I, He II, Li III, Be IV, B V, etc.), characterized by the principal quantum number n, under the action of photoionization and radiative recombination processes. The energy in state n is given by

$$E_n = -\frac{hRZ^2}{n^2}, \tag{6.34}$$

where R is the Rydberg constant, $R = 2\pi^2\mu e^4/h^3$, and μ is the reduced mass of the electron–ion system:

$$\mu = \frac{m_e m_n}{m_e + m_n}. \tag{6.35}$$

For a nucleus with infinite mass, $\mu = m_e$ and

$$R = R_\infty = \frac{2\pi^2 m_e e^4}{h^3} = 3.29 \times 10^{15}\ \mathrm{s}^{-1}. \tag{6.36}$$

In this chapter we will always use $R \simeq \mathrm{R}_\infty$. We see that when $n = 1$, $E_1 = -hRZ^2 = -13.6$ eV for H. For $n \to \infty$, $E_\infty \to 0$ and the continuum states have positive energy. The energy conservation equation for the photoionization and recombination process may be written

$$\frac{1}{2} m_e v^2 = h\nu - \frac{hRZ^2}{n^2}, \tag{6.37}$$

and the bound energy of state n, which is the minimum energy required for an ionization from state n to occur, is

$$\phi_n = E_\infty - E_n = -E_n = \frac{hRZ^2}{n^2}. \tag{6.38}$$

From (6.34) the frequency of a transition between two bound states $m \to n$ $(m > n)$ is

$$\nu_{mn} = -RZ^2\left(\frac{1}{m^2} - \frac{1}{n^2}\right) = RZ^2\left(\frac{1}{n^2} - \frac{1}{m^2}\right). \tag{6.39}$$

We can visualize a bound-free transition where frequency is given by (6.39) replacing m by il, where l is the real quantum number characteristic of the free state, not necessarily an integer. In this case

$$\nu_{\ell n} = \nu = RZ^2\left(\frac{1}{n^2} + \frac{1}{\ell^2}\right). \tag{6.40}$$

Replacing (6.40) in (6.37), we obtain

$$\frac{1}{2} m_e \upsilon^2 = \frac{hRZ^2}{\ell^2} \tag{6.41}$$

$$m_e\, \upsilon\, d\upsilon = h\, d\nu = -\frac{2hRZ^2}{\ell^3} d\ell. \tag{6.42}$$

6.4.1 *Photoionization Cross Section*

Considering the photoionization cross section $\sigma_{\nu \mathrm{ph}}$ given by (6.33), we have

$$\sigma_{\nu \mathrm{ph}} = \frac{\pi e^2}{m_e c}\frac{df_\nu}{d\nu} = \frac{\pi e^2}{m_e c}\frac{df_\nu}{d\ell}\left|\frac{d\ell}{d\nu}\right|. \tag{6.43}$$

Replacing $dl/d\nu$ in (6.42)

$$\sigma_{\nu \mathrm{ph}} = \frac{\pi e^2}{m_e c}\frac{\ell^3}{2RZ^2} f_{n\ell}, \tag{6.44}$$

where we define

$$f_{n\ell} = \frac{df_\nu}{d\ell}. \tag{6.45}$$

The oscillator strength may be calculated quantum mechanically, and we obtain

$$f_{n\ell} = \frac{2^6}{3\sqrt{3}\pi}\frac{1}{2n^2}\frac{1}{(1/n^2 + 1/\ell^2)^3}\left[\frac{1}{n^3}\frac{1}{\ell^3}\right] g_{n\mathrm{ph}}, \tag{6.46}$$

where $g_{n\mathrm{ph}}$ is the Gaunt factor for bound-free transitions, generally of the order of 1. Replacing (6.46) and (6.40) in (6.44), we have

$$\sigma_{\nu\,\mathrm{ph}} = \frac{16}{3\sqrt{3}} \frac{e^2}{m_e c} \frac{R^2 Z^4}{n^5 \nu^3} g_{nf} = \frac{32}{3\sqrt{3}} \frac{\pi^2 e^6}{ch^3} \frac{RZ^4}{n^5 \nu^3} g_{n\,\mathrm{ph}}. \tag{6.47}$$

In logarithmic terms, with the usual units,

$$\log \sigma_{\nu\,\mathrm{ph}} = 29.45 - 3\log\nu - 5\log n + 4\log Z + \log g_{n\mathrm{ph}}. \tag{6.48}$$

Comparing (6.33) and (6.47)

$$\frac{\mathrm{d}f_\nu}{\mathrm{d}\nu} = \frac{16}{3\sqrt{3}\pi}\left(\frac{1}{\nu}\right)^3 R^2 Z^4 \frac{g_{n\mathrm{ph}}}{n^5}. \tag{6.49}$$

Defining a reference frequency ν_1, corresponding to the Lyman limit, from (6.39) with $m \to \infty$ and $n = 1$, we have

$$\nu_1 = RZ^2 = 3.29 \times 10^{15}\,\mathrm{Hz} \tag{6.50}$$

for H, and

$$\begin{aligned} \frac{\mathrm{d}f_\nu}{\mathrm{d}(\nu/\nu_1)} &= \nu_1 \frac{\mathrm{d}f_\nu}{\mathrm{d}\nu} = \frac{16}{3\sqrt{3}\pi} \frac{R^3 Z^6}{\nu^3} \frac{g_{n\mathrm{ph}}}{n^5} \\ &= \frac{16}{3\sqrt{3}\pi}\left(\frac{\nu_1}{\nu}\right)^3 \frac{g_{n\mathrm{ph}}}{n^5}. \end{aligned} \tag{6.51}$$

However, from (6.33), $\sigma_{\nu\mathrm{ph}}$ may be written in terms of $\mathrm{d}f_\nu/\mathrm{d}(\nu/\nu_1)$,

$$\begin{aligned} \sigma_{\nu\,\mathrm{ph}} &= \frac{\pi e^2}{m_e c} \frac{\mathrm{d}f_\nu}{\mathrm{d}(\nu/\nu_1)} \frac{1}{\nu_1} = \frac{\pi e^2}{m_e c R} Z^{-2} \frac{\mathrm{d}f_\nu}{\mathrm{d}(\nu/\nu_1)} \\ &= 8.07 \times 10^{-18} Z^{-2} \frac{\mathrm{d}f_\nu}{\mathrm{d}(\nu/\nu_1)}\,\mathrm{cm}^2. \end{aligned} \tag{6.52}$$

Particular interesting cases are hydrogen, or hydrogen-like atoms in general, ionized from the ground-state level of the neutral atom, $n = 1$. In this case, considering (6.47) and (6.50),

$$\sigma_{\nu\,\mathrm{ph}} = \frac{16\,e^2}{3\sqrt{3} m_e c\, R\, Z^2}\left(\frac{\nu_1}{\nu}\right)^3 g_{1,\mathrm{ph}} \simeq 7.9 \times 10^{-18} Z^{-2}\left(\frac{\nu_1}{\nu}\right)^3 g_{1,\mathrm{ph}}\,\mathrm{cm}^2. \tag{6.53}$$

The Gaunt factor can be approximated by

$$g_{1\,\mathrm{ph}} = 8\pi\sqrt{3}\frac{\nu_1}{\nu} \frac{\exp(-4z\cot^{-1} z)}{1 - \mathrm{e}^{-2\pi z}}, \tag{6.54}$$

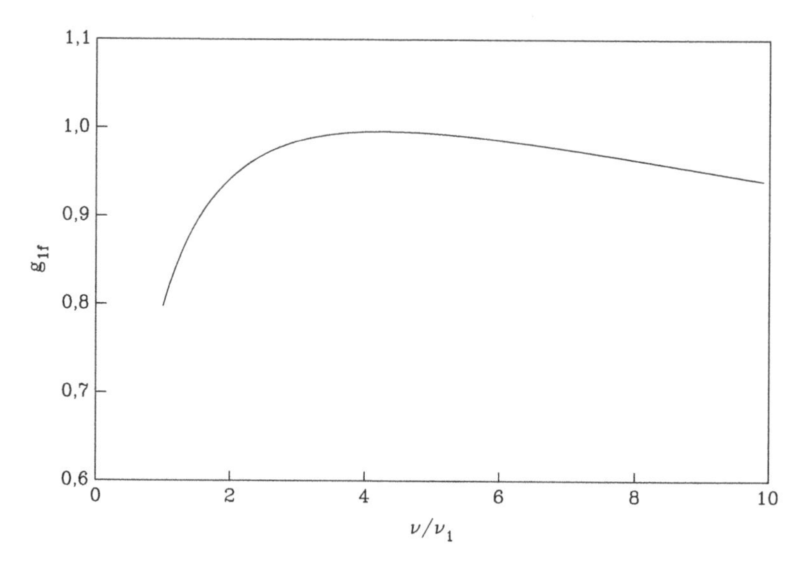

Fig. 6.1 The hydrogen Gaunt factor as a function of the frequency

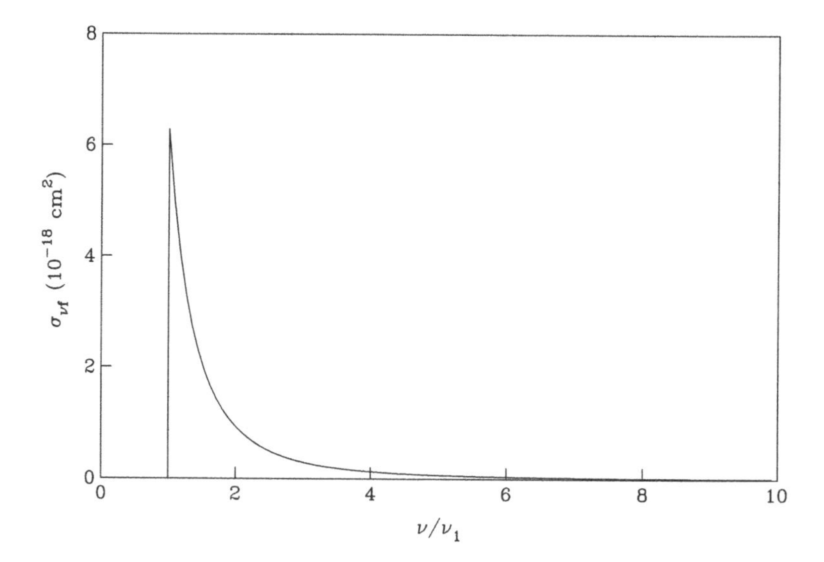

Fig. 6.2 The hydrogen photoionization cross section as a function of the frequency

where $z^2 = \nu_1/(\nu - \nu_1)$. Values of g_{1ph} are given in the literature, and Fig. 6.1 shows this factor for frequencies $\nu/\nu_1 < 10$.

Figure 6.2 shows the H photoionization cross-section variation with frequency, ionized from the first level. We see that the maximum cross section is of the order of $\sigma_{\nu ph} \simeq 6 \times 10^{-18}$ cm^2 for $\nu = \nu_1 = 3.29 \times 10^{15}$ s^{-1} or $\lambda \simeq 912$ Å, with $g_{1ph} \simeq 0.8$.

6.4.2 Radiative Recombination Cross Section

From the Milne relation (6.30) and using (6.51) and (6.52) with $g_n = 2n^2$ and $g_{p,1} = 1$, the H recombination cross section σ_{cn} may be written

$$\sigma_{cn} = \frac{g_n}{g_{p,1}} \frac{h^2 \nu^2}{m_e^2 c^2 v^2} \sigma_{\nu \mathrm{ph}} = A_r \frac{\nu_1}{\nu} \frac{h\nu_1}{(1/2) m_e v^2} \frac{g_{n\mathrm{ph}}}{n^3}, \tag{6.55}$$

where we define the recapture constant as

$$A_r = \frac{2^4}{3\sqrt{3}} \frac{he^2}{m_e^2 c^3} \simeq 2.1 \times 10^{-22}\, \mathrm{cm}^2. \tag{6.56}$$

In the recombination process, $(1/2)m_e v^2$ is the electron energy before capture. By energy conservation, from (6.37) and (6.50), we have

$$h\nu = \frac{1}{2} m_e v^2 + \frac{h\nu_1}{n^2}, \tag{6.57}$$

where $h\nu$ is the emitted photon energy.

6.4.3 Radiative Recombination Coefficient

According to (6.17) and (6.18) the recombination coefficient to state m is

$$\alpha_m = 4\pi \left(\frac{m_e}{2\pi kT} \right)^{3/2} \int_0^\infty \sigma_{cm} v^3 \mathrm{e}^{-m_e v^2/2kT} \mathrm{d}v. \tag{6.58}$$

Let us define the recombination coefficient up to level n by

$$\alpha^{(n)} = \sum_n^\infty \alpha_m, \tag{6.59}$$

so that

$$\alpha^{(2)} = \sum_2^\infty \alpha_m \tag{6.60}$$

$$\alpha^{(1)} = \alpha = \sum_1^\infty \alpha_m = \alpha_1 + \sum_2^\infty \alpha_m = \alpha_1 + \alpha^{(2)} \tag{6.61}$$

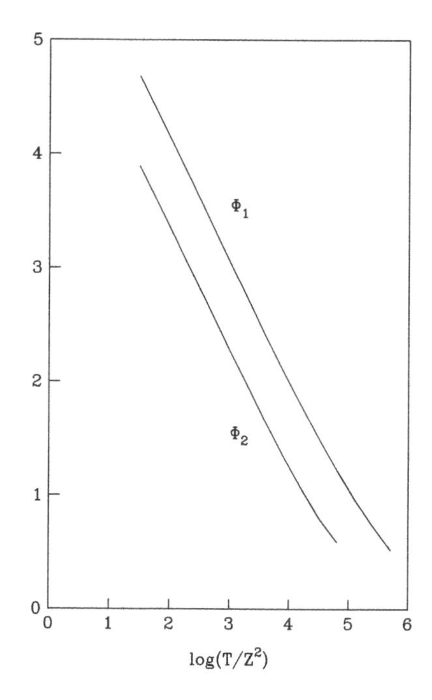

Fig. 6.3 The functions Φ_1 and Φ_2 for some typical temperatures in the interstellar gas

or $\alpha^{(2)} = \alpha - \alpha_1$. From (6.58) and (6.59)

$$\alpha^{(n)} = 4\pi\left(\frac{m_e}{2\pi kT}\right)^{3/2} \sum_n^{\infty} \int_0^{\infty} \sigma_{cm} v^3 \, e^{-m_e v^2/2kT} dv. \tag{6.62}$$

Coefficient $\alpha^{(n)}$ can be written in a more convenient way

$$\alpha^{(n)} = 2A_r \left(\frac{2kT}{\pi m_e}\right)^{1/2} \beta\phi_n(\beta) \simeq 2.1 \times 10^{-11} Z^2 T^{-1/2} \phi_n(\beta) \text{ cm}^3 \text{ s}^{-1}, \tag{6.63}$$

where

$$\beta = \frac{h\nu_1}{kT} = \frac{158,000\, Z^2}{T} \tag{6.64}$$

and $\phi_n(\beta)$ is a function tabulated in the literature. Figure 6.3 shows functions ϕ_1 and ϕ_2 for temperatures observed in the interstellar medium. The approximation errors in the figure are smaller or of the order of 2 % for $T/Z^2 \lesssim$ 16,000 K or log $(T/Z^2) \simeq 4.2$ but can be of the order of 10 % for $T/Z^2 \sim 10^6$ K or log $(T/Z^2) \sim 6$.

A classical example of the H photoionization process occurs in H II regions and planetary nebulae, where a hot star is surrounded by a gas cloud. The high abundance of ionizing ultraviolet photons and the high optical depth of H for these photons result in an ionized region surrounded by an H I region, of neutral hydrogen. These objects will be considered in detail in Chap. 8.

6.5 Ionization of H by Thermal Electrons

Practically all ionizing photons produced in H II regions are absorbed within the region. Thus, outside these regions ionization of H and other elements with ionization potential above 13.6 eV must be accomplished by other mechanisms. Several processes are proposed, applicable to different regions in the interstellar medium, generally based on the interaction between H atoms and energetic particles.

6.5.1 *Thermal Electrons and Coronal Gas*

For temperatures $T \gg 10^4$ K, H and other elements ionization by means of electron collisions may be important. Collision cross sections are generally known quite accurately. In these conditions, the thermal velocities of the electrons are $v \simeq (kT/m_e)^{1/2} \gtrsim 400\ \mathrm{km\ s^{-1}}$, and collisions are very energetic, causing whatever molecules present to dissociate, which in turn simplifies the ionization equilibrium.

Temperatures of the order of 10^4 K or higher are expected in O VI line formation regions, observed in the ultraviolet (1,031.9 Å and 1,037.6 Å). These lines are broad and observed in absorption in the direction of many hot stars, with a Doppler parameter given by

$$b = \left(\frac{2kT}{m}\right)^{1/2} = \frac{c}{\nu_{jk}}\Delta\nu_{\mathrm{D}} = \sqrt{2}\sigma_{\upsilon} \simeq 15 - 50\ \mathrm{km\ s^{-1}}. \tag{6.65}$$

Here, m is the absorbing atom mass, $\Delta\nu_{\mathrm{D}}$ is the Doppler width, and $\sigma_{\nu} = (kT/m)^{½}$ is the radial velocity dispersion of the region that formed the line (see Chap. 3). Assuming pure thermal scattering and letting A be the atomic mass of the absorbing atom, we have

$$b = 1.3 \times 10^4 \left(\frac{T}{A}\right)^{1/2}\ \mathrm{cm\ s^{-1}}, \tag{6.66}$$

where $T \simeq 2.2 \times 10^5 - 2.4 \times 10^6$ K.

The O VI mean density estimated from these lines gives $n(\mathrm{O\ VI}) \simeq 1.7 \times 10^{-8}\ \mathrm{cm^{-3}}$. However, traces of N V were observed and the N V/O VI ratio is smaller than 0.025 for several cases. These, as well as other similar line observations, imply a limitation in the gas temperature and density. Thus, from the above results $T \gtrsim 10^5$ K and from line widths $T \lesssim 10^6$ K. If the O abundance is normal, or cosmic, the H density will be

$$n_{\mathrm{H}} \simeq \frac{n(\mathrm{O})}{a(\mathrm{O})} > \frac{n(\mathrm{O\,VI})}{a(\mathrm{O})} \simeq \frac{10^{-8}}{10^{-4}} \sim 10^{-4}\ \mathrm{cm^{-3}}. \tag{6.67}$$

The nature of this gas is not very well understood. Part of it must occupy the central parts of H II regions, and another part is produced in supernovae explosions, where heating is triggered by shock waves. Part of the gas has intergalactic origin, as it was detected by the FUSE satellite in the direction of active galactic nuclei. We estimate that only about 0.1 % of the galactic disk mass in the solar neighborhood is in this diffuse and ionized form. This gas is usually identified as *coronal gas*, referring to a suggestion from L. Spitzer. It comprises a hot and diffuse region, which occupies a large part of the galactic disk volume. In these conditions, the ionization equilibrium is essentially determined by *collisional ionizations* balanced by *radiative and dielectronic recombinations*.

Let γ_{jph} (X^r) be the ionization rate of element X in ionization state r by means of collisions with thermal electrons. In this case, as in (6.2), the ionization equilibrium equation is

$$n_e \sum_j n_j(X^r)\gamma_{jph} = n_e n(X^{r+1}) \sum_j \alpha_j. \tag{6.68}$$

Note that, in the more general case, the recombination coefficient α_j should include radiative recombinations, where emission of a photon occurs, and dielectronic recombinations where, after recombination, two electrons occupy excited levels without emission of radiation. There is emission only when the atom decays to the ground level by cascading.

6.5.2 H Ionization

For hydrogen the ionization equilibrium equation (6.68) becomes

$$\sum_n n_m(H^0)\gamma_{mph} = n_p \sum_{m=1}^{\infty} \alpha_m = n_p \alpha^{(1)}. \tag{6.69}$$

The $n = 1$ level recombinations produce photons in the Lyman continuum, with $\lambda \leq 912$ Å, if n(H I) is high enough. In this case, we should consider $\alpha^{(2)}$ instead of $\alpha^{(1)}$ in the last equation. However, in the case of coronal gas, n(H I) is low and the Lyman continuum photons end up escaping the considered region.

Similarly to (6.3), let us just consider ionizations from the ground level. From (6.69)

$$n(H^0)\gamma_{1ph} = n_p \alpha^{(1)}. \tag{6.70}$$

Introducing the degree of ionization x [see (6.6) and (6.7)],

$$x = \frac{n_p}{n(H)}, \tag{6.71}$$

$$\frac{x}{1-x} \simeq \frac{n_p}{n(H^0)}, \tag{6.72}$$

where we neglect the formation of molecules involving H. From (6.70) and (6.72) follows

$$(1-x)\gamma_{1ph} = x\,\alpha^{(1)} \tag{6.73}$$

that corresponds to (6.9). We also have the relations:

$$(1-x) = 1 - \frac{n_p}{n(H^0)}(1-x), \tag{6.74}$$

$$1 - x = \frac{1}{1 + \gamma_{1ph}/\alpha^{(1)}}, \tag{6.75}$$

$$x = \frac{1}{1 + \alpha^{(1)}/\gamma_{1ph}}. \tag{6.76}$$

The collisional rates γ_{1ph} for several atoms can be theoretically determined or estimated from observations. Coefficients $\alpha^{(1)}$ can be, for instance, calculated from (6.63), and for $T \sim 10^6$ K results show that

$$\frac{\gamma_{1ph}}{\alpha^{(1)}} \gtrsim 10^6, \tag{6.77}$$

that is, $x \simeq 1$ and all H is essentially ionized.

6.6 Ionization of H: Cosmic Rays and X-Rays

The presence of cosmic rays—protons, electrons, and heavy nuclei—is well known from high atmosphere balloon and satellite observations. Suprathermal particles permeate the whole interstellar medium, moving at high speeds and triggering highly energetic collisions with atoms in the medium. Part of these collisions can ionize H, resulting in a total ionization rate ζ_H, which includes *primary ionization*, triggered by cosmic rays, and *secondary ionization*, triggered by energetic electrons coming from primary ionization. H ionization by cosmic rays can be particularly important in H I clouds, where energetic photons cannot penetrate and the temperature is not high enough for electron collisions to be important. Considering ionization by cosmic rays and radiative recombinations, the ionization equilibrium equation becomes

$$(1 - x)\varsigma_H = x\, n_e\, \alpha^{(2)}, \tag{6.78}$$

where again we consider ionizations from the ground level alone and replace α by $\alpha^{(2)}$, since Lyman continuum photons are quickly reabsorbed by the medium. Replacing (6.7) in (6.78) gives

$$\frac{n_p}{n(H^0)} n_e\, \alpha^{(2)} - \varsigma_H = 0. \tag{6.79}$$

Considering $n_e = n_p + n_i$, where n_i is the density of positively charged ions heavier than He, supposed neutral, we obtain

$$\frac{\alpha^{(2)}}{n(H^0)} n_p^2 + \frac{\alpha^{(2)} n_i}{n(H^0)} n_p - \varsigma_H = 0, \tag{6.80}$$

that can be solved as

$$n_p = \frac{1}{2} n_i \left\{ \left[1 + \frac{4\zeta_H n(H^0)}{\alpha^{(2)} n_i^2} \right]^{1/2} - 1 \right\}. \tag{6.81}$$

The value of ζ_H for cosmic protons and secondary electrons is highly uncertain, because the solar wind affects the low-energy particles energy spectrum (less or of the order of 10^9 eV per nucleon), which are the ones that have a larger cross section for H ionization. From extrapolations of the cosmic ray flux observed from the ground, we may estimate $\zeta_H \simeq 7 \times 10^{-18}$ s^{-1}. This value is sometimes considered a lower limit, and higher values have been proposed, such as $\zeta_H \simeq 10^{-15}$ s^{-1}, though evidence in favor of this value is scarce. More recent works suggest that this rate can be considered as an upper limit for diffuse clouds. In clouds with higher density and lower temperature, (6.81) is not valid because there is molecule formation that will alter the ionization equilibrium. Besides that, interstellar gas ionization triggers a series of chemical reactions involving atoms and molecules, whose observation leads to the determination of rates $\zeta_H \lesssim 10^{-17}$ s^{-1}.

Let us consider a low-density cloud with $n(H^0) \lesssim 10\,\mathrm{cm}^{-3}$. In this case $n(H^0) \simeq n_H$ and for $T \simeq 6{,}000$ K the recombination coefficient is $\alpha^{(2)} \simeq 4 \times 10^{-13}\,\mathrm{cm}^3\,\mathrm{s}^{-1}$. Since $n_e = n_p + n_i$, we have $n_e/n_H = n_p/n_H + n_i/n_H$, $n_p/n_H \ll 1$, and $n_i/n_H \simeq n_e/n_H \sim 10^{-4}$. Replacing $n_i \simeq 10^{-4} n_H$ and $\zeta_H \simeq 7 \times 10^{-18}$ s^{-1}, $\alpha^{(2)} \simeq 4 \times 10^{-13}$ cm^3 s^{-1} and $n(H^0) \simeq n_H$, we obtain

$$n_p \simeq 0.5 \times 10^{-4} n_H \left\{ \left[1 + \frac{4 \times 7 \times 10^{-18} n_H}{4 \times 10^{-13} \times 10^{-8} n_H^2} \right]^{1/2} - 1 \right\}, \tag{6.82}$$

Table 6.1 The hydrogen degree of ionization for cosmic ray ionization

T (K)	n_H (cm^{-3})	ζ_H	n_p (cm^{-3})	x (%)
6,000	0.1	Min.	1.3×10^{-3}	1.3
6,000	0.1	Max.	1.6×10^{-2}	16
6,000	0.01	Min.	4.2×10^{-4}	4.2
6,000	0.01	Max.	5.0×10^{-3}	50
100	0.1	Min.	3.1×10^{-4}	0.1
100	0.1	Max.	3.8×10^{-3}	3.8
100	0.01	Min.	1.0×10^{-4}	1.0
100	0.01	Max.	1.2×10^{-3}	12

$$n_p \simeq 4.2 \times 10^{-3} n_H^{1/2}. \tag{6.83}$$

For values of n_H of the order of 0.01–0.1 cm^{-3}, $n_p \simeq 1.3 \times 10^{-3}$–$4.2 \times 10^{-4}$ cm^{-3} $\ll n_H$, $x = n_p/n_H \simeq 0.013$–0.042, or $x \simeq 1.3$–4.2 %. Using the highest value of the ionization rate, $\zeta_H \sim 10^{-15}$ s^{-1} and $n_p \simeq 5 \times 10^{-2}\, n_H{}^{1/2}$.

In the above conditions $n_p \simeq 1.6 \times 10^{-2}$–$5 \times 10^{-3}$ cm^{-3} $< n_H$, $x \simeq 0.16$–0.50, or $x \simeq 16$–50 %. For lower temperatures, α is higher and the degree of ionization decreases, if the other conditions remain the same. In Table 6.1 we show a relation between the obtained values of n_p and x(%) for $T = 6{,}000$ K and $T = 100$ K, using $n_H = 0.1$ and 0.01 cm^{-3} and $\zeta_H = 7 \times 10^{-18}$ and 10^{-15} s^{-1}, with $\alpha^{(2)} \sim 6.9 \times 10^{-12}$ cm^3 s^{-1} for $T = 100$ K.

As we have seen in Chap. 2, there is an observed background X-ray radiation, especially soft X-rays, with energies between 100 and 250 eV. These X-rays can ionize H and the estimated ionization rate is of the order of $\zeta_H{}^X \lesssim 10^{-15}$ s^{-1}. More recent results show lower rates, $\zeta_H{}^X \lesssim 10^{-17}$ s^{-1}, though rates of the order of $\zeta_H{}^X \lesssim 10^{-16}$ s^{-1} have been suggested for soft X-rays. This rate is lower or at least comparable to the H ionization rate by cosmic rays. However, measured fluxes apparently indicate much lower rates than the upper limit. Besides stationary X-rays, transient X-ray emission processes may be important, such as emission coming from supernovae explosions. These bursts can also trigger the ionization of H and heavy elements.

6.7 Ionization of Heavy Elements

Basically, the same physical processes that trigger the ionization of H can also ionize heavy elements, in some cases a competition existing between the two. Other processes, such as ionization triggered by photons with energy $h\nu < 13.6$ eV, will of course only affect heavy elements and even so just those with low ionization potential, such as Si I and S I. Besides radiative processes—photoionization and recombination—other processes may be important, like dielectronic recombination; collisional ionization triggered by collisions with atoms, ions, electrons, molecules and high-energy particles; as well as charge exchange reactions. We will now consider some of these processes.

6.7.1 Photoionization

The heavy elements ionization equilibrium equation, when photoionization balances radiative or dielectronic recombination, can be written in a similar way as (6.9)

$$(1-x)\beta = x\, n_{\rm e}\, \alpha, \tag{6.84}$$

where we omit the ionization rate subscript. The recombination coefficient α is much more complex now, and we generally cannot use approximation (6.63) for hydrogen-like atoms. Values of α were numerically calculated from photoionization cross sections for the ions of several heavy elements. The radiative recombination coefficient may be written

$$\alpha_{\rm R}({\rm X}^r) = \alpha_{n{\rm v}}({\rm X}^r) + \sum_{n>n_{\rm v}} \alpha_n({\rm X}^r), \tag{6.85}$$

where $\alpha_{n{\rm v}}$ refers to the recombination coefficient for the ${\rm X}^r$ ion ground level, being $n_{\rm v}$ the principal quantum number of the valence state. In a general way, excited levels behave more or less as hydrogen-like atoms, which does not happen with the ground-state level. Calculated recombination coefficients can be approximated by

$$\alpha_{\rm R}({\rm X}^r) = A_{\rm R}({\rm X}^r)\left[\frac{T}{10^4\,{\rm K}}\right]^{-\eta({\rm X}^r)}, \tag{6.86}$$

and parameters $A_{\rm R}({\rm X}^r)$ and $\eta({\rm X}^r)$ can be found in the literature. The error introduced by using (6.86) depends on the ion and also on the electron temperature and is of the order of 10 %. For dielectronic recombination, generally important for $T \gtrsim 10^4$ K, the recombination coefficient can be approximated by the expression

$$\alpha_{\rm D}({\rm X}^r) = A_{\rm D}({\rm X}^r) T^{-3/2} \exp\left[\frac{-T_0({\rm X}^r)}{T}\right]\left\{1 + B_{\rm D}({\rm X}^r)\exp\left[\frac{-T_1({\rm X}^r)}{T}\right]\right\}, \tag{6.87}$$

with a mean accuracy of the order of 15 %, with parameters $A_{\rm D}({\rm X}^r)$, $B_{\rm D}({\rm X}^r)$, $T_0({\rm X}^r)$, and $T_1({\rm X}^r)$ also found in the literature. The recombination coefficients for the main ions of elements C, N, O, etc., are of the order of 10^{-16}–10^{-9} cm^3 s^{-1} and vary rapidly with temperature. Figure 6.4 shows the variation of the recombination coefficients of N ions as a function of electron temperature, using (6.86) and (6.87). The total recombination coefficient is essentially equal to the radiative one, except for log $T > 4.5$ when dielectronic recombinations introduce an important correction. Note that the dielectronic recombination coefficient varies more rapidly with temperature than the radiative recombination coefficient, according to (6.86) and (6.87).

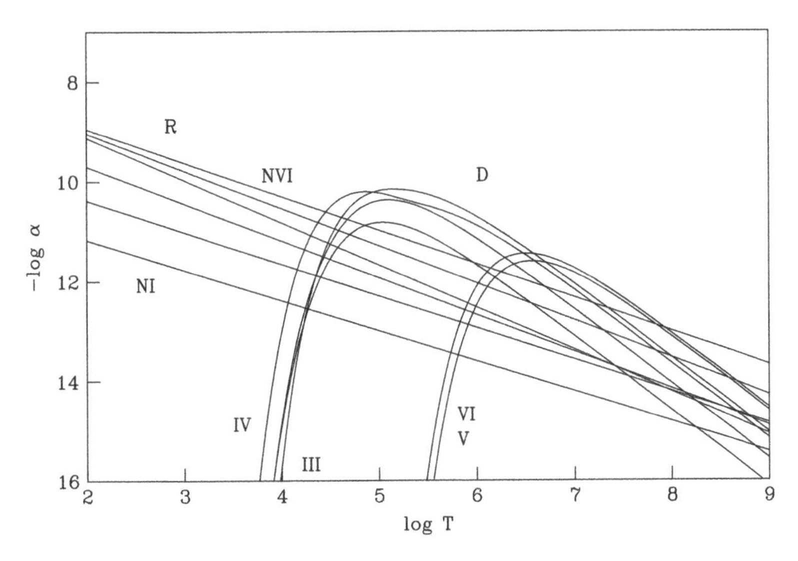

Fig. 6.4 The nitrogen radiative and dielectronic recombination coefficients

Table 6.2 Typical photoionization rates and ionization potentials for some important ions in HI clouds

Ion	ϕ (eV)	β (s^{-1})
C I	11.26	3.1×10^{-10}
Na I	5.14	2.1×10^{-11}
Mg I	7.65	8.1×10^{-11}
Al I	5.99	1.0×10^{-9}
Si I	8.15	3.8×10^{-9}
S I	10.36	2.1×10^{-9}
K I	4.34	6.1×10^{-11}
Ca I	6.11	3.8×10^{-10}
Ca II	11.87	4.0×10^{-12}
Ti I	6.82	5.0×10^{-10}
Mn I	7.44	1.1×10^{-10}
Fe I	7.87	3.7×10^{-10}

The photoionization rate β that appears in (6.84) may be calculated from the same expression (6.15). For a typical H I region, outside the influence of near hot stars, U_ν is given by the integrated stellar radiation field added to scattered radiation (Chap. 2). The upper integration limit is $\nu = c/912$ Å, because the ionizing photons are used to ionize H atoms near the sources. Typical values of β are given in Table 6.2, besides the ion ionization potential (ϕ).

In principle, (6.84) can have several applications. For instance, if the α and β coefficients of a certain ion are known, the degree of ionization can be determined as a function of electron density. Alternatively, comparing determinations of the degrees of ionization from direct observation of the present ions, we can determine the electron density of the observed region.

Table 6.3 Ionization potentials for the first Ca and Na ions

Ion	I	II	III
Na	5.14	47.29	71.64
Ca	6.11	11.87	50.91

Finally, (6.84) may be used to determine the ionization rate and subsequently the radiation field in the relevant spectral region. The ζ Oph direction is the most studied line of sight with respect to ionization conditions. Results show electron densities typically of the order of $n_e \sim 0.06$–$0.25\ \mathrm{cm}^{-3}$ corresponding to two components, diffuse ($n_H \simeq 500\ \mathrm{cm}^{-3}$, $T \simeq 110$ K) and dense ($n_H \simeq 2{,}500\ \mathrm{cm}^{-3}$, $T \simeq 20$ K). For other stars less detailed results have been obtained from models of one isothermal component. In this case, $n_e \simeq 0.1\ \mathrm{cm}^{-3}$, $n_H \simeq 10^3\ \mathrm{cm}^{-3}$, and $T \simeq 70$ K.

The study of the Na and Ca ionization equilibrium in interstellar clouds allows us to check the depletion of Ca, mentioned earlier. In Table 6.3 the ionization potentials (eV) for the first ions of these elements are listed.

Therefore, for H I regions under normal conditions, the Na ionization equilibrium will be determined between Na I and Na II, though the great majority will be Na II atoms. Similarly, Ca ionization equilibrium will be determined between Ca II and Ca III, the major part being Ca III. Applying (6.10) [or (6.84)] to Na, we may write

$$\frac{n(\mathrm{Na\,I})}{n(\mathrm{Na\,II})} = \frac{\alpha(\mathrm{Na\,I})}{\beta(\mathrm{Na\,I})} n_e. \tag{6.88}$$

And by analogy, for Ca, we may write

$$\frac{n(\mathrm{Ca\,II})}{n(\mathrm{Ca\,III})} = \frac{\alpha(\mathrm{Ca\,II})}{\beta(\mathrm{Ca\,II})} n_e \tag{6.89}$$

and thus

$$\frac{n(\mathrm{Na\,I})}{n(\mathrm{Ca\,II})} = \frac{\alpha(\mathrm{Na\,I})}{\alpha(\mathrm{Ca\,II})} \frac{\beta(\mathrm{Ca\,II})}{\beta(\mathrm{Na\,I})} \frac{n(\mathrm{Na\,II})}{n(\mathrm{Ca\,III})}. \tag{6.90}$$

Since we have $n(\mathrm{Na\,II}) \simeq n(\mathrm{Na}) = a(\mathrm{Na})n(\mathrm{H})$ and $n(\mathrm{Ca\,III}) \simeq n(\mathrm{Ca}) = a(\mathrm{Ca})n(\mathrm{H})$, then it follows

$$\frac{n(\mathrm{Na\,I})}{n(\mathrm{Ca\,II})} = \frac{\alpha(\mathrm{Na\,I})\beta(\mathrm{Ca\,II})}{\alpha(\mathrm{Ca\,II})\beta(\mathrm{Na\,I})} \frac{a(\mathrm{Na})}{a(\mathrm{Ca})}. \tag{6.91}$$

Using the a(Na) and a(Ca) values according to cosmic abundance and radiation field mean values, (6.91) predicts a ratio $n(\mathrm{Na\,I})/n(\mathrm{Ca\,II}) \simeq 0.01$ for temperatures $T \simeq 80$ K. Results of analysis performed with the methods explained in Sect. 4.7 show that for low-velocity normal interstellar clouds, ratio $n(\mathrm{Na\,I})/n(\mathrm{Ca\,II}) \simeq 1$–$10$.

Supposing that the gas spatial distribution does not affect the Na I/Ca II ratio, we see that the Ca abundance relative to the Na must be of a hundred to a thousand times lower in the interstellar medium than the cosmic value. Similar depletion factors are obtained for other elements, such as Mg, Mn, and Fe, and their consequences for interstellar dust grain formation will be examined in Chap. 9.

6.7.2 *Collisional Ionization by Thermal Electrons*

By analogy to H, for high temperatures ($T \gtrsim 10^4$ K), collisional ionization may be important for heavy elements, with the recombination being radiative or dielectronic. In these conditions, the ionization equilibrium equation is again (6.68), which may be written

$$\sum_j n_j(\mathrm{X}^r)\gamma_{j\mathrm{ph}} = n(\mathrm{X}^{r+1})\sum_j \alpha_j, \tag{6.92}$$

$$n(\mathrm{X}^r)\gamma = n(\mathrm{X}^{r+1})\alpha, \tag{6.93}$$

where we again include just the ionizations coming from the ground-state level, we omit the subscript in γ and we use

$$\alpha = \sum_j \alpha_j = \alpha^{(1)}. \tag{6.94}$$

Therefore,

$$\frac{n(\mathrm{X}^{r+1})}{n(\mathrm{X}^r)} = \frac{\gamma}{\alpha} \tag{6.95}$$

and the ratio $n(\mathrm{X}^{r+1})/n(\mathrm{X}^r)$ does not depend on the electron density. Considering once more the degree of ionization, with $n(\mathrm{X}) \simeq n(\mathrm{X}^r) + n(\mathrm{X}^{r+1})$, we have relations:

$$x = \frac{n(\mathrm{X}^{r+1})}{n(\mathrm{X})}, \tag{6.96}$$

$$1 - x \simeq \frac{n(\mathrm{X}^r)}{n(\mathrm{X})}, \tag{6.97}$$

$$\frac{x}{(1-x)} \simeq \frac{n(\mathrm{X}^{r+1})}{n(\mathrm{X}^r)} \tag{6.98}$$

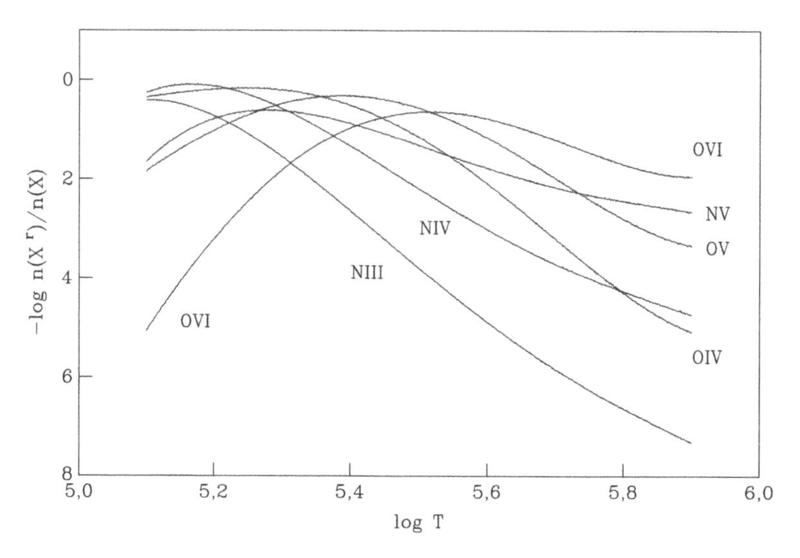

Fig. 6.5 Ratio of N and O ion abundances to the total element abundances as a function of the temperature

so that, by analogy with (6.75),

$$1 - x = \frac{n(X^r)}{n(X)} = \frac{1}{1 + \gamma/\alpha}. \tag{6.99}$$

Figure 6.5 shows the $n(X^r)/n(X)$ ratio for N and O ions for temperatures $T \gtrsim 10^5$ K. For instance, for $\log T \simeq 5.8$, $T \simeq 6.3 \times 10^5$ K we have $n(\text{O VI})/n(\text{O}) \simeq 2.3 \times 10^{-2}$ and $n(\text{O V})/n(\text{O}) \simeq 1.8 \times 10^{-3}$ so that $1 - x = n(\text{O V})/n(\text{O}) \simeq 1.8 \times 10^{-3}$ and $x \simeq 0.998$. We see that, according to Fig. 6.5, for this temperature, the major part of the oxygen is in the form of O VI. From (6.99) we also have $\gamma/\alpha \simeq 550$. Using recombination coefficients from Aldrovandi and Pequignot (1973), $\alpha \simeq 2 \times 10^{-11}$ cm^3 s^{-1}, we have $\gamma \simeq 10^{-8}$ s^{-1}, which is higher than the photoionization rates given by Table 6.2.

Based on the results of Fig. 6.5, the observed ratio $N(\text{N V})/N(\text{O VI}) < 0.025$ mentioned in Sect. 6.5 can be interpreted in terms of gas temperature. We have

$$\begin{aligned}\frac{N(\text{N V})}{N(\text{O VI})} &= \frac{[N(\text{N V})/N(\text{N})][N(\text{N})/N(\text{H})]}{[N(\text{O VI})/N(\text{O})][N(\text{O})/N(\text{H})]} \\ &= \frac{N(\text{N V})/N(\text{N})}{N(\text{O VI})/N(\text{O})}\,\frac{a(\text{N})}{a(\text{O})}.\end{aligned} \tag{6.100}$$

Cosmic abundances are approximately given by $a(\text{N}) = N(\text{N})/N(\text{H}) \simeq 10^{-4}$ and $a(\text{O}) = N(\text{O})/N(\text{H}) \simeq 10^{-3.2}$. Therefore,

$$\log\left[\frac{N(\mathrm{N\,V})}{N(\mathrm{O\,VI})}\right] = \log\left[\frac{N(\mathrm{N\,V})}{N(\mathrm{N})}\right] - \log\left[\frac{N(\mathrm{O\,VI})}{N(\mathrm{O})}\right] + \log\left[\frac{a(\mathrm{N})}{a(\mathrm{O})}\right]. \tag{6.101}$$

The observed ratio implies

$$\log\left[\frac{N(\mathrm{O\,VI})}{N(\mathrm{O})}\right] > \log\left[\frac{N(\mathrm{N\,V})}{N(\mathrm{N})}\right] + 0.8. \tag{6.102}$$

According to Fig. 6.5 this occurs for $T \gtrsim 4 \times 10^5$ K or $\log T \gtrsim 5.6$. The obtained lower limits for ratio N(S IV)/N(Si IV) are similar to the above (6.102) value. The H mean density for this gas can be estimated from

$$a(\mathrm{O}) = \frac{N(\mathrm{O})}{N(H)} \simeq \frac{n(\mathrm{O})}{n(\mathrm{H})} \simeq \frac{n(\mathrm{O})}{n_\mathrm{H}}, \tag{6.103}$$

$$n_\mathrm{H} \simeq \frac{n(\mathrm{O})}{a(\mathrm{O})} \simeq \frac{n(\mathrm{O})}{n(\mathrm{O\,VI})}\frac{n(\mathrm{O\,VI})}{a(\mathrm{O})} = \frac{n(\mathrm{O\,VI})/a(\mathrm{O})}{n(\mathrm{O\,VI})/n(\mathrm{O})}. \tag{6.104}$$

Since n(O VI)/n(O) has a maximum in $\log n(\mathrm{O\,VI})/n(\mathrm{O}) \simeq -0.59$ for $\log T = 5.5$ or $T \simeq 3.2 \times 10^5$ K, $n(\mathrm{O\,VI}) \simeq 1.7 \times 10^{-8}$ cm^{-3} (cf. Sect. 6.5) and $a(\mathrm{O}) = 10^{-3.2}$, we obtain $n_\mathrm{H} \gtrsim 1.1 \times 10^{-4}$ cm^{-3}.

6.7.3 Ionization by Cosmic Rays and X-Rays

H collisional ionization processes by cosmic rays or X-rays can in principle be important for heavy element ionization in regions where electron temperature is low, $T_\mathrm{e} \lesssim 1{,}000$ K, and for heavy elements whose ionization potential $\phi(\mathrm{X}^r)$ is higher than 13.6 eV. For other cases, ionization by the stellar radiation field is probably more important. If $\zeta(\mathrm{X}^r)$ is the ionization rate of heavy element X^r produced by collisions with cosmic rays or X-rays, we have, in a very rough way,

$$\frac{\zeta(\mathrm{X}^r)}{\zeta_\mathrm{H}} = \frac{\sigma(\mathrm{X}^r)}{\sigma_\mathrm{H}} \simeq \left[\frac{\phi_\mathrm{H}}{\phi(\mathrm{X}^r)}\right]^2 \xi(\mathrm{X}^r), \tag{6.105}$$

where σ_H and $\sigma(\mathrm{X}^r)$ are the ionization collisional cross sections of H and element X^r, ϕ_H and $\phi(\mathrm{X}^r)$ are the ionization potentials of H and X^r, and $\xi(\mathrm{X}^r)$ is the electron number in the last level of element X^r. This relation essentially reflects Thomson's classical relation and may be improved by considering the ionization cross section of an arbitrary target by particles of a certain mass, charge, and energy. For instance, for N I with 7 electrons, $\phi(\mathrm{N\,I}) = 14.5$ eV, $\xi(\mathrm{N\,I}) = 5$ (2 electrons in level K and 5 in level L), and $\zeta(\mathrm{N\,I})/\zeta(\mathrm{H}) \simeq 4.4$.

6.7.4 Charge Exchange Reactions

The relative populations of the different ionization states of the elements in an interstellar cloud or ionized nebula may in general be affected by charge exchange reactions, as well as by chemical reactions involving atoms, ions, or molecules. In fact, studies have been made these last years concerning several charge exchange reactions and their rates. They have revealed that these reactions are important in the formation and destruction of several ions. Let us consider, for instance, the reaction

$$\mathrm{O}^+ + \mathrm{H} \rightarrow \mathrm{O} + \mathrm{H}^+ + 0.20\,\mathrm{eV}. \tag{6.106}$$

Neglecting the inverse reaction, the number of O atoms formed per unit volume per unit time is

$$\frac{\mathrm{d}n_\mathrm{O}}{\mathrm{d}t} = k\,n(\mathrm{O}^+)n_\mathrm{H}, \tag{6.107}$$

where we introduce the charge exchange rate or coefficient k measured in cm^3 s^{-1}. Therefore, the ionization equilibrium equation between O I and O II is modified to

$$\begin{aligned} n(\mathrm{O\,I})\beta(\mathrm{O\,I}) &= n(\mathrm{O\,II})n_\mathrm{e}\alpha(\mathrm{O\,I}) + n(\mathrm{O\,II})n_\mathrm{H}k \\ &= n(\mathrm{O\,II})[n_\mathrm{e}\alpha(\mathrm{O\,I}) + n_\mathrm{H}k] \end{aligned} \tag{6.108}$$

or

$$\frac{n(\mathrm{O\,II})}{n(\mathrm{O\,I})} = \frac{\beta(\mathrm{O\,I})}{n_\mathrm{e}\alpha(\mathrm{O\,I}) + n_\mathrm{H}k}. \tag{6.109}$$

Assuming $T = 300$ K, $k \simeq 0.4 \times 10^{-9}$ cm^3 s^{-1}, and $\alpha(\mathrm{O\ I}) \simeq 3.3 \times 10^{-12}$ cm^3 s^{-1}, we obtain

$$\alpha(\mathrm{O\,I})\frac{n_\mathrm{e}}{n_\mathrm{H}} + k = 3.3 \times 10^{-12}\frac{n_\mathrm{e}}{n_\mathrm{H}} + 4.0 \times 10^{-10} \tag{6.110}$$

so that charge exchange is usually more important for O I than radiative recombination, because $n_\mathrm{e}/n_\mathrm{H} < 1$. Besides this reaction, others may be important for the ionization equilibrium of heavy elements, though many of the rates are not well known.

6.7.5 Example 1: Element with Two Ionization States

Let us write the ionization equilibrium equation for a heavy element X considering photoionization and ionization by nonthermal particles or cosmic rays. Let us assume that the element has just two states of ionization, X I and X II, and let β(X I) be the photoionization rate, ζ(X I) the collisional ionization rate, and α(X I) the total recombination coefficient (radiative + dielectronic) of ion X I. Defining the total ionization rate by Γ(X I) $= \beta$(X I) + ζ(X I), the ionization equilibrium equation is

$$n(\mathrm{X\,I})\Gamma(\mathrm{X\,I}) = n(\mathrm{X\,II})n_e\alpha(\mathrm{X\,I}). \tag{6.111}$$

We also have that

$$n(\mathrm{X}) = n(\mathrm{X\,I}) + n(\mathrm{X\,II}). \tag{6.112}$$

We see that there are two equations and three unknowns, n(X I), n(X II), and n(X), that can be determined if we know Γ, α and n_e. Let us define the degrees of ionization $x_{\mathrm{I}} = n(\mathrm{X\ I})/n(\mathrm{X})$, $x_{\mathrm{II}} = n(\mathrm{X\ II})/n(\mathrm{X})$, and also the ratio between densities in the two levels, or relative degree of ionization:

$$R_{\mathrm{II,I}} = \frac{n(\mathrm{X\,II})}{n(\mathrm{X\,I})}. \tag{6.113}$$

We obtain $R_{\mathrm{II,I}} = \Gamma(\mathrm{X\ I})/n_e\alpha(\mathrm{X\ I})$ and the equation system

$$\frac{x_{\mathrm{II}}}{x_{\mathrm{I}}} = R_{\mathrm{II,I}}. \tag{6.114}$$

$$x_{\mathrm{I}} + x_{\mathrm{II}} = 1. \tag{6.115}$$

Now we have two equations and two unknowns (x_{I} and x_{II}) that can be solved knowing R:

$$x_{\mathrm{I}} = \frac{1}{1 + R_{\mathrm{II,I}}}, \tag{6.116}$$

$$x_{\mathrm{II}} = \frac{1}{1 + 1/R_{\mathrm{II,I}}}. \tag{6.117}$$

If $R_{\mathrm{II,I}} \gg 1$ we have $x_{\mathrm{I}} \rightarrow 0$ and $x_{\mathrm{II}} \rightarrow 1$. If $R_{\mathrm{II,I}} \ll 1$ we have $x_{\mathrm{I}} \rightarrow 1$ and $x_{\mathrm{II}} \rightarrow 0$. Let us consider, for instance, Si in a diffuse H I cloud where $T \simeq 100$ K, $n_{\mathrm{H}} \simeq 10$ cm^{-3}, $\zeta_{\mathrm{H}} \simeq 10^{-15}\,\mathrm{s}^{-1}$, $n_e \simeq 10^{-4} n_{\mathrm{H}} = 10^{-3}\,\mathrm{cm}^{-3}$, $\beta(\mathrm{Si\,I}) \simeq 3.8 \times 10^{-9}\,\mathrm{s}^{-1}$, $\zeta(\mathrm{Si\,I}) \simeq 1.1 \times 10^{-14}\,\mathrm{s}^{-1}$, and α(Si I) $\simeq 9.4 \times 10^{-12}\,\mathrm{cm}^3\,\mathrm{s}^{-1}$. We obtain Γ(Si I) $\simeq 3.8 \times 10^{-9}\,\mathrm{s}^{-1}$ and

$$R_{\mathrm{II,I}} = \frac{\Gamma(\mathrm{Si\,I})}{n_e\alpha(\mathrm{Si\,I})} \simeq 4.0 \times 10^5 \gg 1. \tag{6.118}$$

Therefore,

$$x_{\rm I} = x({\rm Si\,I}) = \frac{n({\rm Si\,I})}{n({\rm Si})} \simeq 2.5 \times 10^{-6} \ll 1 \tag{6.119}$$

$$x_{\rm II} = \frac{n({\rm Si\,II})}{n({\rm Si})} \simeq 1.0. \tag{6.120}$$

6.7.6 Example 2: Element with Three Ionization States

Let us now consider element X in the first three ionization state levels according to the same hypotheses of the previous section. In this case, we have

$$n({\rm X\,I})\Gamma({\rm X\,I}) = n({\rm X\,II})n_{\rm e}\alpha({\rm X\,I}), \tag{6.121}$$

$$n({\rm X\,II})\Gamma({\rm X\,II}) = n({\rm X\,III})n_{\rm e}\alpha({\rm X\,II}), \tag{6.122}$$

$$n({\rm X}) = n({\rm X\,I}) + n({\rm X\,II}) + n({\rm X\,III}). \tag{6.123}$$

We can define the rates

$$R_{\rm II,I} = \frac{n({\rm X\,II})}{n({\rm X\,I})} = \frac{\Gamma({\rm X\,I})}{n_{\rm e}\alpha({\rm X\,I})}, \tag{6.124}$$

$$R_{\rm III,II} = \frac{n({\rm X\,III})}{n({\rm X\,II})} = \frac{\Gamma({\rm X\,II})}{n_{\rm e}\alpha({\rm X\,II})}, \tag{6.125}$$

$$R_{\rm III,I} = \frac{n({\rm X\,III})}{n({\rm X\,I})} = \frac{n({\rm X\,III})}{n({\rm X\,II})}\frac{n({\rm X\,II})}{n({\rm X\,I})} = R_{\rm III,II}R_{\rm II,I}. \tag{6.126}$$

And introducing as before the degrees of ionization $x_{\rm I} = n({\rm X\ I})/n({\rm X})$, $x_{\rm II} = n({\rm X\ II})/n({\rm X})$ and $x_{\rm III} = n({\rm X\ III})/n({\rm X})$, we have relations

$$\frac{x_{\rm II}}{x_{\rm I}} = R_{\rm II,I}, \tag{6.127}$$

$$\frac{x_{\rm III}}{x_{\rm II}} = R_{\rm III,II}, \tag{6.128}$$

$$\frac{x_{\rm III}}{x_{\rm I}} = R_{\rm III,I} = R_{\rm III,II}R_{\rm II,I}, \tag{6.129}$$

$$x_{\rm I} + x_{\rm II} + x_{\rm III} = 1. \tag{6.130}$$

Since only rates $R_{\mathrm{II,I}}$ and $R_{\mathrm{III,II}}$ are independent, the system has three equations and three unknowns, x_{I}, x_{II}, and x_{III}, that can be determined as a function of ratios R. Solving the system we obtain

$$x_{\mathrm{I}} = \frac{1}{1 + R_{\mathrm{II,I}} + R_{\mathrm{III,II}} R_{\mathrm{II,I}}} = \frac{1}{1 + R_{\mathrm{II,I}} + R_{\mathrm{III,I}}}, \tag{6.131}$$

$$x_{\mathrm{II}} = \frac{1}{1 + 1/R_{\mathrm{II,I}} + R_{\mathrm{III,II}}} = \frac{1}{1 + 1/R_{\mathrm{II,I}} + R_{\mathrm{III,I}}/R_{\mathrm{II,I}}}, \tag{6.132}$$

$$x_{\mathrm{III}} = \frac{1}{1 + 1/(R_{\mathrm{III,II}} R_{\mathrm{II,I}}) + 1/R_{\mathrm{III,II}}} = \frac{1}{1 + 1/R_{\mathrm{III,I}} + 1/R_{\mathrm{III,II}}}. \tag{6.133}$$

If $R_{\mathrm{III,II}} \to 0$, we can neglect the third ionization state level, and from (6.131) we find again (6.116). From (6.133) we also have $x_{\mathrm{III}} \to 0$. Let us again consider the case of Si, supposing now ionization states Si I, Si II, and Si III. We have $\beta(\mathrm{Si\,II}) = 0$ because $\phi(\mathrm{Si\ II}) = 16.3 > 13.6$ eV, $\zeta(\mathrm{Si\ II}) \simeq 2.1 \times 10^{-15}$ s^{-1}, $\alpha(\mathrm{Si\ II}) \simeq 3.7 \times 10^{-11}$ cm^3 s^{-1}, and $\Gamma(\mathrm{Si\ II}) \simeq 2.1 \times 10^{-15}$ s^{-1}. In this case we obtain

$$R_{\mathrm{III,II}} \simeq \frac{\Gamma(\mathrm{Si\,II})}{n_{\mathrm{e}} \alpha(\mathrm{Si\,II})} \simeq 5.7 \times 10^{-2} \ll 1. \tag{6.134}$$

$R_{\mathrm{II,I}} \simeq 4.0 \times 10^5$ does not change and $R_{\mathrm{III,I}} \simeq R_{\mathrm{III,II}}\, R_{\mathrm{II,I}} \simeq 2.3 \times 10^4$. From (6.131)–(6.134), the degrees of ionization are $x_{\mathrm{I}} \simeq 2.4 \times 10^{-6} \ll 1$, which hardly changes relatively to the previous example, $x_{\mathrm{II}} \simeq 0.95$ and $x_{\mathrm{III}} \simeq 0.05$.

Exercises

6.1 (a) Show that the photoionization cross section from level n for hydrogen-like atoms may be written in the form (6.48). (b) Calculate the maximum cross section for bound-free transitions in the H Balmer continuum. To what wavelength does this transition correspond? What is the error introduced by this approximation, considering $g_{n\mathrm{ph}} \simeq 1$?

6.2 Estimate the fraction of ions Ca I, Ca II, and Ca III in an interstellar cloud with $T = 100$ K and $n_{\mathrm{e}} = 10^{-4}$ cm^{-3}, assuming only photoionization and radiative recombination. Given: $\beta(\mathrm{CaI}) = 3.8 \times 10^{-10}$ s^{-1}, $\beta(\mathrm{CaII}) = 4.0 \times 10^{-12}$ s^{-1}, $\alpha_{\mathrm{R}}(\mathrm{CaI}) = 5.1 \times 10^{-12}$ cm^3 s^{-1}, and $\alpha_{\mathrm{R}}(\mathrm{CaII}) = 2.6 \times 10^{-11}$ cm^3 s^{-1}.

6.3 The galactic-disk-integrated radiation spectra in the optical and ultraviolet regions show a steep decrease in the observed flux for $\lambda \lesssim 1{,}000$ Å. Explain why this is so.

6.4 Consider an atom X that can occupy ionization states X^r and X^{r+1}, with densities $n(\mathrm{X}^r)$ and $n(\mathrm{X}^{r+1})$, respectively. If $\beta(\mathrm{X}^r)$ is the ionization rate and

$\alpha(X^r)$ is the total recombination coefficient, write the atom ionization equilibrium. What is the physical significance of this equation?

6.5 H atoms in an interstellar cloud with $T = 100$ K are ionized by high-energy particles (cosmic rays and X-rays), according to rate $\zeta_H = 10^{-16}$ s^{-1}. Assume a density of H atoms in the interval $0.01 \lesssim n_H\,(\mathrm{cm}^{-3}) \lesssim 1$ and calculate (a) the corresponding proton density, n_p; (b) the electron density n_e; and (c) the degree of ionization of H.

Bibliography

Aldrovandi, S.M.V., Pequignot, D.: Radiative and dielectronic recombination coefficients for complex ions. Astron. Astrophys. **25**, 137 (1973). Detailed calculations of the recombination coefficients of the principal heavy elements. Figure 6.4 uses values taken from this reference. See also Rev. Bras. Fis. vol. 4, p.491, 1974; Astron. Astrophys. vol. 47, p.321, 1976 and vol. 252, p.680, 1991; and also S. N. Nahar. Astrophys. J. Suppl. vol. 101, p.423, 1995

Hollenbach, D.J., Thronson, H.A. (eds.): Interstellar Processes. Reidel, Dordrecht (1987). Collection of articles concerning the main physical processes in the interstellar medium, including ionization and recombination

Menzel, D.H., Pekeris, C.L.: Absorption coefficients and hydrogen line intensities. Mon. Notices Roy. Astron. Soc. **96**, 77 (1935). Classical discussion on hydrogen ionization, with reference values tabulated. See also D. H. Menzel. Selected Papers on Physical Processes in Ionized Plasmas, New York, Dover, 1962 e W.J. Karzas and R. Latter. Astrophys. J. Suppl. vol. 6, p.167, 1961

Osterbrock, D.: Astrophysics of Gaseous Nebulae and Active Galactic Nuclei. University Science Books, Mill Valley (1989). Referred to in Chapter 1. Excellent discussion on the ionization conditions in interstellar photoionized nebulae

Spitzer, L.: Physical Processes in the Interstellar Medium. Wiley, New York (1978). Referred to in Chapter 1. Includes discussions on ionization equilibrium in several situations applied to the interstellar medium. It also includes numerical values for Gaunt factors, recombination coefficients, and other important parameters for the ionization equilibrium. Figure 6.5 is based on this reference. See also Ann. Rev. Astron. Astrophys. vol. 28, p.71, 1990

Chapter 7
Interstellar Gas Heating

7.1 Introduction

In this chapter, we will consider the main processes that determine the interstellar gas temperature, in particular neutral H regions and intercloud medium. H II regions and planetary nebulae will be treated in Chap. 8.

From the first law of thermodynamics, when a system experiences an interaction in an infinitesimal process, we have

$$\mathrm{d}E = \mathrm{d}Q - \mathrm{d}W, \tag{7.1}$$

where $\mathrm{d}E$ is the internal energy variation, $\mathrm{d}Q$ is the system absorbed heat, and $\mathrm{d}W$ is the work done by the system, measured, for instance, in ergs. If we have a quasi-steady-state system, the absorbed heat $\mathrm{d}Q$ may be put in terms of the system's entropy variation $\mathrm{d}S$ (second law of thermodynamics):

$$\mathrm{d}Q = \mathrm{T}\ \mathrm{d}S, \tag{7.2}$$

where $\mathrm{d}S$ is measured in erg K^{-1}. Let $\mathrm{d}V$ be the volume variation of the system, then we have

$$\mathrm{d}W = p\,\mathrm{d}V, \tag{7.3}$$

where p is the gas pressure (dyne cm^{-2}). Replacing (7.2) and (7.3) in (7.1) gives

$$T\,\mathrm{d}S = \mathrm{d}E + p\,\mathrm{d}V. \tag{7.4}$$

For an ideal gas, the internal energy only depends on temperature, $E = E(T)$, and the equation of state is $p = nkT$, where k is the Boltzmann constant and n is the number of particles in the gas per unit volume or $n = N/V$, N being the total

W.J. Maciel, *Astrophysics of the Interstellar Medium*,
DOI 10.1007/978-1-4614-3767-3_7, © Springer Science+Business Media New York 2013

number of gas particles in volume V. This way, the absorbed heat per unit volume (erg cm^{-3}) may be written

$$\frac{dQ}{V} = n\,d\left(\frac{3}{2}kT\right) - kT\,\mathrm{d}n, \tag{7.5}$$

where we use the fact that the internal energy of an ideal and monatomic gas is its kinetic energy, $E \simeq (3/2)NkT$.

7.2 Equilibrium Temperature

Let us now consider the problem of thermal equilibrium in the interstellar medium. In the more general case, heating and cooling processes depend on time, and from (7.5), the total energy input per unit volume per unit time is

$$\varDelta = n\frac{\mathrm{d}}{\mathrm{d}t}\left(\frac{3}{2}kT\right) - kT\frac{\mathrm{d}n}{\mathrm{d}t}, \tag{7.6}$$

given in erg cm^{-3} s^{-1}. Let $\Gamma_{\xi\eta}$ be the energy per unit volume per unit time added to the gas by the interaction of particles ξ and η. Similarly, let $A_{\xi\eta}$ be the contribution of these particles to the gas cooling. Let us then define the *heating function* Γ as

$$\Gamma = \sum_{\xi\eta} \Gamma_{\xi\eta}. \tag{7.7}$$

Similarly, the *cooling function* Λ is defined by

$$\Lambda = \sum_{\xi\eta} \Lambda_{\xi\eta}. \tag{7.8}$$

According to this definition,

$$\varDelta = \Gamma - \Lambda = \sum_{\xi\eta} (\Gamma_{\xi\eta} - \Lambda_{\xi\eta}). \tag{7.9}$$

Therefore, in the general case, temperature T and density n vary with time, the rate of variation being determined by the heating and cooling functions.

In (7.6) or (7.9), we neglect thermal conduction, which is not important for typical interstellar temperatures, $T \lesssim 2 \times 10^4$ K. For higher temperatures, such as in the coronal gas, the thermal conductivity K is high, and (7.6) should be written as

$$\varDelta = n\frac{\mathrm{d}}{\mathrm{d}t}\left(\frac{3}{2}kT\right) - kT\frac{\mathrm{d}n}{\mathrm{d}t} + \nabla\cdot(K\nabla T). \tag{7.10}$$

The presence of a magnetic field in the high-temperature region may affect the thermal conductivity K, changing even further (7.10).

In the steady state, dT/dt and dn/dt are null, and from (7.6) and (7.9), we have

$$\Delta = 0 \quad \text{or} \quad \Gamma = \Lambda. \tag{7.11}$$

In general, functions Γ and Λ depend on temperature. Thus, it is possible to define a steady-state equilibrium temperature T_E, corresponding to the temperature where (7.11) is fulfilled. In these conditions, to determine the equilibrium temperature in an interstellar medium region, we need to know all heating processes (which determine Γ) and all cooling processes (which determine Λ) and then apply condition (7.11). In this chapter, we will only consider steady-state heating and cooling processes of the interstellar gas. Meanwhile, transient processes can eventually be important, heating the gas on a relatively short timescale, whereas cooling and recombination take longer.

7.3 Cooling Timescale

Let us consider a process where temperature varies from value T to equilibrium value T_E and where dT/dt is the variation rate. In this case, we can define the gas cooling time t_T as

$$t_T = -\frac{T - T_E}{dT/dt} = -\frac{\Delta T}{dT/dt}. \tag{7.12}$$

If we have cooling, $\Delta T > 0$, $dT/dt < 0$, and $t_T > 0$; if we have heating, $\Delta T < 0$, $dT/dt > 0$, and $t_T > 0$. We may write

$$\frac{d}{dt}\left(\frac{3}{2}kT\right) = -\frac{3\,k(T - T_E)}{2\,t_T}, \tag{7.13}$$

that is, the cooling time corresponds to the ratio between the gas excess energy relative to the equilibrium value and the net heating function Γ–Λ. If t_T and T_E are constant in time, then

$$\frac{dT}{T - T_E} = -\frac{dt}{t_T}. \tag{7.14}$$

Considering $T = T_0$ in $t = 0$, we obtain the relations

$$\int_{T_0}^{T} \frac{dT}{T - T_E} = -\frac{1}{t_T}\int_0^t dt, \tag{7.15}$$

$$T - T_E = (T_0 - T_E)\, e^{-t/t_T}, \tag{7.16}$$

that is, if t_T and T_E are constants, T will tend to T_E according to the function exp $(-t/t_T)$. For $t/t_T \to 0$, $T \to T_0$; for $t/t_T \simeq 1$, $T - T_E \simeq (T_0 - T_E)/e$. Generally, the cooling time t_T defined by (7.12) is positive. However, this time may be negative for certain regions, where instabilities occur. We will see this later on, when we discuss some interstellar cloud formation processes triggered by interstellar gas instabilities. Some H I region cooling time estimates will be presented in Sect. 7.6, and H II regions will be considered in Chap. 8.

7.4 Photoionization of Neutral Atoms

One of the most important interstellar gas heating mechanisms comes from photo-ionization of neutral atoms. In this process, a photon with energy $h\nu$ gives rise to an electron with energy E_2. This electron may then collide with other gas particles, redistributing energy excess and thus inducing heating. Naturally, the gas energy input in this process is less than or equal to E_2 because part of the energy is lost by electron–ion recombination. The ionization equilibrium equation between photoionizations and radiative recombinations is

$$\sum_j n_j(\mathrm{X}^r)\beta_{jph} = \sum_j n(\mathrm{X}^{r+1})\, n_e\, \alpha_j \tag{7.17}$$

[see (6.2)]. Replacing $n(\mathrm{X}^{r+1})$ by n_i, the density of ionized atoms, assumed identical and in the ground state, and remembering relations $\alpha_j = \langle \sigma_{cj} v \rangle$ and $\alpha = \sum_j \alpha_j$ for the recombination coefficient of level j, where we take the mean relatively to a Maxwellian electron velocity distribution, we may write

$$\Gamma_{ei} = n_e\, n_i \sum_j \left[\langle \sigma_{cj}\, v \rangle\, \bar{E}_2 - \langle \sigma_{cj}\, v\, E_1 \rangle \right], \tag{7.18}$$

where E_1 is the kinetic energy lost by the electron in the recombination. In this equation, Γ_{ei} (measured in erg cm^{-3} s^{-1}) is the heating function for interaction between particles ξ = e (electrons) and η = i (ions), and $\bar{E}_2$ is the photoelectron mean energy, calculated with respect to the ionizing photons flux (Sect. 7.7). Assuming that photoionizations come essentially from the ground-state level, we obtain

$$\begin{aligned} \Gamma_{ei} &= n_e\, n_i \left[\bar{E}_2 \sum_j \langle \sigma_{cj}\, v \rangle - \sum_j \left\langle \sigma_{cj}\, v\, \frac{1}{2}\, m_e\, v^2 \right\rangle \right] \\ &= n_e\, n_i \left[\alpha\, \bar{E}_2 - \frac{1}{2} m_e \sum_j \langle \sigma_{cj}\, v^3 \rangle \right]. \end{aligned} \tag{7.19}$$

In (7.18) and (7.19), Γ_{ei} does not depend on β_{jph} or U_ν, because according to (7.17), in the steady state, the total number of photoionizations must balance the total number of recombinations per unit volume per unit time.

7.5 Electron–Ion Collisional Excitation

The most important cooling mechanisms in the interstellar medium involve collisions between particles (electrons, ions, and neutral atoms) with excitation of energy levels close to the ground level. Once excited, the atom tends to return to the ground state by emitting radiation, which may escape the region, corresponding thus to energy loss and cooling of the gas. Let us consider a collision between electrons (density n_e) and ions. Let n_{ij} be the number of ions i in level j per unit volume. In this case, following Chap. 5 notation, the number of excitations $j \rightarrow k$ per cm^3 per second is

$$\frac{\text{number of excitations. } j \rightarrow k}{\text{cm}^3 \text{ s}} = n_e \, n_{ij} \, \gamma_{jk}, \tag{7.20}$$

where $\gamma_{jk} = \langle u\sigma_{jk} \rangle$ is once again the collision coefficient or rate given in cm^3 s^{-1}. The energy lost per electron in the collisional excitation process is simply $E_k - E_j = E_{jk}$. In the collisional, de-excitation process part of this energy is recovered so that the cooling function for electron–ion interactions is

$$\Lambda_{ei} = n_e \sum_j \sum_{k>j} E_{jk} \left(n_{ij} \, \gamma_{jk} - n_{ik} \, \gamma_{kj} \right). \tag{7.21}$$

If all ions are in the ground state ($j = 1$), the j summation is eliminated, and we obtain

$$\begin{aligned} \Lambda_{ei} &= n_e \sum_{k>1} E_{1k} \left(n_{i1} \, \gamma_{1k} - n_{ik} \, \gamma_{k1} \right) \\ &= n_e \sum_{k>1} E_{1k} \left(n_i \, \gamma_{1k} - n_{ik} \, \gamma_{k1} \right). \end{aligned} \tag{7.22}$$

7.6 The Cooling Function in H I Regions

In H II regions, the principal energy source is known, that is, the central star itself. In H I regions, this is not the case, and several heating and cooling mechanisms should be analyzed. Let us consider some of the main cooling mechanisms in these regions.

7.6.1 Cooling by Electron–Ion Collisional Excitation

In this process, function Λ_{ei} is given by (7.21). In this case, the important ions are C II, Si II, O I, Fe II, N II, C I, etc. For instance, for C II transition $^2P_{1/2}-^2P_{3/2}$, we have $E_{jk} = 0.0079$ eV $= 1.3 \times 10^{-14}$ erg, corresponding to a temperature $E_{jk}/k \simeq 90$ K, of the order of the kinetic temperature found in diffuse interstellar clouds. Other transitions may be Si II ($^2P_{1/2}-^2P_{3/2}$) with $E_{jk}/k \simeq 400$ K and O I ($^3P_2-^3P_{1,0}$) with $E_{jk}/k \simeq 200$–300 K.

The collisional rates were discussed in Chap. 5. We saw that coefficient γ_{kj} can be related to the collision strength $\Omega(j,k)$ by (5.32) and the relation between excitation and de-excitation rates is given by (5.13). Let us write (7.21) for a $j \rightarrow k$ transition, neglecting the correction for energy input, $E_{jk}\, n_{ik}\, \gamma_{kj}$, because $n_{ij}\, \gamma_{jk} > n_{ik}\, \gamma_{kj}$ holds in general. In this case, we have

$$\begin{aligned}
\Lambda_{ei} &\simeq n_e\, E_{jk}\, n_{ij}\, \gamma_{jk} = n_e\, n_{ij}\, E_{jk} \frac{g_k}{g_j} \frac{h^2\, \Omega\,(j,k)}{g_k\,(2\pi m_e)^{3/2} (kT)^{1/2}} \mathrm{e}^{-E_{jk}/kT} \\
&= \frac{h^2}{(2\pi m_e)^{3/2}\, k^{1/2}} n_e\, n_{ij}\, T^{-1/2}\, \mathrm{e}^{-E_{jk}/kT} \frac{E_{jk}\, \Omega\,(j,k)}{g_j} \\
&= 8.6 \times 10^{-6} n_e\, n_{ij}\, T^{-1/2}\, \mathrm{e}^{-E_{jk}/kT} \frac{E_{jk}\, \Omega(j,k)}{g_j}. \qquad (7.23)
\end{aligned}$$

In the last relation, the cooling function is given in erg cm^{-3} s^{-1}. Results of the contribution to cooling from different ions are shown in Fig. 7.1 for $T \lesssim 10^4$ K, where we have plotted on the y-axis the $\Lambda_{ei}/n_e n_H$ ratio, given in erg cm^3 s^{-1}. The abundances used in this figure are similar to the values already mentioned in Chap. 4. We see that for $T \lesssim 100$ K, the most important contribution comes from C II, followed by Si II and Fe II. We can estimate the C II cooling function if we consider a cloud with $T \simeq 100$ K and $n_i \simeq 10^{-4}\, n_H$. For transition $^2P_{1/2}-^2P_{3/2}$, we have $E_{jk} \simeq 0.0079$ eV, $\Omega(j,k) \simeq 1.33$, and $g_J = 2$ so that (7.23) gives $\Lambda_{ei}/n_e n_H \simeq 2.9 \times 10^{-25}$ erg cm^3 s^{-1} or $\log(\Lambda_{ei}/n_e n_H) \simeq -24.5$, in good agreement with Fig. 7.1.

7.6.2 Cooling by Electron–H Collisional Excitation

For high temperatures, $T \gtrsim 10^3$ K, loss of energy by excitation of neutral H levels may be important, especially for $n = 2$, due to collisions with thermal electrons. For 12,000 $\gtrsim T(\mathrm{K}) \gtrsim$ 4,000, the cooling function may be approximated by

$$\Lambda_{eH} \simeq 7.3 \times 10^{-19} n_e\, n_H\, \mathrm{e}^{-118{,}400/T}, \qquad (7.24)$$

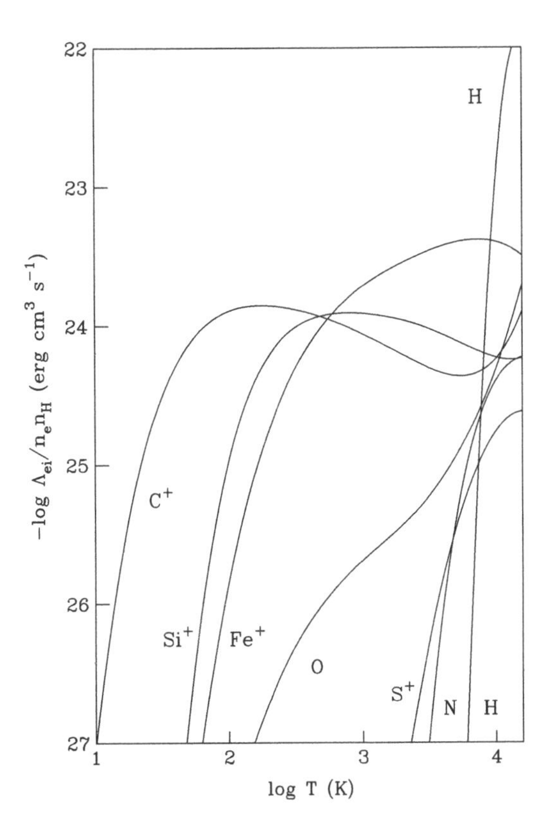

Fig. 7.1 The cooling function for some important ions in the interstellar gas

given in erg cm^{-3} s^{-1}, where the exponential term essentially takes into account excitations to the second level. More detailed calculations differ from (7.24) by 3 % maximum. After excitation to level $n = 2$, the atom will return to the ground level by emitting a quantum L_α that will probably be absorbed by dust grains. In Fig. 7.1, we have also indicated the variation of $\Lambda_{\rm eH}/n_{\rm e}n_{\rm H}$ with temperature. For $T \simeq$ 10,500 K, cooling is essentially controlled by H. For this temperature, we obtain $\log(\Lambda_{\rm eH}/n_{\rm e}n_{\rm H})$ $\simeq -23.0$ from (7.24).

7.6.3 Cooling by H–Ion Collisional Excitation

Collisional excitation of ions such as C I, C II, and Fe II by neutral H can be one of the most important cooling processes in the interstellar medium, mainly when the fraction ionization $n_{\rm e}/n_{\rm H}$ is low. In a general way, the cooling function is given by (7.21) replacing $n_{\rm e}$ by $n_{\rm H}$:

$$\Lambda_{\rm Hi} = n_{\rm H} \sum_j \sum_{k>j} E_{jk} \left(n_{ij}\, \gamma_{jk} - n_{ik}\, \gamma_{kj}\right). \qquad (7.25)$$

Collisional rates for several ions may be found in the literature. We will now obtain the cooling function for the C II ion excitation, which is probably the most abundant positively charged ion in H I regions. The excitation process may be written

$$\mathrm{H} + \mathrm{C\,II}(^2P_{1/2}) \rightarrow \mathrm{H} + \mathrm{C\,II}(^2P_{3/2}). \tag{7.26}$$

In this case and neglecting the de-excitation correction, (7.25) becomes

$$\Lambda_{\mathrm{H,C\,II}} \simeq n_{\mathrm{H}} \sum_{j} \sum_{k>j} E_{jk}\, n_{\mathrm{C\,II}_j}\, \gamma_{jk}. \tag{7.27}$$

Supposing that excitations occur from the ground state, we can eliminate the j summation:

$$\Lambda_{\mathrm{H,C\,II}} \simeq n_{\mathrm{H}} \sum_{k>j_g} E_{j_g k}\, n_{\mathrm{C\,II}}\, \gamma_{j_g k}, \tag{7.28}$$

where we represent the ground-state level by j_g. Let us also eliminate the k summation, considering only one transition, from level $^2P_{1/2}$ to level $^2P_{3/2}$. In this case,

$$\begin{aligned}\Lambda_{\mathrm{H,C\,II}} &\simeq n_{\mathrm{H}}\, n_{\mathrm{CII}}\, E(^2P_{3/2} - {}^2P_{1/2})\gamma(^2P_{1/2} - {}^2P_{3/2}) \\ &= n_{\mathrm{H}}\, n_{\mathrm{CII}}\, E_{jk}\, \gamma_{jk}.\end{aligned} \tag{7.29}$$

Supposing that practically all atoms are in C II form, because the ionization potentials are $\phi(\mathrm{C\ I}) = 11.26$ eV and $\phi(\mathrm{C\ II}) = 24.38$ eV and letting $a(\mathrm{C})$ be the C abundance, we have

$$n_{\mathrm{CII}} \simeq n_{\mathrm{C}} = a(\mathrm{C})\, n(\mathrm{H}) \simeq a(\mathrm{C})\, n_{\mathrm{H}}, \tag{7.30}$$

that is, we assume that the H atoms density is essentially equal to the H nuclei density. Taking into account carbon depletion, characterized by parameter d_{C}, defined by

$$d_{\mathrm{C}} = \frac{a(\mathrm{C})}{a_{\mathrm{c}}(\mathrm{C})} = 10^{f_{\mathrm{d}}(C)}, \tag{7.31}$$

where $a_{\mathrm{c}}(\mathrm{C})$ is the cosmic abundance of the carbon [see (4.68)], we have

$$n_{\mathrm{CII}} = d_{\mathrm{C}}\, a_{\mathrm{c}}(\mathrm{C})\, n_{\mathrm{H}}. \tag{7.32}$$

Replacing in (7.29),

$$\Lambda_{\mathrm{H,CII}} = n_{\mathrm{H}}^2\, d_{\mathrm{C}}\, a_{\mathrm{c}}\, (\mathrm{C})\, E_{jk}\, \gamma_{jk}. \tag{7.33}$$

The carbon abundance cosmic value is defined as log $a_c(C) = \epsilon_c(C) - \epsilon_c(H)$ (cf. Table 4.3) so that $a_c(C) \simeq 10^{-3.4} = 4.0 \times 10^{-4}$. The energy difference of levels $^2P_{1/2}$ and $^2P_{3/2}$ is $E_{jk} = 0.0079$ eV, as we saw. For temperatures $T < 100$ K, we have $\gamma_{kj} \simeq 7.8 \times 10^{-10}$ cm^3 s^{-1}, and so

$$\gamma_{jk} = \frac{g_k}{g_j} \gamma_{kj} \, e^{-Ejk/kT} \simeq 1.6 \times 10^{-9} \, e^{-92.0/T} \, \text{cm}^3 \, \text{s}^{-1}. \tag{7.34}$$

Replacing $a_c(C)$, E_{jk}, and γ_{jk} in (7.33),

$$\Lambda_{\text{H,CII}} \simeq 7.9 \times 10^{-27} \, n_{\text{H}}^2 \, d_{\text{C}} \, e^{-92.0/T} \, \text{erg cm}^{-3} \, \text{s}^{-1}. \tag{7.35}$$

For $T \simeq 80$ K,

$$\Lambda_{\text{H,CII}} \simeq 2.5 \times 10^{-27} \, n_{\text{H}}^2 \, d_{\text{C}} \; \text{erg cm}^{-3} \, \text{s}^{-1}. \tag{7.36}$$

We have seen in Chap. 4 that for the line of sight of ζ Oph, depletion is of the order of $d_C \simeq 10^{-0.7} \simeq 0.2$. Other stars with less reddening have $d_C \simeq 1$. The H−C II collisional de-excitation rates were determined as a function of temperature, as well as for other ions, such as Fe II, C I, and O I.

7.6.4 Cooling by H–H$_2$ Collisional Excitation

In dense interstellar clouds, the H_2 molecule may be a cooling source by means of collisional excitation of H rotational levels $J = 0 \rightarrow J = 2$ (corresponding to temperature $E/k \simeq 500$ K) and $J = 1 \rightarrow J = 3$ (electric quadrupole transitions). On the other hand, collisional de-excitations produce energy input. If the higher rotational levels are populated by radiation, this input can be intensified. The cooling function is again (7.25), now replacing n_i by n_H or simply n. Considering only the transition from rotational level J to rotational level $J + 2$,

$$\Lambda_{\text{H,H}_2} = n_{\text{H}} \sum_J E_{J,J+2} \big[n(J) \, \gamma_{J,J+2} - n(J+2) \, \gamma_{J+2,J} \big]. \tag{7.37}$$

Using again (7.34),

$$\Lambda_{\text{H,H}_2} = n_{\text{H}} \sum_J E_{J,J+2} \left[n(J) \frac{g_{J+2}}{g_J} \gamma_{J+2,J} \, e^{-E_{J,J+2}/kT} - n(J+2) \, \gamma_{J+2,J} \right]. \tag{7.38}$$

Using the excitation equation and considering the thermodynamic equilibrium deviation coefficients,

$$\frac{n(J+2)}{n(J)} = \frac{b_{J+2}}{b_J} \frac{g_{J+2}}{g_J} \, e^{-E_{J,J+2}/kT}, \tag{7.39}$$

we finally obtain

$$\Lambda_{\mathrm{H,H_2}} = n_{\mathrm{H}}^2 \sum_J \frac{n(J+2)}{n_{\mathrm{H}}} E_{J,J+2}\, \gamma_{J+2,\,J} \left[\frac{b_J}{b_{J+2}} - 1 \right]. \tag{7.40}$$

A typical rate for an interstellar cloud with $T \simeq 100$ K is $\Lambda_{\mathrm{H,H_2}}/n_{\mathrm{H_2}} \simeq 3 \times 10^{-26}$ erg s^{-1}. In a very dense cloud with $n_{\mathrm{H_2}} \simeq 10^4\ \mathrm{cm}^{-3}$, $\Lambda_{\mathrm{H,H_2}}/n_{\mathrm{H_2}}^2 \simeq 3 \times 10^{-26}$ erg cm^3 s^{-1}.

For the H_2 molecule, the lifetime of rotational levels for quadrupole radiation is relatively high, of the order of 10^{10} s, whereas populations obtained by collisions have lower timescales, in particular for denser clouds. This leads to a population typical of the gas kinetic temperature, and the radiation represents a small perturbation of the population of the levels. In the case where optical pumping is important, the population of the higher levels ($J \neq 2$) will be larger than the one of the lower levels relatively to TE values or

$$\frac{b_J}{b_{J+2}} = \frac{n_J/n_J^*}{n_{J+2}/n_{J+2}^*} < 1 \tag{7.41}$$

and $\Lambda_{\mathrm{H,H2}}$ is negative, corresponding to an energy input. Besides H_2, the contribution of the HD molecule can also be important for collisional cooling, if the temperature is low enough, $T < 80$ K. In this case, the molecule's small asymmetry produces a certain electric dipole moment triggering transitions with $\Delta J = \pm 1$. The cooling per molecule is thus more efficient than in the previous case, though relative abundance (~10^{-5}) is too low to have any effect on the cooling function.

7.6.5 *Other Processes*

Other physical processes may also be important in the cooling of the interstellar medium. Among them, it is worth mentioning (1) cooling by interstellar dust grains (Sect. 7.7); (2) cooling by other molecules besides H_2 and HD, in particular CO, which has typical abundances of $n_{\mathrm{CO}} \sim 10^{-4}\, n_{\mathrm{H2}}$, OH, and H_2O; and (3) cooling by collisional excitation between protons and ions.

7.6.6 *The Cooling Function*

The total cooling function, including electron–ion and H–ion collisional processes, is plotted in Fig. 7.2 as a function of temperature.

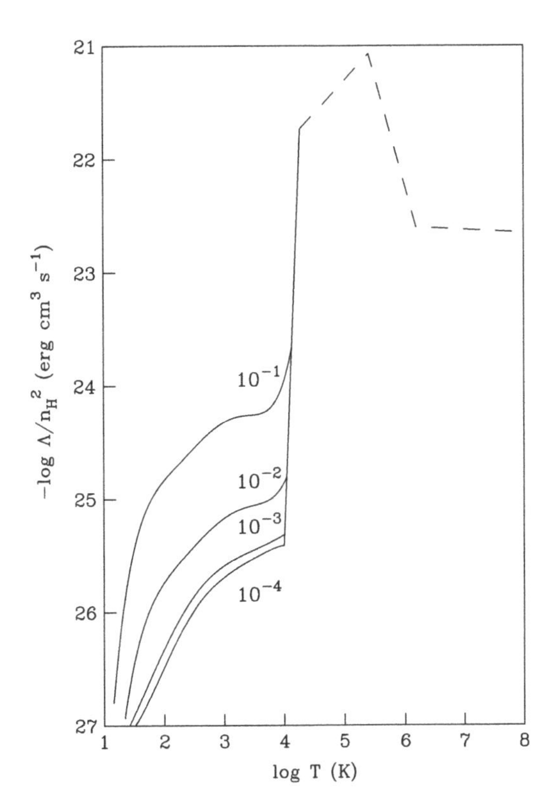

Fig. 7.2 The total cooling function including collisional electron-ion and H-ion interactions in the interstellar gas

On the y-axis, we have function Λ/n_{H}^2, given in erg cm^3 s^{-1}. The different curves refer to different fractional ionization values, $x = n_e/n_{\mathrm{H}}$. In this figure, we do not include processes that involve dust grains or molecules. For $T \lesssim 10^4$ K, cooling depends mainly on fractional ionization n_e/n_{H}. For $n_e/n_{\mathrm{H}} \lesssim 10^{-3}$, the cooling function is essentially due to the impact of H atoms and does not depend on the n_e/n_{H} ratio. Hotter regions are indicated in an approximated way by a dashed line. As we shall see in Sect. 7.9, the cooling function behavior is critical in the study of the interstellar medium thermal stability.

7.6.7 The Cooling Time

The cooling time t_{T} may be estimated from Fig. 7.1 for the case where $T \gg T_{\mathrm{E}}$. From (7.13), we have

$$\begin{aligned} t_{\mathrm{T}} &\simeq -\frac{\frac{3}{2}k(T - T_{\mathrm{E}})}{\frac{\mathrm{d}}{\mathrm{d}t}\left(\frac{3}{2}kT\right)} \simeq -\frac{(3/2)kT}{(\Gamma - \Lambda)/n} \\ &\simeq \frac{3}{2}\frac{n\,k\,T}{\Lambda} \quad (\Gamma \ll \Lambda), \end{aligned} \tag{7.42}$$

where we use (7.6) and (7.9) with n constant and the assumption that for $T \gg T_E$, we have $\Gamma \ll \Lambda$. For $n_e/n_H \ll 1$, we can also write

$$t_T \simeq \frac{3\,kT}{2}\frac{n/n_H}{(\Lambda/n_H^2)n_H} \simeq \frac{3\,kT}{2}\frac{1+n_e/n_H}{(\Lambda/n_H^2)n_H}$$
$$\simeq \frac{3\,kT}{2}\frac{1}{n_H\,(\Lambda/n_H^2)} \quad (n_e/n_H \ll 1). \tag{7.43}$$

Considering $T \simeq 100$ K and $n_e/n_H \simeq 10^{-4}$, we obtain from Fig. 7.2 $\Lambda/n_H^2 \simeq 3 \times 10^{-27}$ erg cm^3 s^{-1} so that

$$t_T \simeq \frac{2.2 \times 10^5}{n_H}\,\text{year}. \tag{7.44}$$

With $n_H \simeq 10$ cm^{-3}, it corresponds to $t_T \simeq 2 \times 10^4$ years, which is a small value relative to the interstellar clouds lifetime scales, $t_{ic} \gtrsim 10^6$ years. Similarly, for a hotter region, with $T \sim 10^4$ K and $n_e/n_H \sim 10^{-1}$, we have $\Lambda/n_H^2 \simeq 6 \times 10^{-25}$ erg c m^3 s^{-1} and $t_T \simeq 1.1 \times 10^5/n_H$ years. The accuracy of these estimates is of the order of 30 % for densities $1 \lesssim n_H\,(\text{cm}^{-3}) \lesssim 300$ and temperatures $50 \lesssim T(\text{K}) \lesssim 600$.

7.7 The Heating Function in H I Regions

7.7.1 Heating by Stellar Radiation

In the process of photoionization of heavy elements by the integrated stellar radiation field, an electron is produced with a certain kinetic energy. Electrons transfer this energy to the surrounding gas by collision. Of course, part of this energy is reused in the recombination process, but the energy distribution by collisions occurs quite rapidly so that thermalization is reached before electron removal by recombination.

We have seen that the radiation field above 13.6 eV is absorbed near the sources. Therefore, the electron energy is always much lower than this value, being of the order of 2 eV (which corresponds to an ionization potential of the order of 11 eV, similar to the ionization potential of C I) for photoionization of a gas with cosmic abundance. The heating function is given by (7.19). The photoelectron mean energy is essentially the ratio between the photoelectron energy per second and the number of absorbed photons per second or

$$\bar{E} = \frac{\displaystyle\int_{\nu_0}^{\infty} \frac{h(\nu-\nu_0)\,\sigma_\nu\,c\,U_\nu\,d\nu}{h\nu}}{\displaystyle\int_{\nu_0}^{\infty} \frac{\sigma_\nu\,c\,U_\nu\,d\nu}{h\nu}}, \tag{7.45}$$

where ν_0 is the threshold frequency for the considered ion, σ_ν is the photoabsorption cross section, and the upper limit is formally ∞, though energy density $U_\nu \to 0$ for frequencies above $\nu_1 = 13.6$ eV h^{-1}. For C I and other stellar radiation field ions, we obtain $\bar{E}_2 \simeq 2$ eV.

The correction for the energy lost by the electron in the recombination process, that is, the non-thermalized energy, depends on the summation $\sum_j \langle \sigma_{cj} v^3 \rangle$ that appears in the second term of (7.19). In this case, the capture cross section depends on the radiation frequency or on the electron velocity in a complex way, similarly to the recombination coefficient. In general, we can reduce this summation to a relatively simple temperature function of the form

$$\sum_{j=k}^{\infty} \langle \sigma_{cj} v^3 \rangle = f(T)\, T^{1/2} \chi_k(T), \tag{7.46}$$

where $f(T)$ is a function that varies slowly with temperature and $\chi_k(T)$ are functions tabulated in the literature. We will now obtain an expression for the heating function by neglecting the above correction. From (7.19),

$$\Gamma_{eX^r} \simeq n_e\, n(X^r)\, \alpha(X^r)\, \bar{E}_2 \simeq n_e\, n(X^r)\, A(X^r) \left[\frac{T}{10^4\,\mathrm{K}}\right]^{-\eta(X^r)} \bar{E}_2, \tag{7.47}$$

where we use approximation (6.86) for the recombination coefficient of element X^r. Assuming $\eta(X^r) \simeq 1/2, A(X^r) \simeq 10^{-13}$ cm^3 s^{-1}, and $\bar{E}_2 \simeq 2\,\mathrm{eV} = 3.2 \times 10^{-12}$ erg, we have

$$\Gamma_{eX^r} \simeq 3.2 \times 10^{-23}\, n_e\, n(X^r) T^{-1/2}, \tag{7.48}$$

given in erg cm^{-3} s^{-1}. Assuming now that $n(X^r) = a(X^r) n(H) \simeq a(X^r) n_H$ with $a(X^r) \simeq 4 \times 10^{-4}$, which is suitable for C, we have

$$\Gamma_{eX^r} \simeq 1.3 \times 10^{-26}\, n_e\, n_H\, T^{-1/2}\, \mathrm{erg\,cm^{-3}\,s^{-1}} \tag{7.49}$$

or in terms of fractional ionization $x = n_e/n_H$,

$$\Gamma_{eX^r} \simeq 1.3 \times 10^{-26}\, x\, n_H^2\, T^{-1/2}\, \mathrm{erg\,cm^{-3}\,s^{-1}}. \tag{7.50}$$

The equilibrium temperature can be determined by making the cooling function obtained in Sect. 7.6 equal to the heating function estimated by (7.50). This is represented in Fig. 7.3, where the two upper curves correspond to the cooling function of Fig. 7.2 for $x = n_e/n_H = 0.1$ and 0.01, whereas the lower left corner lines represent function Γ_{eX_r} given by (7.50). We see that even for relatively high values ($x \simeq 0.1$) of fractional ionization, equilibrium temperatures are low, of the order of $T_E \simeq 13$ K. This value is still an upper limit because we have neglected energy loss by recombination. We have already seen that the analysis of spectral lines in the radio or in the optical and ultraviolet points to the presence of diffuse clouds in the interstellar medium, with temperatures of the order of 80–100 K, in addition to an intercloud medium of low density and temperature $T \gtrsim 1{,}000$ K. Therefore, besides photoionization, other heating mechanisms are necessary.

7.7.2 *Heating by Cosmic Rays*

As we have seen in Chap. 6, cosmic rays can ionize H (and also He) producing energetic electrons. Part of the energy of these electrons is thermalized, resulting in a heating function $\Gamma_{H,CR}$ or $\Gamma_{He,CR}$.

The ejected electron energy naturally depends on the cosmic particle energy. In the case of 2 MeV protons, the electron mean energy is of the order of 32 eV. For 10 MeV protons, this energy is 36 eV. Naturally, only part of this energy can be used for heating. In a faintly ionized gas, where the degree of ionization $x = n_e/n_H \ll 1$, the major part of the electron energy is used for excitations and new ionizations, and detailed calculations show that an energy $\bar{E}$ between 3.4 eV and 8.5 eV is available for gas heating by means of collisions. Considering again the cosmic ray ionization rate ζ_H, the number of ejected electron per cm^3 per second is $\zeta_H n_H$, and the heating function may be written

$$\Gamma_{H,CR} \simeq n_H \, \zeta_H \, \bar{E}. \tag{7.51}$$

Using the ζ_H limits mentioned before (Sect. 6.6) and the above $\bar{E}$ limits, we obtain

$$\Gamma_{H,CR}(\text{min.}) \simeq 3.8 \times 10^{-29} \, n_H \tag{7.52a}$$

$$\Gamma_{H,CR}(\text{max.}) \simeq 1.4 \times 10^{-26} \, n_H \tag{7.52b}$$

in units of erg cm^{-3} s^{-1}, with n_H in cm^{-3}. For instance, for $n_H \simeq 1 - 10^2$ cm^{-3}, this process produces maximum rates of $10^{-26} - 10^{-24}$ erg cm^{-3} s^{-1}, corresponding to values of the heating function higher than the ones shown in Fig. 7.3. Note that we have not included He ionization by cosmic rays, which may increase the above rate. Comparing (7.52) with the cooling functions of Figs. 7.2 and 7.3, we can see that if the cosmic ray ionization rate is near the upper limit $\zeta_H \simeq 10^{-15}$ s^{-1}, the equilibrium temperature may reach values of the order of 100 K, for clouds with $n_H \gtrsim 10$ cm^{-3}. Nevertheless, several problems exist relative to this mechanism, in particular the difficulty in obtaining a realistic rate near the upper limit as well as difficulties encountered in the study of cosmic ray propagation through the interstellar medium.

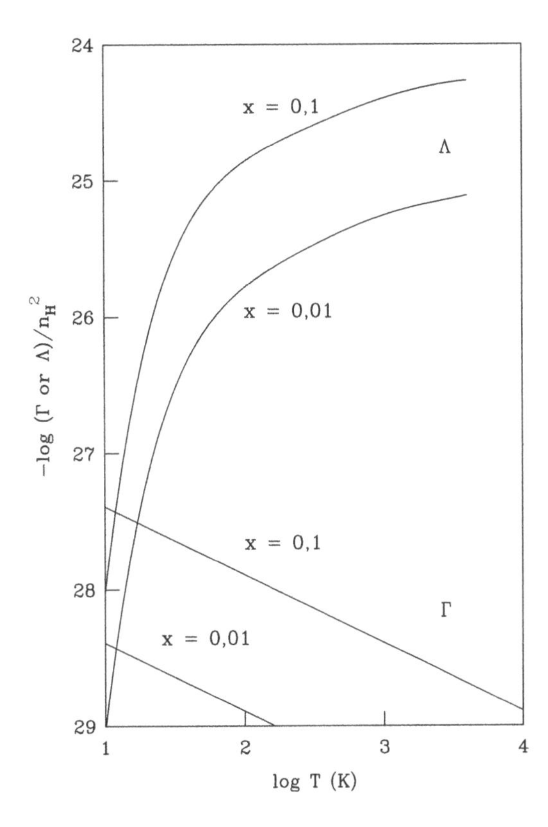

Fig. 7.3 Determination of the equilibrium temperature

Besides cosmic rays, heating of the interstellar medium by X-rays was also proposed. Mainly due to its relatively low flux, this mechanism presents the same difficulties as heating by cosmic rays.

7.7.3 *Heating by H_2 Molecules Formation*

H_2 is the most abundant molecule in dense interstellar clouds, where the temperature is relatively low and the opacity is high enough so that energetic ultraviolet photons are unable to penetrate, which would increase the destruction rate of this molecule (Chap. 10). Being a homopolar molecule, its rovibrational transitions for electric dipole radiation are forbidden, and radiative association of two H atoms to form the molecule has a low probability. There is a general consensus that this molecule is formed by association of two H atoms on the surface of a solid grain, which works as a catalyst. Additional gas heating can be produced when these molecules are present, by means of photodissociation of H_2, and the distribution of the kinetic energy of the produced atoms, as well as during the formation process, when energy is released.

In the process of H_2 molecule formation, there is an energy release of $\bar{E}(H_2) \simeq 4.5$ eV, corresponding to the bound energy of the H atoms. The energy distribution between the heating of the dust grain, excitation of the produced molecule, adsorption energy compensation at the surface of the grain, and kinetic energy of the molecule can be quite complex. Let z_H be the energy fraction released in the form of the H_2 molecule kinetic energy, that is, the energy that is available for gas heating. Also let n(HI) be the neutral H particle density and $(R_{fj})_d$ the probability per unit time of the formation of an H_2 molecule in state j by collision between an H atom and a grain. The number of molecules formed per cm^3 per second is

$$\frac{\text{number of H}_2\text{ molecules formed}}{\text{cm}^3\text{ s}} = n(\text{HI}) \sum_j (R_{fj})_\text{d}. \tag{7.53}$$

The total probability, given by the above summation, can be written in the form

$$\sum_j (R_{fj})_\alpha = \frac{1}{2} \langle \gamma \, v_\text{H} \rangle \Sigma_d n_\text{H} = R \, n_\text{H}, \tag{7.54}$$

where Σ_d is the projected area of the grains per H nucleus in the gas, that is, $n_H \Sigma_d$ is the total projected area per unit volume, n_H being the total number of H nuclei per cm^3. For instance, if the gas contains only protons, electrons, H atoms, and H_2 molecules, we have $n_H = n(\text{HI}) + n_p + 2n(H_2)$. v_H is the velocity of the H atoms, and γ represents the fraction of colliding atoms that effectively produce an H_2 molecule, that is, the process efficiency. Finally, the factor 1/2 takes into account the fact that two H atoms are needed to produce one H_2 molecule. Defining coefficient R by (7.54), we may write

$$R \, n_\text{H} \, n(\text{HI}) = \frac{\text{number of H}_2\text{ molecules}}{\text{cm}^3\text{ s}}. \tag{7.55}$$

Therefore, the heating function due to this process is

$$\begin{aligned} \Gamma_{\text{Hd}} &= R \, n_\text{H} \, n(\text{HI}) \; z_\text{H} \, \bar{E}(\text{H}_2) \\ &\simeq 7.2 \times 10^{-12} R \, z_\text{H} \, n_\text{H} \, n(\text{HI}), \end{aligned} \tag{7.56}$$

given in erg cm^{-3} s^{-1}. Coefficient R can be determined from H_2 molecule observations, and a mean value for $R \simeq 3 \times 10^{-17}$ cm^3 s^{-1} is found. In this case,

$$\Gamma_{\text{Hd}} = 2.2 \times 10^{-28} \, z_\text{H} \, n_\text{H} \, n(\text{HI}) \, \text{erg cm}^{-3}\,\text{s}^{-1}. \tag{7.57}$$

Assuming $z_H \simeq 1$ and $n_H \simeq n(\text{HI}) \simeq 1{-}10^2$ cm^{-3}, we have $\Gamma_{Hd} \sim 10^{-28}{-}10^{-24}$ erg cm^{-3} s^{-1} or $\Gamma_{Hd}/n^2_H \sim 10^{-28}$ erg cm^3 s^{-1}, again of the order of the values of the heating function shown in Fig. 7.3. Thus, Γ_{Hd} may eventually balance cooling by collisional excitation of C II by H [(7.35) or (7.36) for $T = 80\,\mathrm{K}$], if $d_C \simeq 0.1, z_H \simeq 1$, and $n(\mathrm{HI}) \simeq n_H$. Meanwhile, the values of parameters d_C and z_H are not well determined, and more accurate calculations show that this mechanism can maintain temperatures of the order of 80 K only in very restricted situations.

7.7.4 Heating by Photoelectrons Ejected by Dust Grains

When a photon with relatively high energy is absorbed by a dust grain, an electron with a certain kinetic energy may be ejected (photoelectric effect), thus heating the interstellar gas. Neglecting the energy lost by the electron in recombination [term $\langle \sigma_{cj} v E_I \rangle$ of (7.18)], the energy transferred to the gas per unit volume per unit time is

$$\Gamma_{ed} = \int \left[\frac{c U_\lambda \, d\lambda}{h\nu} \sigma_d \, n_d \right] y_e \, E_2, \tag{7.58}$$

where σ_d is the photoabsorption cross section of a dust grain and n_d is the number of dust grains per unit volume. The product of the parameters between square brackets, integrated over a certain wavelength interval, is the number of absorbed photons by dust grains per cm^3 per second. Multiplying by y_e, the quantum yield, that is, the ratio between the number of ejected electrons and the number of absorbed photons, we obtain the number of ejected electrons per cm^3 per second. Finally, we multiply by the photoelectron energy E_2 to obtain the heating function in erg $cm^{-3}\,s^{-1}$. We also have

$$\sigma_d = \sigma_g \, Q_a(\lambda), \tag{7.59}$$

where σ_g is the grain geometric section and $Q_a(\lambda)$ is the grain efficiency factor for radiation absorption with wavelength λ (see Chap. 9). Let Σ_d be, once again, the ratio between the total projected area of the grains and the number of H nuclei. Then we have

$$\Sigma_d = \frac{\sigma_g \, n_d}{n_H}, \tag{7.60}$$

so that

$$\sigma_d \, n_d = n_H \, \Sigma_d Q_a(\lambda) \tag{7.61}$$

and (7.58) becomes

$$\Gamma_{ed} = n_H \Sigma_d \int \frac{c \, Q_a(\lambda) \, y_e \, U_\lambda \, E_2}{h\nu} \, d\lambda. \tag{7.62}$$

We can consider the integral in (7.62) as $\bar{E}_2 F_e$, where we consider the photoelectron mean energy and define the flux of photoelectrons produced by grains per projected area of the grains, F_e ($cm^{-2}\,s^{-1}$):

$$\Gamma_{ed} = n_H \Sigma_d \bar{E}_2 F_e. \tag{7.63}$$

The quantum yield y_e is not well known and depends on the grain size and nature, as well as on the absorbed photon frequency. For ultraviolet radiation, y_e can be of the order of 0.1–1.0, being generally negligible for longer wavelengths. Using the integrated stellar radiation field, the photoelectrons flux is of the order of

$$F_e \simeq 2 \times 10^7\, y_e\, Q_a\, e^{-\tau}\, cm^{-2}\, s^{-1}, \tag{7.64}$$

where the exponential term takes into account the radiation attenuation inside the cloud. Adopting typical values for dust grains in interstellar clouds (Chap. 9), $Q_a \simeq 1$, $\Sigma_d \simeq 1.1 \times 10^{-21}$ cm^2 per H atom, and $\bar{E}_2 \simeq 5$ eV, we obtain

$$\Gamma_{ed} \simeq 1.8 \times 10^{-25}\, y_e\, n_H\, e^{-\tau}\, erg\, cm^{-3}\, s^{-1}. \tag{7.65}$$

For $\tau \ll 1$ and $y_e \simeq 1$, the heating function Γ_{ed} can balance $\Lambda_{H,CII}$ for the equilibrium temperature $T_E \simeq 80$ K, even if $d_C \simeq 1$ ($n_H \simeq 100$ cm^{-3}). For instance, with $y_e \sim 1, \tau \ll 1$, and $n_H \simeq 1–10^2\, cm^{-3}$, we have $\Gamma_{ed} \sim 10^{-25}–10^{-23}\, erg\, cm^{-3}\, s^{-1}$ and $\Gamma_{ed}/n_H^2 \sim 10^{-25} - 10^{-27}\, erg\, cm^{-3}\, s^{-1}$, which may be compared to values seen in Fig. 7.3. In this case, the photoelectron flux is $F_e \simeq 2 \times 10^7\, cm^{-2}\, s^{-1}$. If y_e is lower or the optical depth is higher, this mechanism loses importance, though it is still one of the most important processes to interstellar gas heating.

Recent models for dense clouds and photodissociation regions consider photoelectric ejection by grains containing polycyclic aromatic hydrocarbon (PAH, see Chap. 9) molecules. Regions with temperatures in the interval $10 \lesssim T(K) \lesssim 10^4$ and electron densities $10^{-3} \lesssim n_e(cm^{-3}) \lesssim 10^2$ are heated by grains with dimensions between 15 and 100 Å. For $T \lesssim 10^4$ K, the heating rate depends on the incident ultraviolet flux, temperature, and cloud density. Typical values for the heating function are $\Gamma_{ed}/n_H \sim 10^{-26}–10^{-25}\, erg\, s^{-1}$ or $\Gamma_{ed}/n_H^2 \sim 10^{-28} - 10^{-25}$ $erg\, cm^{-3}\, s^{-1}$ for $n_H \sim 1–10^2$ cm^{-3}. The cooling corresponding to electron recombination was also obtained, being of the order of $\Lambda_{ed}/n_e n_H \sim 10^{-27}$ -10^{-24} erg $cm^3\, s^{-1}$ for $10^2 \lesssim T(K) \lesssim 10^4$. The obtained quantum yield is lower than 0.1, typically $y_e \simeq 0.03$.

In reality, the role of dust grains on interstellar cloud temperature determination is much more complex than shown above. Besides energy lost in the electron–grain recombination process, grains can also act as coolers. H atoms may collide with dust grains and suffer coalescence, transferring their kinetic energy to the grain, which will in turn be emitted as infrared radiation. Typical rates $\Lambda_{dH} \sim 10^{-31} n_H^2\, erg\, cm^{-3}\, s^{-1}$ are obtained in clouds with $T \simeq 100$ K.

7.7.5 *Other Processes*

Other heating processes may be important in localized regions, such as (1) heating by ionizing photons absorption in H II regions (Chap. 8) and (2) heating by shock waves triggered by supernovae interaction with the interstellar medium.

7.8 Heating of the Intercloud Medium

We saw in Chaps. 4 and 6 that there is evidence of a hot and diluted gas, identified as the intercloud medium. Heating of this gas cannot obviously be accomplished by photoionization of heavy elements or H_2 molecule formation. From Fig. 7.2, we have that, for $T \simeq 6{,}000$ K ($\log T \simeq 3.8$),

$\Lambda/n_H^2 \gtrsim 3 \times 10^{-26}\,\mathrm{erg\,cm^{-3}\,s^{-1}}$ and (7.57) cannot reach this value, even if $z_H \simeq 1$. Heating by cosmic rays may explain these temperatures if the rate is very high, $\zeta_H \simeq 10^{-15}\,\mathrm{s^{-1}}$, and the H atoms density is relatively low, $n_H \simeq 0.2\,\mathrm{cm^{-3}}$. In this case, if the available photoelectrons energy is of the order of 4 eV, from (7.51),

$$\frac{\Gamma_{H,CR}}{n_H^2} \simeq 3.2 \times 10^{-26}\,\mathrm{erg\,cm^3\,s^{-1}}, \tag{7.66}$$

which is a higher value than the ones shown in Fig. 7.3. Meanwhile, evidence for such a high rate is scarce, as we have already mentioned.

Photoelectric heating by grains can maintain temperatures of the order of 10^3 K, for values of $n_H \simeq 0.1\ \mathrm{cm^{-3}}$ and with a considerable depletion of some of the cooling agents. In this case, we need heating by cosmic or X-rays, with a rate lower than $10^{-15}\ \mathrm{s^{-1}}$, in order to balance part of the gas cooling. Another mechanism considered is magnetohydrodynamic (MHD) waves dissipation associated with supernovae explosions.

The photoelectric heating model is applicable to a relatively wide temperature interval, being able to reproduce typical features of the intercloud medium, where the equilibrium temperature between the photoelectric heating rate and the recombination cooling rate has values of the order of $T \simeq 15{,}000$–$20{,}000$ K.

7.9 Interstellar Gas Instabilities

In this chapter, we consider steady-state processes for interstellar gas heating and cooling. One of the consequences of this kind of treatment is the occurrence of two different gas phases, that is, a cool and dense region identified with *interstellar clouds* and a hot and diluted phase identified with the *intercloud medium*, as illustrated in Fig. 7.4.

These phases can be understood in terms of thermal instabilities of the interstellar medium. For a certain set of heating and cooling mechanisms, we may have variation of the gas temperature T (Fig. 7.4a) and of the gas pressure P/k (Fig. 7.4b) with the total density of particles n. We see that there are two different regions: (1) a low-density ($n \lesssim 1$ cm^{-3}) and high-temperature ($T \simeq 10^4$ K) region, identified as the *intercloud medium*, and (2) a higher density ($n > 1$ cm^{-3}) and lower temperature ($T \lesssim 10^2$ K) region, identified as the *interstellar clouds*. Between the two regions, the temperature rapidly decreases with an increase in density, which is a consequence of the higher efficiency of gas cooling. When the gas is compressed (n increases), energy losses increase proportionally to n^2 [see, for instance (7.33)], whereas energy input increases less dramatically because $\Gamma \propto n$ [see, for instance (7.51)]. Therefore, the gas temperature decreases. In the intermediate region, the temperature drops off so quickly that it is unable to balance the density increase and thus pressure decreases because $p \propto nT$. This can be seen in Fig. 7.4b, where we show that the phases identified as clouds and intercloud medium are in pressure equilibrium. In the region characterized by point A (intercloud medium), pressure increases with density and the same happens in the region characterized by point C (clouds). However, in the intermediate region, for densities above the characteristic value of point B, pressure *decreases* with density increase, characterizing a *thermal instability* region. Therefore, we can interpret Fig. 7.4b in a qualitative way, initially considering a diluted and hot gas with $T \simeq 10^4$ K. This gas may suffer a compression and maintain thermal equilibrium until reaching a region where temperature drops off so quickly that it cannot balance density increase, and so pressure drops off. For still higher densities, temperature reaches a new equilibrium region, where pressure increases again.

Basic works on thermal instabilities and interstellar clouds formation from the intercloud medium were mainly developed from the 1960s onward. Let us briefly consider the principal physical aspects related to the interpretation of instabilities in an infinite, uniform, and static interstellar medium, with density ρ_0 and temperature T_0. In thermal equilibrium, we may write

$$\mathcal{L}(\rho_0, T_0) = 0, \tag{7.67}$$

where $\mathcal{L}$ is the generalized loss–input function, defined as the energy net loss per gram of material and per second [see (7.11)]. To obtain this function, we may consider processes such as photoionization, heating by cosmic rays, and collisional excitation cooling. As we saw, this equilibrium may be unstable for density or temperature perturbations. Considering an isovolumetric temperature perturbation, T', we have

$$T = T_0 + T' \tag{7.68}$$

$$\mathrm{d}Q = c_V \, \mathrm{d}T \tag{7.69}$$

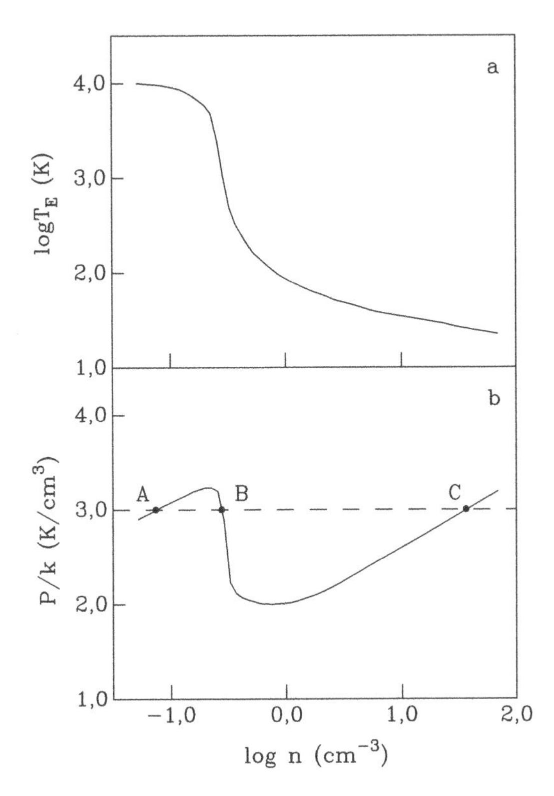

Fig. 7.4 Phases of the interstellar gas

$$c_V \frac{dT}{dt} = -\mathcal{L}, \tag{7.70}$$

where dQ is the absorbed heat per gram in an infinitesimal process, c_V is the specific heat at constant volume, and d/dt represents the total derivative, $d/dt = \partial/\partial t + (\mathbf{v} \cdot \Delta)$. Since $\mathbf{v}_0 = 0$, we have

$$c_V \frac{\partial T'}{\partial t} = -\left(\frac{\partial \mathcal{L}}{\partial T}\right)_\rho T' = -\mathcal{L}_T T'. \tag{7.71}$$

The gas will be unstable if the $\mathcal{L}$ variation has an opposite sign relative to the gas entropy variation. The *instability criterion for isovolumetric temperature perturbations* is

$$\mathcal{L}_T = \left(\frac{\partial \mathcal{L}}{\partial T}\right)_\rho < 0. \tag{7.72}$$

For instance, for $T' > 0$, there will be instability if $\partial T'/\partial t > 0$.

Let us now consider an *isobaric* perturbation. Let c_{p} be the specific heat at constant pressure, then we have

$$dQ = c_{\mathrm{p}}\, dT, \tag{7.73}$$

$$c_{\mathrm{p}} = \frac{\partial T'}{\partial t} = -\left(\frac{\partial \mathcal{L}}{\partial T}\right)_{\mathrm{p}} T'. \tag{7.74}$$

There will be *instability for an isobaric perturbation* if

$$\left(\frac{\partial \mathcal{L}}{\partial T}\right)_p < 0. \tag{7.75}$$

But

$$\left(\frac{\partial \mathcal{L}}{\partial T}\right)_p = \left(\frac{\partial \mathcal{L}}{\partial T}\right)_\rho + \left(\frac{\partial \mathcal{L}}{\partial \rho}\right)_T \left(\frac{\partial \rho}{\partial T}\right)_p. \tag{7.76}$$

Introducing the ideal gas law, the instability criterion for an isobaric perturbation takes the form

$$\left(\frac{\partial \mathcal{L}}{\partial T}\right)_\rho - \frac{\rho_o}{T_o}\left(\frac{\partial \mathcal{L}}{\partial \rho}\right)_T < 0. \tag{7.77}$$

Using the equation of state and condition (7.67), we obtain

$$\left(\frac{\partial \mathcal{L}}{\partial T}\right)_\rho \left(\frac{\partial \log p}{\partial \log \rho}\right)_{\mathcal{L}=0} < 0. \tag{7.78}$$

Supposing that $(\partial \mathcal{L}/\partial T)_\rho > 0$, we obtain

$$\left(\frac{\partial \log p}{\partial \log \rho}\right)_{\mathcal{L}=0} < 0, \tag{7.79}$$

which is the condition for thermal instability for an isobaric perturbation. This relation helps us to understand the result shown in Fig. 7.4, relative to plane $p \times \rho$ or $p \times n$.

Besides steady-state heating and cooling processes, the interstellar medium suffers other dynamical nonsteady-state processes that lead to gas heating, frequently with temperatures higher than the one of the intercloud medium or $T \gtrsim 10^4$ K. These processes include H II regions expansion, supernovae explosions, and stellar winds associated with hot stars. Some of these involve high-velocity shocks, where mechanical energy is transformed into thermal energy, with temperatures reaching values of the order of 10^6 K and where O VI lines, among others, and X-ray emission may be produced.

Exercises

7.1 (a) Estimate the cooling time for an H I cloud with $T = 100$ K, $n_H = 10$ cm^{-3}, and $n_e/n_H = 10^{-3}$. (b) Estimate the recombination time for radiative capture of an electron by an heavy element X^r, defined as $1/t_r = n_e\alpha(X^r)$, where $\alpha(X^r)$ is the radiative recombination coefficient. Compare the two timescales.

7.2 (a) Consider an interstellar cloud with $n_H = 20$ cm^{-3} heated by cosmic particles coming from H ionization, with rate $\zeta_H = 10^{-15}$ s^{-1}. Estimate the energy per cm^3 per second transferred to the gas, supposing that the mean energy of the electrons ejected by the cosmic rays is 3.4 eV. (b) Suppose that the interstellar cloud is cooled by collisional excitation of C II by H atoms. Consider a depletion parameter $d_C = 0.2$ and obtain the cloud equilibrium temperature.

7.3 Suppose that solid dust grains of an interstellar cloud are spherical with a radius $a = 100$ Å and internal density $s = 3$ g cm^{-3}. (a) What is the geometric cross section of the grains? (b) What is the mass of the grains relative to the H atom mass? (c) Estimate the grains' projected area per hydrogen nucleus Σ_d, supposing that the ratio between the total mass of the grains and the total mass of the gas (grain-to-gas ratio) is of the order of 1/200. (d) Estimate the energy provided to cloud heating by photoelectric emission, considering a cloud with $n_H = 1$ cm^{-3}. Assume that the photoelectrons flux is $F_e = 2 \times 10^6$ cm^{-2} s^{-1} and the photoelectron mean energy is 5 eV.

7.4 An interstellar cloud is heated by two processes: (1) H ionization by cosmic rays at a rate of 5×10^{-16} s^{-1}, corresponding to photoelectrons with mean energy of 5 eV, and (2) stellar radiation, by means of carbon photoionization. The cloud cooling is exclusively accomplished by C collisional excitation due to electrons. The cloud has a density $n_H = 1$ cm^{-3} and a fractional ionization $n_e/n_H = 0.1$. Assume that all carbon atoms are ionized, the carbon abundance is 4×10^{-4} n_H, and also that 75 % of the C atoms are contained in interstellar dust grains. (a) Estimate the heating function (erg cm^{-3} s^{-1}) by cosmic rays. (b) Estimate the heating function by stellar radiation for typical cloud temperatures. Which one of these processes dominates? (c) Estimate the cooling function by C ions. (d) Estimate the cloud temperature.

7.5 Assume that the cooling function for typical temperatures of the intercloud medium is given by $\Lambda/n_H^2 \simeq 3 \times 10^{-26}$ erg cm^3 s^{-1}. Bakes and Tielens model (1994) predicts a heating rate by hydrogen atoms of the order of 7×10^{-27} erg s^{-1}. What is the density of this interstellar region?

Bibliography

Bakes, E.L.O., Tielens, A.G.G.M.: The photoelectric heating mechanism for very small graphitic grains and polycyclic aromatic hydrocarbons. Astrophys. J. **427**, 822 (1994). Updated discussion on dense clouds and photodissociation regions and on the principal heating processes involving dust grains. (See also Annu. Rev. Astron. Astrophys. vol. 35, p.179, 1997)

Bowers, R.L., Deeming, T.: Astrophysics II. Jones and Bartlett, Boston (1984). Includes a chapter about interstellar gas heating processes and temperature determination in interstellar clouds

Dalgarno, A., McCray, R.A.: Heating and ionization of HI regions. Annu. Rev. Astron. Astrophys. **10**, 375 (1972). Excellent overview article, with a detailed discussion on the principal heating and cooling processes of the interstellar gas. Figures 7.1 and 7.2 are based on this reference

Dyson, J., Williams, D.A.: The Physics of the Interstellar Medium. Institute of Physics Publishing, London (1997). Referred to in Chapter 1. Accessible discussion on interstellar gas heating and cooling processes and the role of interstellar grains

Field, G.B.: In: Habing, H.J. (ed.) IAU Symposium 39, p. 51. Reidel, Dordrecht (1970). Discussion on thermal instability processes in the interstellar medium by one of the leading scientists of this study (See also Astrophys. J. vol. 142, p.531, 1965 and Astrophys. J. Lett. vol. 155, p.49, 1969)

Kaplan, S.A., Pikelner, S.B.: The Interstellar Medium. Harvard University Press, Cambridge (1970). Referred to in Chapter 1. Includes a discussion on temperatures of the interstellar gas and some heating processes

Scheffler, H., Elsässer, H.: Physics of the Galaxy and Interstellar Matter. Springer, Berlin (1988). Referred to in Chapter 1. Includes an analysis of interstellar gas temperature determination and of heating and cooling processes, besides instability processes and references. Figure 7.4 is based on this reference

Spitzer, L.: Physical Processes in the Interstellar Medium. Wiley, New York (1978). Referred to in Chapter 1. Includes a discussion on the principal heating and cooling processes of the interstellar gas and numerical estimates of the main contributions to gas heating, as well as on the functions used to equilibrium temperature calculation and references (See also Astrophys. J. vol. 107, p.6, 1948; Annu. Rev. Astron. Astrophys. vol. 13, p.133, 1975; and Annu. Rev. Astron. Astrophys. vol. 28, p.71, 1990)

Chapter 8
Interstellar Ionized Nebulae

8.1 Introduction

In this chapter, we will consider interstellar ionized gaseous nebulae, which include H II regions, planetary nebulae, and supernovae remnants. The first two are essentially gas and dust clouds associated with very hot stars that cause gas photoionization. Supernovae remnants are also composed of ionized gas, though in this case ionization has collisional origin.

8.1.1 H II Regions

H II regions or diffuse nebulae are essentially gas clouds associated with OB stars, such as the Orion Nebula (NGC 1976 = M42, Fig. 8.1), Rosette (NGC 2237), or the Trifid Nebula (NGC 6514 = M20). They are mainly composed of hydrogen, with some helium and traces of heavy elements. H II region complexes may contain several hot stars and are associated with dark and dense clouds, having molecules such as H_2 and CO.

Due to the large number of ultraviolet photons coming from the central star, H is ionized, as well as He and the heavy elements. Table 8.1 compares some of the main H II regions properties with the ones of planetary nebulae.

H II regions can be detected at several wavelengths, from X-rays to radio. The optical region of the spectrum is characterized by strong emission lines superimposed to a faint continuum background, similar to planetary nebulae, as we have seen in the spectrum of Fig. 5.2. Some lines are typical of ions of elements O, N, S, etc., such as the [O III]4959/5007 Å lines and the [N II] 5754 Å line. These are forbidden lines, meaning that the radiative de-excitation probability is very low, and so in laboratory conditions, de-excitation is accomplished collisionally. In nebulae, however, densities are low (see Tables 1.1 and 8.1), so de-excitation is radiative, with emission of an optical photon. Lines such as these are basically excited through *collisions* between electrons and ions.

W.J. Maciel, *Astrophysics of the Interstellar Medium*,
DOI 10.1007/978-1-4614-3767-3_8, © Springer Science+Business Media New York 2013

Fig. 8.1 Orion Nebula (Rodrigo Prates Campos, LNA)

Table 8.1 Properties of photoionized nebulae

	H II regions	Planetary nebulae
Spectral type	O, B	O, W
Effective temperature	30,000–50,000 K	30,000–300,000 K
Population	I young	I old II
Electron temperature	10^4 K	10^4 K
Electron density	10–10^2 cm^{-3}	10^2–10^4 cm^{-3}
Total mass	10^2–10^4 $M_\odot$	0.01–1 $M_\odot$
Typical dimension	10 pc	$\lesssim$0.5 pc
H state	H^+	H^+
He state	He^+	He^+, He^{++}
Heavy elements	Ionized	Ionized
Typical velocity	10 km s^{-1} (thermal)	25 km s^{-1} (expansion)

In the H recombination process, there is also line emission, such as the ones of the Balmer series (Hα 6563 Å, Hβ 4861 Å, Hγ 4340 Å). The recombination spectrum is due to electron–proton and electron–ion recombination, also involving ions such as He^+, C^+, N^+, and O^+.

Emission-line spectra also include *fluorescence*-excited lines, that is, lines formed from energy level excitations by photons originated in the nebular emission itself. For that, it is necessary that the photon energy coincides more or less with the line excitation energy, as it happens with the O^{++} line 303.799 Å excitation by photons which come from the He^+ recombination at 303.780 Å.

Besides the emission-line spectrum, a continuum spectrum may be observed at radio, infrared, and optical wavelengths. This radiation is partially produced by free–free radiation (electrons thermal bremsstrahlung) at radio wavelengths. Superimposed to this continuum, radio recombination lines are observed, essentially H and He lines.

In the infrared, there is emission from dust grains (Chap. 9). For not very hot central stars (spectral type B1 or later), starlight reflection by dust grains may be observed. In this case, we will have a continuum spectrum similar to the one of the star, and the nebula will be a *reflection nebula*, like in the case of the Pleiades (M 45).

H II regions present a wide interval of physical properties. Objects with dimensions of the order of 10 pc are *classical H II regions*. Larger objects, with dimensions of the order of or higher than 100 pc, are *giant H II regions*, and those with dimensions below 1 pc are *compact H II regions*, whose densities can reach about 10^6 cm^{-3}.

Besides ionized gas, there is evidence for neutral gas associated with nebulae, grains, and molecular line emission. In H II regions, comet-tail or elephant-trunk structures are frequently observed. Such features are generally associated with ionization fronts near the nebula edges. Nebulae geometry is usually complex, involving interaction between a group of young hot stars and relatively dense gas regions, reaching maybe 10^5 cm^{-3}.

H II regions are particularly important for mapping the spiral structure of the Galaxy and of other galaxies. As population I objects, H II regions are concentrated in the galactic plane, at low galactic latitudes, and are associated with the spiral arms. Their rotation velocities define the Galaxy's rotation curve quite accurately, which may in turn be used to estimate kinematic distances and the mass of the galactic system. Finally, these objects are particularly important in the study of the Galaxy's chemical evolution because their abundances are normally representative of the ones of the younger galactic objects.

8.1.2 Planetary Nebulae

Planetary nebulae are shells of gas ejected by an intermediate mass star (0.8–8 $M_{\odot}$) already in its last evolutionary stages, between red giants and white dwarfs. They appear as gas clouds surrounding a very hot star, frequently with spherical or bipolar symmetry.

Table 8.1 shows some planetary nebulae properties. Assuming a typical expansion velocity of the order of 25 km s^{-1} and a mean radius of the order of 0.4 pc, we see that the timescale for the planetary phase is about 16,000 years, which is rather short compared with the main timescales of stellar evolution. The protoplanetary phase, between the asymptotic giant branch (AGB) and the beginning of photoionization, is even shorter by a factor of the order of 10.

In spite of different origin and evolution, planetary nebulae have much in common with H II regions, since both consist of a hot star surrounded by a gas cloud. Planetary nebulae are relatively distant objects, with the closest ones at a distance of about 200 pc. Some examples are NGC 7293 (Helix) in the Aquarius constellation, NGC 7027, and NGC 7009.

We have seen a typical spectrum of a planetary nebula in Fig. 5.2. The spectral energy distribution of these objects, from the ultraviolet to the far infrared, includes infrared continuum radiation produced by dust grains embedded in the nebula, apart from bright emission lines at ultraviolet and optical wavelengths. In many cases, the central star can be observed, and models for the whole spectrum, including the central star radiation, the emission nebular lines, and the infrared emission of the dust grains, can be obtained. The planetary nebulae central stars are, in fact, the hottest stars in the Galaxy, able to attain effective temperatures above 200,000 K.

Planetary nebulae originate from main sequence stars in the mass interval 1–8 $M_\odot$, approximately. The more massive stars are in general younger, so planetary nebulae include objects of different populations, a fact that makes them particularly attractive for studies of the chemical and dynamical evolution of the Galaxy.

8.1.3 Supernovae Remnants

The most massive stars ($\gtrsim 10\ M_\odot$) evolve through more violent final stages, exploding as supernovae. A rarer class of interstellar ionized nebulae, supernova remnants, are frequently ionized gas filaments moving at speeds of the order of 300–6,000 km s^{-1}, around the region where the explosion took place. The best known example is the Crab Nebula (NGC 1952 = M1), a supernova remnant that exploded in 1054 AD and was registered by Chinese astronomers. Other examples are Vela X, with an age of the order of 13,000 years, and the Cygnus Ring with 17,000 years. After explosion, a shock wave is produced, which heats up the gas to temperatures higher than or of the order of 10^5 K. For this temperature, collisions between atoms and electrons are highly energetic, ionizing H, He, and heavy elements and producing X-rays.

The main observational characteristic of these objects is the presence of intense radiation at radio frequencies, coming from nonthermal mechanisms. In the Crab Nebula, this emission extends to optical and ultraviolet wavelengths, mostly triggered by the *synchrotron* mechanism (see Chap. 2). For the other cases, the major part of the energy is produced by conversion into heat of the gas particles kinetic energy, that is, by means of filaments collisions with interstellar gas. The energy coming from this collisional process ionizes and heats up the gas and may eventually be converted into radiation, including X-rays. We see that, contrary to H II regions and planetary nebulae which are photoionized, supernova remnants are *collisionally* ionized.

8.2 Transition Between H II and H I Regions

Due to the abundance of ionizing photons, the region around hot stars contains essentially ionized gas, whereas in the general interstellar medium, the major elements (H, He) are neutral, particularly in interstellar clouds. We may roughly estimate the dimensions of the transition region between H II and H I regions, assuming only interstellar gas opacity.

8.2.1 Transition Region Thickness

We saw in Chap. 6 that the photoionization cross section for an hydrogen-like atom in level $n = 1$ is

$$\sigma_\nu \simeq 7.9 \times 10^{-18} Z^{-2} \left(\frac{\nu_1}{\nu}\right)^3 g_1 \, \mathrm{cm}^2, \tag{8.1}$$

where $\nu_1 = R\,Z^2 = 3.29 \times 10^{15}$ Hz for H and g_1 in the Gaunt factor [see (6.53)]. For frequencies near the threshold ν_1, we have $g_1 \simeq 0.8$ and cross section $\sigma_{\nu 1} \simeq 6.3 \times 10^{-18}$ cm^2 for H. The mean free path for an ionizing photon is

$$L_1 \simeq \frac{1}{n(\mathrm{HI})\sigma_{\nu_1}} \simeq 0.05\,\mathrm{pc}, \tag{8.2}$$

where we assume $n(\mathrm{HI}) \simeq 1$ cm^{-3} for the neutral hydrogen atom density. For frequencies above ν_1, cross section (8.1) decreases and the mean free path can reach $L_{180} \simeq 5$ pc for $\lambda = 180$ Å ($\nu \simeq 5\ \nu_1$), where $g_1 \simeq 0.99$. The mean free path is always smaller than or of the order of 1 pc because for frequencies near 180 Å, there is less radiation, unless the central star is extremely hot. Naturally, for n (HI) > 1 cm^{-3}, length L decreases even more. In the next section, we shall see that the ionized sphere dimensions are relatively large, typically of the order of a few pc (see Table 8.1), much larger than the mean free path. Therefore, a consequence of the interstellar neutral gas high opacity is that *the transition region between an H II region and an H I region must be very thin*, less than 1 pc, typically. In other words, the ionized region, assumed spherical, can be characterized by a relatively well-defined dimension, the Strömgren radius, thus called in reference to the first analysis done by Bengt Strömgren in 1939.

8.2.2 Strömgren Radius

Let us estimate the dimensions of the ionized H region around a hot central OB star. In Chap. 6, we have seen that the probability per second of H photoionization from level 1 is

$$\beta_{1\mathrm{ph}} = \int_{\nu_1}^{\infty} \frac{\sigma_\nu\, c\, U_\nu\, \mathrm{d}\nu}{h\nu} \tag{8.3}$$

(Equation 6.15). The energy density U_ν has now two components: U_ν^*, the stellar radiation field contribution, and U_ν^{D}, the "diffuse" radiation field contribution, that is, the Lyman continuum photons emitted in the recombination process of electrons to H level $n = 1$. For a star with luminosity $L_\nu d\nu$ in the frequency interval $d\nu$, where L_ν is the luminosity per frequency interval, the contribution U_ν^* at a distance r from the star is

$$U_\nu^* = \frac{L_\nu \mathrm{e}^{-\tau_\nu}}{4\pi r^2 c}, \tag{8.4}$$

where we introduce the attenuation factor $\mathrm{e}^{-\tau\nu}$. The optical depth may be defined by $\mathrm{d}\tau_\nu = k_\nu\, \mathrm{d}r$, where k_ν is the absorption coefficient per volume, measured in cm^{-1}. Near the star we have $r = R_*$ and for $r \gg R_*$ we have

$$\tau_\nu = \tau_{\nu r} = \int_0^r k_\nu\, \mathrm{d}r. \tag{8.5}$$

Let $n(\mathrm{H\,I})$ be the atomic H density and n_{H} the total number of H nuclei per cm^3. In a general way, $n_{\mathrm{H}} = n(\mathrm{H\,I}) + n_{\mathrm{p}}$ or

$$1 - x = \frac{n_{\mathrm{H}} - n_{\mathrm{p}}}{n_{\mathrm{H}}} = \frac{n(\mathrm{H\,I})}{n_{\mathrm{H}}}, \tag{8.6}$$

where $x = n_{\mathrm{p}}/n_{\mathrm{H}}$ is once again the degree of ionization of H and we neglect molecule formation. Since $k_\nu = n(\mathrm{H\,I})\, \sigma_\nu$, we have

$$\mathrm{d}\tau_\nu = n(\mathrm{H\,I})\, \sigma_\nu\, \mathrm{d}r = (1 - x) n_{\mathrm{H}}\, \sigma_\nu\, \mathrm{d}r. \tag{8.7}$$

Multiplying (8.4) by $4\pi r^2 c$ and differentiating, we obtain

$$\frac{1}{r^2}\frac{\mathrm{d}}{\mathrm{d}r}\left(r^2 c U_\nu^*\right) = -(1 - x) n_{\mathrm{H}}\, \sigma_\nu\, c\, U_\nu^*. \tag{8.8}$$

Let us now consider the diffuse radiation field. If F_ν^{D} is the diffuse flux, we can show that

$$\frac{1}{r^2}\frac{\mathrm{d}}{\mathrm{d}r}\left(r^2 F_\nu^{\mathrm{D}}\right) = -(1 - x) n_{\mathrm{H}}\, \sigma_\nu\, c\, U_\nu^{\mathrm{D}} + 4\pi j_\nu^{\mathrm{D}}, \tag{8.9}$$

where j_ν^{D} is the diffuse emissivity in the Lyman continuum measured in $\mathrm{erg\ cm^{-3}\ s^{-1}\ Hz^{-1}\ sr^{-1}}$. Since $U_\nu = U_\nu^* + U_\nu^{\mathrm{D}}$, (8.8) and (8.9) allow us to write

$$\frac{1}{r^2}\frac{\mathrm{d}}{\mathrm{d}r}(r^2 cU_\nu^* + r^2 F_\nu^{\mathrm{D}}) = -(1-x)n_{\mathrm{H}}\,\sigma_\nu\, c\, U_\nu + 4\pi j_\nu^{\mathrm{D}}. \tag{8.10}$$

The ionization equilibrium equation may be written as $(1-x)\beta_{1\mathrm{ph}} = xn_{\mathrm{e}}\alpha$ [see (6.9)] so that

$$\frac{1}{r^2}\frac{\mathrm{d}}{\mathrm{d}r}\left[r^2\left(cU_\nu^* + F_\nu^{\mathrm{D}}\right)\right] = -\frac{xn_{\mathrm{e}}\alpha}{\beta_{1\mathrm{ph}}}n_{\mathrm{H}}\,\sigma_\nu\, cU_\nu + 4\pi j_\nu^{\mathrm{D}}. \tag{8.11}$$

Multiplying by $\mathrm{d}\nu/h\nu$ and integrating over ν, we obtain

$$\frac{1}{r^2}\frac{\mathrm{d}}{\mathrm{d}r}\left[r^2\int_{\nu_1}^{\infty}\left(cU_\nu^* + F_\nu^{\mathrm{D}}\right)\frac{\mathrm{d}\nu}{h\nu}\right] = -\frac{x\,n_{\mathrm{e}}\,\alpha\, n_{\mathrm{H}}}{\beta_{1\,\mathrm{ph}}}\int_{\nu_1}^{\infty}\frac{\sigma_\nu\, c\, U_\nu\,\mathrm{d}\nu}{h\,\nu} + 4\pi\int_{\nu_1}^{\infty} j_\nu^{\mathrm{D}}\frac{\mathrm{d}\nu}{h\,\nu}. \tag{8.12}$$

The U_ν integral is exactly the same as $\beta_{1\mathrm{ph}}$ (8.3). Multiplying by $4\pi r^2$ and rearranging, we obtain

$$\frac{\mathrm{d}}{\mathrm{d}r}\left[4\pi r^2\int_{\nu_1}^{\infty}\left(c\,U_\nu^* + F_\nu^{\mathrm{D}}\right)\frac{\mathrm{d}\nu}{h\nu}\right] = -4\pi r^2\left[x\,n_{\mathrm{e}}\,n_{\mathrm{H}}\,\alpha - 4\pi\int_{\nu_1}^{\infty} j_\nu^{\mathrm{D}}\frac{\mathrm{d}\nu}{h\,\nu}\right]. \tag{8.13}$$

Recalling that $xn_{\mathrm{H}} = n_{\mathrm{p}}$ and thus term $xn_{\mathrm{e}}n_{\mathrm{H}}\alpha$ simply becomes $n_{\mathrm{p}}n_{\mathrm{e}}\alpha$, the total number of recombinations per cm^3 per second in each point r. The second term on the right $\left(4\pi\int j_\nu^{\mathrm{D}}\mathrm{d}\nu/h\nu\right)$ is the number of recaptures to the ground level per cm^3 per second. Therefore, the difference between square brackets in the second member of (8.13) is simply $xn_{\mathrm{e}}n_{\mathrm{H}}\alpha^{(2)}$ (6.60), and we may write

$$\frac{\mathrm{d}}{\mathrm{d}r}\left[4\pi r^2\int_{\nu_1}^{\infty}\left(c\,U_\nu^* + F_\nu^{\mathrm{D}}\right)\frac{\mathrm{d}\nu}{h\nu}\right] = -4\pi r^2\, x\, n_{\mathrm{e}}\, n_{\mathrm{H}}\,\alpha^{(2)}. \tag{8.14}$$

In this equation, U_ν^* is the energy due to stellar radiation per cm^3 per Hz in point r. Thus, cU_ν^* is the energy per cm^2 per Hz per second or the radiation flux, and cU_ν^* $+F_\nu^{\mathrm{D}}$ gives the total radiation. Dividing by $h\nu$ and integrating along ν, we simply obtain the total number of ionizing photons per cm^2 per second. Multiplying by the area of a sphere with radius r, we obtain the total number of ionizing photons per second in position r, which we will call $Q(r)$. Equation (8.14) then becomes

$$\frac{\mathrm{d}Q(r)}{\mathrm{d}r} = -4\pi\, r^2\, x\, n_{\mathrm{e}}\, n_{\mathrm{H}}\,\alpha^{(2)}. \tag{8.15}$$

This equation may be interpreted in the following way: As we move away from the star, r increases and $\mathrm{d}Q/\mathrm{d}r < 0$, and more ionizing photons are destroyed by ionizations followed by recombinations to levels above the ground level, $n \geq 2$. Therefore, the variation per second of the number of ionizing photons is proportional to the volume of the layer and to the recombination efficiency for levels above the ground level, measured by the recombination coefficient $\alpha^{(2)}$. Naturally, direct recombinations to the ground level produce a Lyman continuum photon that will ionize another H atom and will therefore not be counted as "lost" in the $Q(r)$ calculation.

Let us now integrate (8.15) from $r = 0$ (we saw that $r \gg R_*$) to a certain radius r_s where $Q(r) = 0$. We obtain

$$\int_{Q(r)}^{0} \mathrm{d}Q(r) = -\int_{0}^{r_s} 4\pi r^2 x\, n_e\, n_H\, \alpha^{(2)} \mathrm{d}r, \tag{8.16}$$

$$Q(0) = 4\pi \int_{0}^{r_s} r^2 x\, n_e\, n_H\, \alpha^{(2)} \mathrm{d}r = 4\pi \int_{0}^{r_s} r^2\, n_e\, n_p\, \alpha^{(2)} \mathrm{d}r. \tag{8.17}$$

Assuming that $x \simeq 1$ (completely ionized gas) and that $n_e n_H \alpha^{(2)}$ is r independent, that is, neglecting H II region heterogeneity,

$$Q(0) = \frac{4}{3}\pi r_s^3\, n_e\, n_H\, \alpha^{(2)}. \tag{8.18}$$

For $r \to 0$ (or more exactly $r \to R_*$), F_ν^D is negligible relative to the stellar component, and the total number of photons per second $Q(0)$ is identical to the number of ionizing photons emitted by the star per second, Q_*:

$$Q_* = \frac{4}{3}\pi\, r_s^3\, n_e\, n_H\, \alpha^{(2)}. \tag{8.19}$$

Assuming the star to be a blackbody or adopting an atmosphere model for the central star characterized, for instance, by its spectral type, effective temperature, and radius, we may calculate the number of ionizing photons per second:

$$Q_* = \int_{\nu_1}^{\infty} \frac{L_\nu}{h\nu} \mathrm{d}\nu = 4\pi R_*^2 \int_{\nu_1}^{\infty} \frac{\pi F_\nu(R_*)}{h\,\nu} \mathrm{d}\nu, \tag{8.20}$$

where F_ν is the radiative flux at the surface of the star, given in erg cm^{-2} s^{-1} Hz^{-1}. Knowing $\alpha^{(2)}$ as a function of temperature, which may be estimated by different methods, we can obtain $r_s(n_e n_H)^{1/3}$ or even r_s, the Strömgren radius, if we assume some hypothesis for $n_e n_H$. From (8.19),

$$r_s(n_e\, n_H)^{1/3} = \left[\frac{3Q_*}{4\pi\, \alpha^{(2)}}\right]^{1/3}, \tag{8.21a}$$

Table 8.2 Parameters of H II regions surrounding stars of spectral types O5 to B1

Spectral type	$T_{\rm eff}$ (K)	R_* ($R_\odot$)	Q_* (10^{48} s^{-1})	$r_s\,(n_e\,n_H)^{1/3}$ (pc cm^{-2})	$\tau_{sd}/n_H^{1/3}$ (cm)
O5	47,000	13.8	51	110	0.69
O6	42,000	11.5	17.4	77	0.48
O7	38,500	9.6	7.2	57	0.36
O8	36,500	8.5	3.9	47	0.29
O9	34,500	7.9	2.1	38	0.24
B0	30,900	7.6	0.43	22	0.14
B1	22,600	6.2	0.0033	4.4	0.028

$$r_s = \left[\frac{3Q_*}{4\pi\, n_e\, n_H\, \alpha^{(2)}}\right]^{1/3}. \tag{8.21b}$$

In Table 8.2, we give the values of $r_s(n_e n_H)^{1/3}$ and Q_* for H II regions around main sequence stars of several spectral types. Calculations were done for $T \simeq 8{,}000$K so that $\alpha^{(2)} \simeq 3.1 \times 10^{-13}$ cm^3 s^{-1}. The mean temperature may be estimated from methods such as radio recombination line analysis (see Sect. 8.6). We see that for an O5 star, supposing $n_e \simeq n_H \simeq 10$–100 cm^{-3}, $r_s \simeq 5$–20 pc.

For supergiant stars of luminosity class I, the Strömgren radius increases by a factor 1.5–2 for O7–B0 stars. In H II regions, $n_e \simeq n_H$, neglecting n(HI) contribution, and $n_e \simeq n_p$, neglecting He and heavy elements contribution. Therefore,

$$r_s = (n_e\, n_H)^{1/3} \simeq r_s\, n_e^{2/3}. \tag{8.22}$$

For a given central star, Q_* is given by Table 8.2 and $r_s \propto Q_*^{1/3} n_e^{-2/3}$, that is, for denser regions, the ionized sphere is smaller.

In Fig. 8.2, we show the variation of r_s with n_e for the stars shown in Table 8.2, using approximation (8.22). Naturally, n_H is not exactly the same as n_p, because there is at least some H in atomic form and n_e is generally higher than n_p, taking into account the heavy elements and He contribution, with a decrease in r_s. However, we must take into account that frequently denser H II regions ($n_e \gtrsim 100$ cm^{-3}) have several exciting stars that tend to increase the ionized sphere, forming giant H II regions or H II region complexes.

8.3 Degree of Ionization of Hydrogen

In the previous section, we assumed $x \simeq 1$ for a homogeneous H II region. Let us now roughly estimate this parameter. The H ionization equilibrium equation is

$$n(\mathrm{H\,I})\beta = n_e\, n_p\, \alpha. \tag{8.23}$$

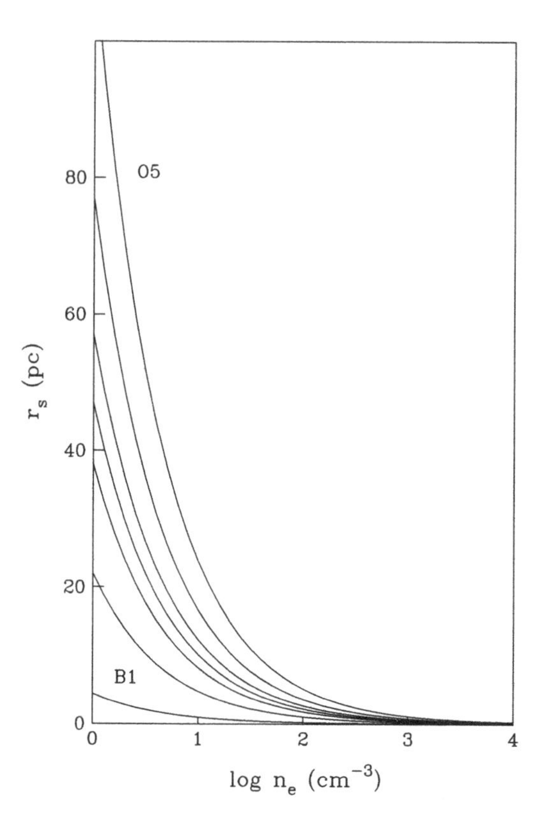

Fig. 8.2 The Stromgrem radius as a function of the electron density for the stars of Table 8.2

Remembering that $x = n_p/n_H$ and $n_H = n(\mathrm{HI}) + n_p$ and considering (8.6), the ionization equilibrium equation allows us to write

$$x = \frac{1}{1 + \frac{n_e \alpha}{\beta}}, \tag{8.24}$$

that is,

$$x \simeq 1 \quad \text{if} \quad \frac{n_e \alpha}{\beta} \ll 1. \tag{8.25}$$

The recombination coefficient can be taken to be of the order of $\alpha \simeq 3 \times 10^{-13}$ cm^3 s^{-1} like in the previous section. The ionization rate β given in (8.3) may be written

$$\beta = \int_{\nu_1}^{\infty} \frac{\sigma_\nu \, c \, U_\nu}{h \nu} \mathrm{d}\nu \simeq \int_{\nu_1}^{\infty} \frac{\sigma_\nu \, c \, U_\nu^*}{h \nu} \mathrm{d}\nu, \tag{8.26}$$

if we neglect the diffuse radiation field. Using (8.4) and neglecting the ionizing radiation attenuation ($\tau_\nu \simeq 0$),

$$\beta \simeq \int_{\nu_1}^{\infty} \frac{\sigma_\nu L_\nu \, d\nu}{4\pi r^2 \, h\nu}. \tag{8.27}$$

Let us now approximate σ_ν by the threshold cross section, that is, $\sigma_{\nu 1} \simeq 6 \times 10^{-18}$ cm^2, obtaining thus

$$\beta = \frac{\sigma_{\nu_1}}{4\pi r^2} \int_{\nu_1}^{\infty} \frac{L_\nu \, d\nu}{h\nu}. \tag{8.28}$$

Comparing with (8.20), we see that the above integral is simply Q_*, the number of ionizing photons emitted by the star per second. We have

$$\beta \simeq \frac{\sigma_{\nu_1}}{4\pi r^2} Q_*. \tag{8.29}$$

Let us estimate β for $r \simeq r_s$ in a typical H II region with an O7 star with $Q_* \simeq 7 \times 10^{48}$ s^{-1} and $r_s n_e{}^{2/3} \simeq 60$ pc cm^{-2} (see Table 8.2). In this case, $\beta \simeq 10^{-10}\, n_e{}^{4/3}$ s^{-1} and $n_e\alpha/\beta \simeq 3 \times 10^{-3} n_e{}^{-1/3}$. For a mean value $n_e \simeq 30$ cm^{-3} (see Table 8.1 and Chap. 4), we have $n_e\alpha/\beta \simeq 10^{-3}$, $\beta \simeq 8 \times 10^{-9}$ s^{-1}, $x \simeq 1$, and $1-x \simeq 10^{-3}$, confirming that the gas is ionized in the region inside the Strömgren radius.

8.3.1 *Photoionization of an H Nebula*

We can obtain the detailed ionization structure of an H II region by solving the ionization equilibrium equation in every point of the nebula. Let us assume the simple case of a single star in a homogenous region containing only H with a constant density. Remembering that $U_\nu = 4\pi J_\nu / c$ [cf. (2.40)], we can write the equilibrium equation:

$$n(\mathrm{H\,I}) \int_{\nu_1}^{\infty} \frac{4\pi J_\nu}{h\nu} \sigma_\nu d\nu = n_p n_e \alpha, \tag{8.30}$$

where $\alpha = \alpha^{(1)}$ is the recombination coefficient for all levels [cf. (6.59)–(6.61)]. The ionizing radiative transfer equation in direction s is

$$\frac{dI_\nu}{ds} = j_\nu - n(\mathrm{H\,I}) \sigma_\nu I_\nu. \tag{8.31}$$

By analogy to Sect. 8.2, we consider the stellar and diffuse components so that $I_\nu = I^*_\nu + I^{\rm D}_\nu$. The stellar component decreases as we move away from the star due to geometric dilution and radiation absorption so that we may write the flux at distance r from the star as

$$4\pi J^*_\nu = \pi F^*_\nu(r) = \pi F^*_\nu(R_*)\frac{R^2_*}{r^2}{\rm e}^{-\tau_\nu}, \tag{8.32}$$

where $F^*_\nu(r)$ and $F^*_\nu(R_*)$ represent the fluxes at distance r and at the surface of the star, respectively, and τ_ν is the optical depth at frequency ν. Similarly to (8.7), we have

$$\tau_\nu(r) = \int_0^r n({\rm H\,I})\sigma_\nu {\rm d}r. \tag{8.33}$$

For diffuse radiation, the transfer equation is given by

$$\frac{{\rm d}I^{\rm D}_\nu}{{\rm d}s} = j_\nu - n({\rm H\,I})\sigma_\nu I^{\rm D}_\nu. \tag{8.34}$$

For typical temperatures of the gas in an H II region, the diffuse ionizing radiation source is the recombinations at the ground level $n = 1$, for which the emission coefficient is a function $j_\nu = f(\nu, T, n_{\rm p}, n_{\rm e})$. The number of photons produced by these recombinations can be related to the recombination coefficient at level $n = 1$, designated by α_1 (Sect. 6.4):

$$4\pi\int_{\nu_1}^{\infty}\frac{j_\nu}{h\nu}{\rm d}\nu = n_{\rm p}\, n_{\rm e}\, \alpha_1. \tag{8.35}$$

For optically thin nebulae, we may neglect $J^{\rm D}_\nu$ on a first approximation. On the other hand, for optically thick nebulae, we may suppose that all ionizing photons coming from the diffuse radiation field end up by being absorbed inside the nebula, that is,

$$4\pi\int_{\nu_1}^{\infty}\frac{j_\nu}{h\nu}{\rm d}\nu = 4\pi\int_{\nu_1}^{\infty}\frac{n({\rm H\,I})J^{\rm D}_\nu\sigma_\nu}{h\nu}{\rm d}\nu. \tag{8.36}$$

Assuming also that these photons are essentially absorbed *in the same region that produces them* (*on the spot* approximation), we have

$$j_\nu = n({\rm H\,I})J^{\rm D}_\nu\sigma_\nu. \tag{8.37}$$

Using (8.32), (8.35), and (8.37), we may rewrite the ionization equilibrium (8.30) in the form

Table 8.3 The degree of ionization as function of position in a pure hydrogen H II region

r (pc)	$x = n_p/n_H$	$x_0 = n\,(HI)/n_H$
0.1	1.0	4.5×10^{-7}
1.2	1.0	2.8×10^{-5}
2.2	0.9999	1.0×10^{-4}
3.3	0.9997	2.5×10^{-4}
4.4	0.9995	4.4×10^{-4}
5.5	0.9992	8.0×10^{-4}
6.7	0.9985	1.5×10^{-3}
7.7	0.9973	2.7×10^{-3}
8.8	0.9921	7.9×10^{-3}
9.4	0.977	2.3×10^{-2}
9.7	0.935	6.5×10^{-2}
9.9	0.838	1.6×10^{-1}
10.0	0.000	1.0

$$\frac{n(\mathrm{H\,I})R_*^2}{r^2}\int_{\nu_1}^{\infty}\frac{\pi F_\nu^*(R_*)}{h\nu}\sigma_\nu\,\mathrm{e}^{-\tau_\nu}\mathrm{d}\nu = n_\mathrm{p}\,n_\mathrm{e}\,\alpha^{(2)}, \tag{8.38}$$

where we again use the fact that $\alpha^{(2)} = \alpha - \alpha_1$. Since the density is constant, we also have

$$n_\mathrm{H} = n(\mathrm{H\,I}) + n_\mathrm{p} = \text{constant}. \tag{8.39}$$

Therefore, relations (8.38) and (8.39) allow us to calculate the degree of ionization as a function of position in the nebula, the optical depth being obtained by relation (8.33). The photoionization cross section can be obtained as a function of frequency by a relation such as (6.53), and the recombination coefficient $\alpha^{(2)}$ comes from expression (6.63). Assuming a blackbody flux with $T_* = 40{,}000$ K and a homogeneous nebula with $T = 7{,}500$ K and $n_\mathrm{H} = 10\ \mathrm{cm}^{-3}$, Table 8.3 shows the variation of the degree of ionization with position. These characteristics are approximately representative of a main sequence O6 star. We see that the degree of ionization is essentially equal to 1 inside the nebula, sharply decreasing in the Strömgren radius neighborhood. The neutral H fraction, $x_0 = n(\mathrm{HI})/n_\mathrm{H} = 1 - x$, is also given in Table 8.3.

We see that in the inner part of the H II region, $x_0 \ll 1$, as expected. The value $x_0 \simeq 10^{-3}$ obtained at the beginning of Sect. 8.3 is reached at an average point in the nebula.

A more complete treatment must also include He and heavy elements, which will alter the results shown in Table 8.3. The inclusion of He leads to the determination of two concentric ionized regions, where He is ionized in the more internal part and H remains ionized up to a larger distance from the star, depending on the star temperature. For very hot objects, He^{++} must be taken into account as well. As an example, Fig. 8.3 shows the degree of ionization or the fractional ionization as a function of position, for two regions containing H and He, excited by central stars with temperatures $T_* = 40{,}000$ K and $T_* = 30{,}000$ K, respectively.

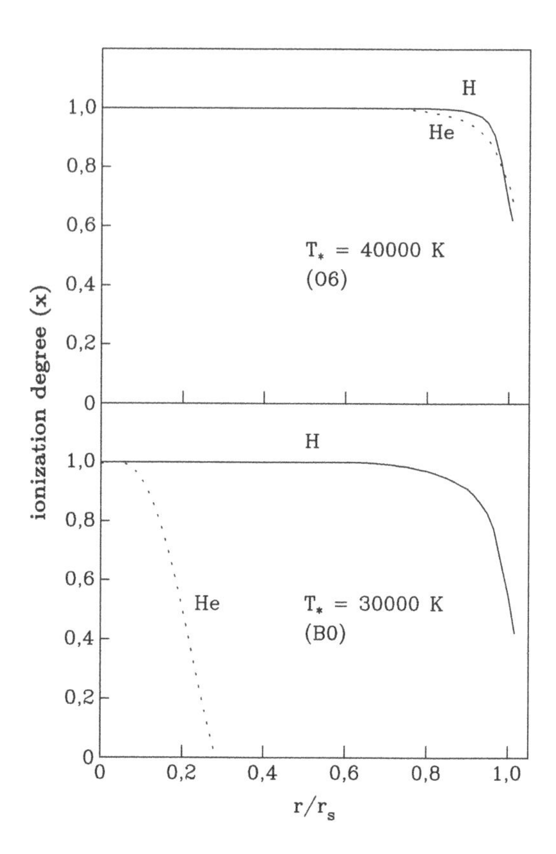

Fig. 8.3 The degree of ionization as function of position in H II region containing H and He

8.4 Dust Grains in H II Regions

8.4.1 Ionizing Photons Absorption by Dust Grains

The presence of dust grains in H II regions is inferred from several observable effects, the main ones being (1) observations of absorption features in various regions of several nebulae, blocking the radiation coming from distant objects or from parts of the nebula; (2) scattering of the central star light, quite evident in reflection nebulae; and (3) continuum infrared emission observed in H II regions, typical of dust grains with temperatures $T_d \lesssim 100$ K (Chap. 9).

The effects of dust grains on the general properties of H II regions are several and subject of intense research. Let us consider essentially the absorption of part of the star's ionizing radiation, with consequent decrease of the ionized region radius. We will get a rough estimate of the optical depth τ_{sd} for the ionizing radiation in the region delimited by r_s. Let r_i be the radius of the effectively ionized region and τ_{id} the optical depth of the dust grains in this region so that $\tau_{id} < \tau_{sd}$. Let τ_{sH} be the optical depth at r_s, if H were completely neutral, taken at the Lyman limit:

$$\tau_{sH} = n_H \sigma_\nu r_s = 6.3 \times 10^{-18} N_H, \tag{8.40}$$

where N_H is the H column density (cm^{-2}). We previously saw that N_H, comprising H^0, H^+, and H_2, linearly correlates with color excess $E(B–V)$. The $N_H/E(B–V)$ ratio (gas-to-grain or gas-to-dust ratio) is more or less constant for H II regions, as well as for H I regions [cf. (4.44)], according to relation

$$N_H \simeq 6 \times 10^{21} E(B-V) \text{ mag}^{-1} \text{ cm}^{-2}. \tag{8.41}$$

If Δm is the magnitude absorption at 912 Å and ψ is the ratio between absorption in this range and color excess, we have

$$\Delta m = 1.086\, \tau_{sd} = \psi\, E(B-V). \tag{8.42}$$

The ψ ratio may be obtained from the study of interstellar extinction (Chap. 9) and is approximately $\psi \simeq 13$. We then have the relations

$$\tau_{sd} \simeq \frac{13E(B-V)}{1.086} \simeq 2.0 \times 10^{-21} N_H, \tag{8.43}$$

$$\frac{\tau_{sd}}{\tau_{sH}} \simeq \frac{1}{3,100}, \tag{8.44}$$

$$\frac{\tau_{sd}}{n_H^{1/3}} \simeq \frac{\sigma_\nu}{3,100} (r_s\, n_H^{2/3}). \tag{8.45}$$

Assuming $n_e \simeq n_H$, $r_s(n_e n_H)^{1/3} \simeq r_s n_H^{2/3}$. Table 8.2 gives some values of $r_s(n_e n_H)^{1/3} \simeq r_s n_H^{2/3}$, with $n_e \simeq n_H$. Therefore, we may obtain $\tau_{sd}/n_H^{1/3}$ from (8.45), and the results are shown in the last column of Table 8.2. Figure 8.4 shows the total optical depth τ_{sd} as a function of effective temperature of the central star, for main sequence stars using data from Table 8.2. The four curves correspond to density values $n_H \simeq n_e = 1, 10, 100,$ and $1,000\ cm^{-3}$.

8.4.2 Ionized Region Radius

Let us estimate the ionized radius r_i with the following approximations: (1) We neglect optical depth variations with frequency using the threshold value in $\nu = \nu_1$, in the same way as previously done. (2) We use again the *on the spot* approximation, that is, we will assume that the Lyman continuum photons emitted after electron capture to level $n = 1$ are absorbed in the same region where the emission took place. Thus, the recombination coefficient is again $\alpha^{(2)}$, and we will simply write α. We will also name τ the total optical depth of gas and dust grains,

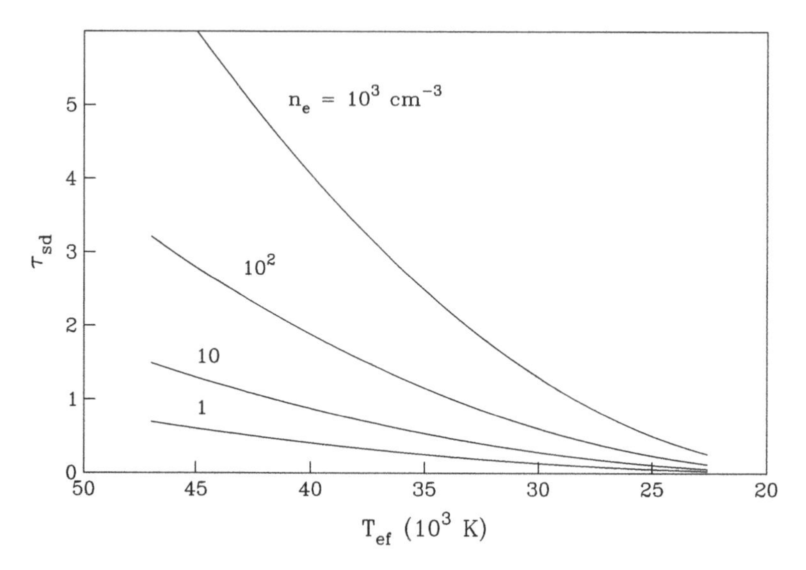

Fig. 8.4 The total dust optical depth as a function of the effective temperature for the stars of Table 8.2

$\tau = \tau_H + \tau_d$. By analogy, to what we have done in Sect. 8.3, but assuming total absorption, we have for the photoionization rate

$$\beta = \int_{\nu_1}^{\infty} \frac{\sigma_\nu \, c \, U_\nu \, d\nu}{h\nu} = \int_{\nu_1}^{\infty} \frac{\sigma_{\nu_1} L_\nu e^{-\tau} d\nu}{4\pi r^2 h\nu}$$
$$= \frac{\sigma_{\nu_1} \, e^{-\tau}}{4\pi \, r^2} \int_{\nu_1}^{\infty} \frac{L_\nu \, d\nu}{h\nu} = \frac{\sigma_{\nu_1} Q_* \, e^{-\tau}}{4\pi r^2}. \tag{8.46}$$

This equation may be compared with (8.29). Defining ratio $f = r/r_s$, we obtain

$$\beta = \frac{\sigma_{\nu_1} \, Q_* \, e^{-\tau}}{4\pi f^2 r_s^2}. \tag{8.47}$$

From the ionization equilibrium equation (see Sect. 8.3), we may write, with $n_e \simeq n_p$,

$$\frac{1-x}{x^2} = \frac{n_e \alpha}{\beta(n_p/n_H)} \simeq \frac{n_H \alpha}{\beta}. \tag{8.48}$$

Replacing β by (8.47) and Q_* by (8.19), taking $n_e n_H \simeq n_H^2$ and using (8.40), we have

$$\frac{1-x}{x^2} \simeq \frac{3f^2 e^{\tau}}{\tau_{\mathrm{sH}}}. \tag{8.49}$$

For a uniform distribution of grains, the optical depth in each point is simply $\tau_{\mathrm{d}} = f\,\tau_{\mathrm{sd}}$. For the gas, the optical depth in each point is

$$\tau_{\mathrm{H}} = n(\mathrm{H\,I})\sigma_{\nu_1} r = (1-x) n_{\mathrm{H}}\, \sigma_{\nu_1} f r_{\mathrm{s}} = (1-x) f \tau_{\mathrm{sH}}, \tag{8.50}$$

where we again use τ_{sH}, the total optical depth of H in r_{s}, if all H were neutral (8.40). From (8.50),

$$\frac{\mathrm{d}\tau_{\mathrm{H}}}{\mathrm{d}f} = (1-x)\tau_{\mathrm{sH}}. \tag{8.51}$$

Replacing (8.49),

$$\frac{\mathrm{d}\tau_{\mathrm{H}}}{\mathrm{d}f} = 3x^2 f^2 e^{\tau} = 3x^2 f^2 e^{\tau_{\mathrm{H}}+\tau_{\mathrm{d}}}. \tag{8.52}$$

Let us integrate (8.52) in an approximate manner, making $x \simeq 1$, which is a very good approximation for $r \ll r_{\mathrm{i}}$. The integration limits are $f = 0$, corresponding to $\tau_{\mathrm{H}} = 0$, and $f = f_{\mathrm{i}} = r_{\mathrm{i}}/r_{\mathrm{s}}$, corresponding to the total optical depth τ_{iH}. We have

$$e^{-\tau_{\mathrm{H}}} \frac{\mathrm{d}\tau_{\mathrm{H}}}{\mathrm{d}f} \simeq 3f^2\, e^{\tau_{\mathrm{d}}}, \tag{8.53}$$

$$\int_0^{\tau_{\mathrm{iH}}} e^{-\tau_{\mathrm{H}}} \mathrm{d}\tau_{\mathrm{H}} = \int_0^{f_{\mathrm{i}}} 3f^2 e^{\tau_{\mathrm{d}}} \mathrm{d}f, \tag{8.54}$$

$$\int_0^{\tau_{\mathrm{iH}}} e^{-\tau_{\mathrm{H}}}\, \mathrm{d}\tau_{\mathrm{H}} = -e^{-\tau_{\mathrm{H}}}\Big]_0^{\tau_{\mathrm{iH}}} \simeq 1 \quad (\tau_{\mathrm{iH}} \gg 1). \tag{8.55}$$

Therefore,

$$\int_0^{f_{\mathrm{i}}} 3f^2\, e^{\tau_{\mathrm{d}}} \mathrm{d}f = 3 \int_0^{f_{\mathrm{i}}} f^2 e^{f\tau_{\mathrm{sd}}} \mathrm{d}f = 1. \tag{8.56}$$

This equation can be integrated, and we obtain the following relation between f_{i} and τ_{sd}:

$$\exp(\tau_{\mathrm{sd}}\, f_{\mathrm{i}}) \left[\frac{f_{\mathrm{i}}^2}{\tau_{\mathrm{sd}}} - \frac{2f_{\mathrm{i}}}{\tau_{\mathrm{sd}}^2} + \frac{2}{\tau_{\mathrm{sd}}^3}\right] - \frac{2}{\tau_{\mathrm{sd}}^3} = \frac{1}{3}. \tag{8.57}$$

Values of $f_i = r_i/r_s$, the radius decreasing factor of the ionized region due to the presence of dust, are given by Spitzer (1978) as a function of optical depth τ_{sd}. For optical depths τ_{sd} varying from 0 to 10, factor f_i decreases from 1.0 to 0.4, approximately. For $\tau_{sd} \simeq 40$, $f_i \simeq 15$ %. We see that for low values of the total optical depth of the grains, τ_{sd}, the ionized region practically does not change, that is, $f_i \simeq 1$. For a larger quantity of grains or a higher ionizing radiation absorption efficiency, τ_{sd} increases and the ionized radius decreases, that is, $f_i < 1$.

With the obtained values of f_i, we can estimate the stellar ionizing photon fraction absorbed by H in the entire ionized region, which varies with f_i^3. For $\tau_{sd} \ll 1$, this fraction is of the order of 1, decreasing to 20% if $\tau_{sd} \simeq 4$. These results are compatible with H II region infrared observations, whose emission is probably caused by dust grains. We shall discuss grain temperature and infrared emission in the next chapter.

8.5 Temperatures in H II Regions

Contrary to H I regions, in HII regions, the main energy source is known: the central star or stars, with the main heating mechanism being photoionization of H, followed by a secondary component due to photoionization of He. Other processes may occur, such as photoelectric heating by dust grains. However, the importance of these processes to the H II region global heating is generally secondary. Let us examine the principal heating and cooling mechanisms in an ionized H region.

8.5.1 *Heating by H Photoionization*

As we have seen in Chap. 7, the heating function or energy provided by H photoionization per unit volume per unit time may be written

$$\Gamma_{ep} = n_e\, n_p \left[\alpha \overline{E}_2 - \frac{1}{2} m_e \sum_j \left\langle \sigma_{cj} v^3 \right\rangle \right], \tag{8.58}$$

where $\overline{E}_2$ is the photoelectron mean energy [cf. (7.45)]:

$$\overline{E}_2 = \frac{\int_{\nu_1}^{\infty} h(\nu - \nu_1)\sigma_\nu\, U_\nu (\mathrm{d}\nu/\nu)}{\int_{\nu_1}^{\infty} \sigma_\nu U_\nu (\mathrm{d}\nu/\nu)}. \tag{8.59}$$

We can define the *input temperature* T_{en} by

$$\overline{E}_2 = \frac{3}{2} k T_{en}, \tag{8.60}$$

Table 8.4 Average photoelectron energy for stars with different color temperatures

T_c (K)	Ψ_0	$\bar{E}_2(\Psi_0)$ (eV)	$\langle\Psi\rangle$	$\langle\bar{E}_2\rangle$ (eV)
4,000	0.977	0.34	1.051	0.36
8,000	0.959	0.66	1.101	0.76
16,000	0.922	1.27	1.199	1.65
32,000	0.864	2.38	1.380	3.81
64,000	0.775	4.28	1.655	9.14

that is, this temperature corresponds to the gas temperature without taking into account any cooling processes. The determination of U_ν is complicated for two reasons: Firstly, U_ν is the sum of two components, one due to stellar radiation U_ν^* and another due to diffuse radiation U_ν^{D}, and secondly, U_ν is attenuated inside the H II region due to gas and dust grain absorption. Therefore, in the more general case, the determination of the photoelectron mean energy $\bar{E}_2$ is quite complex. Let us consider two simple cases in which (i) the photoelectron energy is estimated near the central star. In this case, $U_\nu^* \gg U_\nu^{\mathrm{D}}$ and diffuse radiation may be neglected; (ii) the photoelectron energy is estimated for the whole H II region, meaning that a mean value is estimated $\langle\bar{E}_2\rangle$. Let us assume that the central star emits in the ultraviolet as a blackbody with temperature T_c (color temperature). Energy $\bar{E}_2$ may be written in the form $\bar{E}_2 = \Psi k T_c$, that is,

$$\Psi = \frac{\bar{E}_2}{kT_c}, \tag{8.61}$$

where Ψ is a dimensionless parameter. For case (i), we have $r \to 0$, $\Psi = \Psi_0$, and

$$U_\nu = U_\nu^* = \frac{4\pi B_\nu(T_c)}{c}. \tag{8.62}$$

The solution of (8.59) for these conditions gives the values of Ψ_0 and $\bar{E}_2$, shown in Table 8.4.

For case (ii), the photoelectron mean energy is valid for the entire H II region, being simply the energy per second given to all photoelectrons divided by the total number of photoelectrons produced per second that is equal to the number of ionizing photons emitted per second. In this case, diffuse photons are not considered, because the energy gained by diffuse photon absorption is the same as the one lost in recombination to the ground level. We have thus the relations

$$\bar{E}_2 = \frac{\int h(\nu - \nu_1) U_\nu^* (\mathrm{d}\nu/\nu)}{\int U_\nu^* (\mathrm{d}\nu/\nu)}, \tag{8.63}$$

$$\langle\Psi\rangle = \frac{\langle\bar{E}_2\rangle}{kT_c}. \tag{8.64}$$

In Table 8.4, values of $\langle\Psi\rangle$ and $\langle\bar{E}_2\rangle$ for $U_\nu^* = U_\nu^*(T_c)$ are also given. Therefore, for typical temperatures of H II region central stars, the photoelectron mean energy is of the order of 1–3 eV, a value that should be used in the heating function calculation (8.58).

8.5.2 *Cooling by Electron–Proton Recombination*

The energy lost by photoelectrons in the recombination process is given by the second term of (8.58). Using the H capture cross section and a Maxwellian distribution, we can show that

$$\sum_{j=k}^{\infty} \langle \sigma_{cj} v^3 \rangle = \frac{2A_r}{\sqrt{\pi}} \left(\frac{2kT}{m_e} \right)^{3/2} \frac{h\nu_1}{kT} \chi_k(T), \tag{8.65}$$

where $A_r = 2.1 \times 10^{-22}$ cm^2 is the recapture constant [cf. (6.56)], $h\nu_1 = 13.6$ eV, and $\chi_k(T)$ is a known function of temperature, with a variation interval $3 \gtrsim \chi_2 \gtrsim 1$ and $4 \gtrsim \chi_1 \gtrsim 2$ for $2 \lesssim \log T(\mathrm{K}) \lesssim 4$. Considering (8.58) and (8.65), we obtain

$$\frac{\Gamma_{ep}}{n_e\, n_p} = \frac{2.1 \times 10^{-11}}{T} [\bar{E}_2\, \phi_n(T) - kT\, \chi_n(T)] \tag{8.66}$$

(units, erg cm^3 s^{-1}), where we use

$$\alpha^{(n)} = 2.1 \times 10^{-11} T^{-1/2} \phi_n(T)\, \mathrm{cm^3\, s^{-1}} \tag{8.67}$$

[cf. (6.63)]. For the nearest regions to the star, diffuse radiation is negligible, and $n = 1$, that is, we use ϕ_1 and χ_1, $\bar{E}_2$ being given by (8.61) or Table 8.4. For the case where a photoelectron mean energy value is used (Table 8.4), diffuse radiation is ignored, and we assume in general that the ionizing photons produced in recombination are absorbed near the production site (on the spot approximation). In this case, we consider only recombinations to levels $n \geq 2$ in (8.66).

8.5.3 *Cooling by Electron–Ion Collisional Excitation*

Energy loss in the H recombination process is generally negligible relative to photoionization input. As done in the previous section, these losses are generally introduced as a correction factor to the photoionization input. In an H II region, energy losses are essentially due to excitation of energy levels of C, N, O, and Ne ions, by collisions with electrons. This process is quite efficient despite the low abundance (~10^{-4} n_H) of the ions. If there was no cooling by collisional excitation, then the temperature of H II regions would be of the order of the input temperature or of the order of the central star color temperature or even of the star effective temperature and not of the order of 10^4 K, as indicated by several observational evidence (see Table 8.1).

As we have seen in Chap. 7, the energy lost by excitation is given by

$$\Lambda_{ei} = n_e \sum_j \sum_{k>j} E_{jk}(n_{ij}\gamma_{jk} - n_{ik}\gamma_{kj}) \tag{8.68}$$

[cf. (7.21)]. Let us again consider the simple case of a two-level atom. From (8.68),

$$\begin{aligned}\Lambda_{ei} &= n_e E_{jk}(n_{ij}\gamma_{jk} - n_{ik}\gamma_{kj}) \\ &= n_e n_{ij} E_{jk}\left[\frac{g_k}{g_j}\gamma_{kj}\,e^{-E_{jk}/kT} - \frac{n_{ik}}{n_{ij}}\gamma_{kj}\right] \\ &= n_e n_{ij} E_{jk}\,\gamma_{kj}\frac{g_k}{g_j}\,e^{-E_{jk}/kT}\left(1 - \frac{b_k}{b_j}\right),\end{aligned} \tag{8.69}$$

where we use the relations between γ_{jk} and γ_{kj} and between n_k and n_j. For a two-level atom, considering optical transitions at interstellar conditions, we saw that

$$\frac{b_k}{b_j} = \frac{1}{1 + \frac{A_{kj}}{n_e \gamma_{kj}}} \tag{8.70}$$

[cf. (5.45)]. Therefore, we obtain

$$\Lambda_{ei} = n_e n_{ij} E_{jk} \gamma_{jk} \frac{1}{1 + \frac{n_e \gamma_{kj}}{A_{kj}}}. \tag{8.71}$$

If n_i is the total number of ions per cubic centimeter,

$$n_i = \sum n_{ij} = n_{ij} + n_{ik} = a_i n_H \simeq a_i n_p, \tag{8.72}$$

where a_i is the abundance of element i. Using the degree of ionization $x_j = n_{ij}/n_i$, we may write

$$n_{ij} = n_i x_j = (a_i n_H) x_j \simeq a_i x_j n_p. \tag{8.73}$$

Replacing in (8.71),

$$\frac{\Lambda_{ei}}{n_e n_p} = a_i x_j E_{jk} \gamma_{jk} \frac{1}{1 + \frac{n_e \gamma_{kj}}{A_{kj}}}. \tag{8.74}$$

For the most complex case of atoms with several energy levels, the statistical equilibrium equations must be written considering all possible transitions. The ionization equilibrium must also be solved, in principle, because the cooling function depends on the degree of ionization of the considered ion.

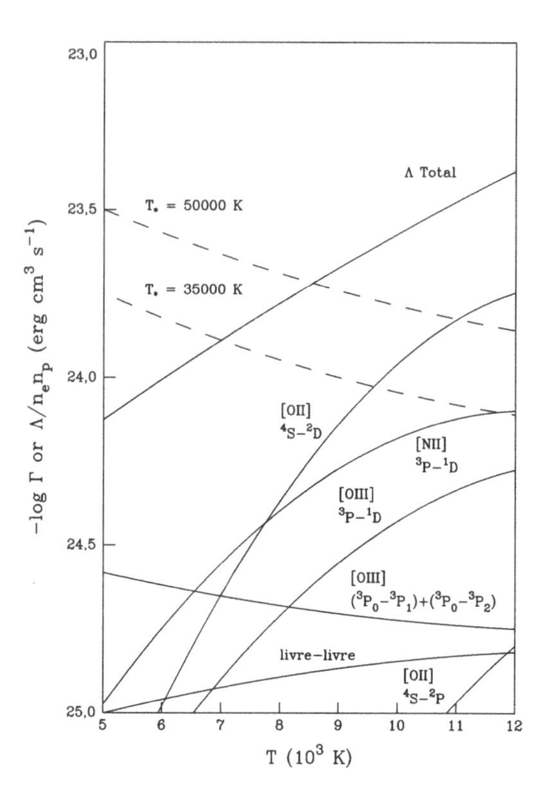

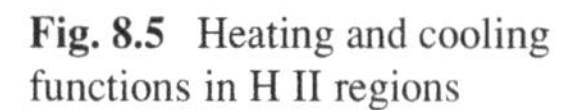
Fig. 8.5 Heating and cooling functions in H II regions

To illustrate the calculation of the cooling function, we will estimate function $\Lambda/n_e n_p$ for the O II ion considering only transitions ${}^4S_{3/2}$–${}^2D_{5/2}$. We will adopt $T \simeq 10^4$ K, $n_e \simeq 10^2$ cm^{-3}, $x_{\text{OII}} = n(\text{O II})/n(\text{O}) \simeq 0.80$, and $a(\text{O}) \simeq 6 \times 10^{-4}$. With $E_{jk} \simeq 3.32$ eV, $g_k = 6$, $g_j = 4$, $A_{kj} \simeq 4.2 \times 10^{-5}$ s^{-1}, and $\Omega_{kj} \simeq 0.88$, we obtain $\gamma_{kj} \simeq 1.3 \times 10^{-8}$ cm^3 s^{-1}, $\gamma_{jk} \simeq 4.2 \times 10^{-10}$ cm^3 s^{-1}, and $\Lambda\ e_{\text{OII}}/n_e n_p \simeq 1.1 \times 10^{-24}$ erg cm^3 s^{-1}. This result can be compared with Fig. 8.5, where we represent functions Γ (for two different stars) and Λ (divided by $n_e n_p$) in typical H II regions. The figure also shows the cooling functions $\Lambda/n_e n_p$ for the main O and N ions in a typical H II region and the total cooling function (solid lines). We also include in the same figure the heating functions $\Gamma/n_e n_p$ for two central stars with temperatures 50,000 K and 30,000 K, respectively (dashed lines). We see that the total cooling by O II at $T \simeq$ 10,000 K is of the order of 10^{-24} erg cm^3 s^{-1}, similar to the above calculated value. In Fig. 8.5, we note that the equilibrium temperature falls in the interval $7{,}000 \lesssim T(\text{K}) \lesssim 9{,}000$.

8.5.4 Photoelectric Heating by Dust Grains

One of the effects of the presence of solid dust grains immersed in H II regions is their influence on the gas electron temperature due to the heating produced by photoelectrons ejected by the grains. This process is similar to H I regions heating (Chap. 7, Sect. 7.7). The heating rate may be written

$$\Gamma_{\rm ed} \simeq \Gamma_{\rm dL\alpha} + \Gamma_{\rm d_*}, \tag{8.75}$$

where $\Gamma_{\rm dL\alpha}$ is the heating produced by liberated photoelectrons after the absorption by the grains of Lyman-α photons produced by H recombination (diffuse field) and $\Gamma_{\rm d*}$ is the heating produced by direct absorption of stellar photons. Component $\Gamma_{\rm dL\alpha}$ may be written in an approximated way

$$\Gamma_{\rm dL\alpha} \simeq f\, n_{\rm p}\, n_{\rm e}\, \alpha E_{\rm L\alpha}\, y_{\rm L\alpha}, \tag{8.76}$$

where $f \simeq 0.7$ is the fraction of recombinations to levels $n \geq 2$ that produce Lyman-α photons, $\alpha \simeq \alpha^{(2)}$ is the total recombination coefficient above level $n = 1$, $E_{\rm L\alpha}$ is the mean energy of the resulting photoelectron that depends on the electric charge of the grains, and $y_{\rm L\alpha}$ is the photoelectric yield, or the process efficiency.

Function $\Gamma_{\rm d*}$ can be divided in two components if we consider the nonionizing photons, for which $h\nu < 13.6$ eV, and the ionizing photons with $h\nu \geq 13.6$ eV, separately. We may write for distances $r \gg R_*$:

$$\Gamma_{\rm d*}(h\nu{<}13.6\,{\rm eV}) \simeq n_{\rm d}\left(\frac{R_*}{r}\right)^2 \int_{\nu t}^{\nu_1} \frac{\pi B_\nu(T_{\rm ef})}{h\nu} \exp[-\tau_{\rm d}(\nu)]\, \sigma_{{\rm d}\nu}\, E_\nu\, y_\nu\, {\rm d}\nu, \tag{8.77}$$

where $n_{\rm d}$ is the grain numerical density; $\tau_{\rm d}(\nu)$ is the grain optical depth at frequency ν; $\sigma_{{\rm d}\nu}$ is the absorption cross section for photons with frequency ν; $E_\nu = h\nu - h\nu_{\rm t}$ is the photoelectron energy, considering the threshold energy $h\nu_{\rm t}$; and y_ν is the process efficiency at frequency ν. In the same way, for the ionizing photons, we have

$$\Gamma_{\rm d*}(h\nu \geq 13.6\,{\rm eV}) \simeq n_{\rm d}\left(\frac{R_*}{r}\right)^2 \int_{\nu_1}^{\infty} \frac{\pi B_\nu(T_{\rm ef})}{h\nu} \exp-[\tau_{\rm d}(\nu) + \tau_{\rm H}(\nu)]\, \sigma_{{\rm d}\nu}\, E_\nu\, y_\nu\, {\rm d}\nu. \tag{8.78}$$

The solution of (8.76)–(8.79) for a simplified model of the H II region and the grain physical properties gives thresholds for the photoelectric heating contribution relative to heating by H photoionization. Results for ratio $\Gamma_{\rm ed}/\Gamma_{\rm H}$ for a star with effective temperature $T_{\rm ef} = 35{,}000$ K and electron density between 0.1 cm^{-3} and 100 cm^{-3} vary typically from $\Gamma_{\rm ed}/\Gamma_{\rm H} \simeq 2$ for $r/r_{\rm s} \simeq 0.3$ to negligible values of $\Gamma_{\rm ed}/\Gamma_{\rm H} \ll 1$ for the nearest parts of the H II region surface. So, in general, we have $\Gamma_{\rm ed}/\Gamma_{\rm H} \lesssim 1$ for the intermediate regions of H II regions, except for very dense nebulae.

The global effect on the temperature may be seen in Fig. 8.5, where we show the total heating and cooling functions, as well as heating functions for the grains by H photoionization in a nebula with electron density $n_{\rm e} = 10$ cm^{-3} in position $r/r_{\rm s} = 0.4$.

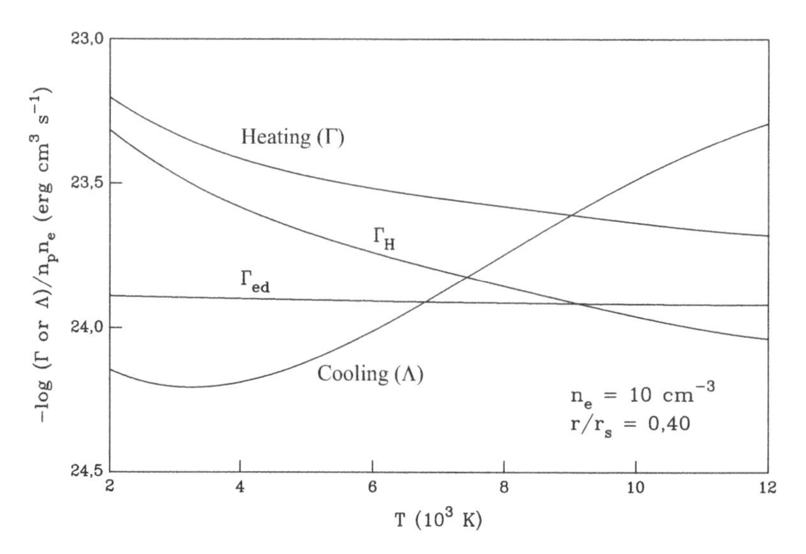

Fig. 8.6 Determination of the equilibrium temperature in H II regions

We may conclude that the effect of dust grains depends on the position in the nebula, being more intense in intermediate regions, between about 20 % and 60 % of the Strömgren radius. In the regions closer to the star, grain survival is hampered by the radiation high density, whereas for very distant regions, stellar radiation is very faint, thus decreasing the radiation fraction absorbed by the grains. Results show that the nebula *global* heating is not much affected by the process, having an upper limit of about 10 %, that is, $\Delta T \lesssim 1{,}000$ K in an H II region with $T \simeq 10^4$ K. *Locally*, however, the dust grain contribution may reach higher values in the intermediate regions of the nebula. Note that in Fig. 8.6 the heating function slightly *decreases* with the increase of electron temperature, thus reflecting the recombination coefficient behavior [cf. (8.58) and (8.67)]. For higher temperatures, the efficiency of the collisional excitation process increases more rapidly, thus increasing the energy losses and the cooling function [cf. (8.69) and (8.74)]. Temperature fluctuations, measured by parameter $t^2 = (\delta T/T)^2$, where δT is the H II region temperature dispersion, may lead to several important consequences for the gas equilibrium, especially if these fluctuations reach values of the order of $t^2 \simeq 0.03$.

8.5.5 *Other Processes*

Other heating and cooling processes may occur in H II regions, though they are less important than the processes already mentioned. Some of these processes are (1) cooling due to free–free emission (Sect. 2.2), (2) cooling by emission in H and He collisionally excited lines, and (3) cooling through grain/gas collisions. An example of the cooling function by free–free emission is also included in Fig. 8.5.

8.5.6 *Cooling Timescale*

The cooling time defined by (7.13) or (7.42) may be estimated for an H II region by

$$t_{\mathrm{T}} \simeq -\frac{(3/2)k(T - T_{\mathrm{E}})}{(\mathrm{d}/\mathrm{d}t)\,(3kT/2)} \simeq -\frac{(3/2)k(T - T_{\mathrm{E}})}{(\Gamma - \Lambda)/n_{\mathrm{e}}}. \tag{8.79}$$

For temperatures close to $T_{\mathrm{E}} \simeq 8{,}000$ K, we obtain

$$t_{\mathrm{T}} \simeq \frac{2 \times 10^4}{n_{\mathrm{p}}}\,\mathrm{year}. \tag{8.80}$$

For a typical electron density $n_{\mathrm{e}} \simeq 100\ \mathrm{cm}^{-3}$, we have $t_{\mathrm{T}} \simeq 200$ years. For instance, for the Orion H II region, we estimate $t_{\mathrm{T}} \simeq 150$ years.

The cooling time may be compared with the recombination characteristic time t_{r}:

$$t_{\mathrm{r}} \simeq \frac{1}{n_{\mathrm{e}}\alpha} \simeq \frac{1}{n_{\mathrm{e}}[2.1 \times 10^{-11} Z^2 \phi_2(T) T^{-1/2}]}, \tag{8.81}$$

where we use (8.67) with $\alpha = \alpha^{(2)}$. Measuring t_{r} in years, we have

$$t_{\mathrm{r}} \simeq \frac{1.5 \times 10^3 T^{1/2}}{Z^2 n_{\mathrm{e}}\, \phi_2(T)}\,\mathrm{year}. \tag{8.82}$$

For $T \simeq 10^4$ K,

$$t_{\mathrm{r}} \simeq \frac{1.2 \times 10^5}{n_{\mathrm{e}}}\,\mathrm{year}. \tag{8.83}$$

We see that $t_{\mathrm{r}} \gtrsim t_{\mathrm{T}}$; for $n_{\mathrm{e}} \simeq n_{\mathrm{p}}$, $t_{\mathrm{r}} \simeq 6\ t_{\mathrm{T}}$. Therefore, the gas is cooling more rapidly than it recombines. For high temperatures, $T \sim 10^5$ K, a similar analysis applied to H I regions, shows that $t_{\mathrm{T}} \gg t_{\mathrm{r}}$ in a general way.

8.6 Radio Recombination Lines

In the H II region electron–proton recombination process, the electron capture may occur at very high H energy levels, $n > 40$. The electron then cascades producing emission lines in the radio domain, the so-called radio recombination lines. The analysis of these lines can lead to the determination of the nebula electron temperature as we shall now see.

8.6.1 High-Energy Level Excitation

For $n > 40$, the statistical equilibrium of the energy levels is simplified because the subshells with different values of the azimuthal quantum number l are populated according with their statistical weight. In this case, l variations occur more frequently than variations of the principal quantum number n, due to collisions with protons. Determinations of the deviation parameter b_n show a continuous variation of this parameter with n, with values of the order of $0.7 \lesssim b_n \leq 1.0$ for $n > 40$ with $T \simeq 10^4$ K and $n_e \sim 10^4$ cm^{-3}. We may use these values to estimate the stimulated emission correction factor χ, defined by (3.67), which may be written

$$\chi = \frac{b_k}{b_j} + \frac{kT}{h\nu_{jk}}\left(1 - \frac{b_k}{b_j}\right) \simeq 1 - \frac{kT}{h\nu_{jk}}\frac{\mathrm{d}\ln b}{\mathrm{d}n}\Delta n, \tag{8.84}$$

where Δn is the n variation for the considered emission line. For $\Delta n = 1, 2$, etc., the lines are named $n\alpha$, $n\beta$, etc. The line frequency is given by

$$\begin{aligned} \nu &= \nu_1 \left[\frac{1}{n^2} - \frac{1}{(n+\Delta n)^2}\right] \\ &\simeq 6.6 \times 10^9 \frac{\Delta n}{n^3}\left[1 - \frac{1.5\,\Delta n}{n} + \ldots\right] \mathrm{MHz}, \end{aligned} \tag{8.85}$$

where $\nu_1 = 3.29 \times 10^{15}$ Hz as we have previously seen. For instance, for H transition, $n = 77 \rightarrow n = 76$, we have line H76α with $\nu = 1.47 \times 10^{10}$ Hz $= 1.47 \times 10^4$ MHz $= 14.7$ GHz or $\lambda \simeq 2$ cm. Values of factor χ may be obtained from (8.84) if we know $b(n)$, including radiative and collisional processes and are given in Fig. 8.7 for $T \simeq 10^4$ K and $n_e \sim 10$ and 1,000 cm^{-3}. For $n = 76$ with $T \simeq 10^4$ K, we have $20 \lesssim -\chi \lesssim 70$, depending on electron density.

We may use these results to obtain the absorption and emission coefficients relative to values in ETE. From (3.8), (3.36), and (3.68),

$$\frac{k_\nu}{k_\nu^*} = b_n \chi, \tag{8.86}$$

and from (3.26),

$$\frac{j_\nu}{j_\nu^*} = b_m, \tag{8.87}$$

where m represents the total quantum number of the upper level and n the one of the lower level. Note that $\chi < 0$, and so *maser* amplification may occur.

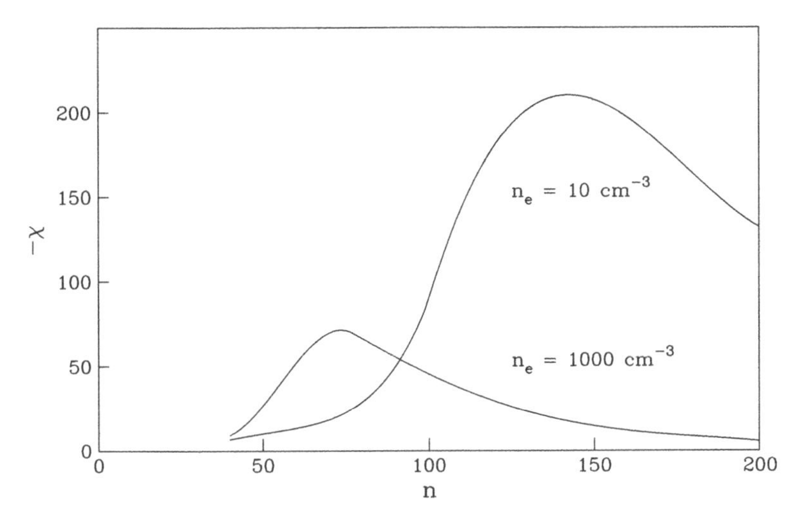

Fig. 8.7 The stimulated emission factor of high energy levels for two values of the electron density

8.6.2 *Temperature and Recombination Lines*

Let us consider radio recombination lines and estimate the electron temperature of the H II region. The radiation intensity may be obtained using the source function (3.72) in the solution of the transfer Equation (3.23), considering that the optical depth $\tau_{\nu r}$ includes both continuum (τ_C) and line (τ_L) absorptions. In ETE, the optical depth τ_L in the line center can be obtained from (3.38) and (3.68) using $\chi = 1$. Considering the Doppler profile (3.44) with $v = 0$ in the line center, relation (3.27), and the Saha Equation (3.13), we obtain

$$\tau_L^* \simeq 27.5 \frac{n_p}{n_e} \frac{n^2 f_{nm}}{b T^{5/2}} \mathrm{EM}, \tag{8.88}$$

where we assume n_e constant and use the emission measure EM given in pc cm^{-6} [see (4.7)]. The oscillator strength f_{mn} may be obtained in terms of n:

$$f_{n,n+\Delta n} = n\, M(\Delta n), \tag{8.89}$$

where $M(1) = 0.191$, $M(2) = 0.0263$, and $M(3) = 0.0081$. The velocity parameter is $b \simeq 10^6$ cm s^{-1} for $T \simeq 10^4$ K [cf. (3.45)]. Using these values for a line with $n \simeq 200$, we have $\tau_L^* \sim 10^{-8}$ EM, with the emission measure in pc cm^{-6}. Outside TE, relation (8.86) shows that $\tau_L/\tau_L^* \simeq \chi$ because $b_n \simeq 1$ as we have seen. From Fig. 8.7, $\chi \simeq -100$ for $T \simeq 10^4$ K and $n \simeq 200$ so that $\tau_L \simeq 1$ for EM $\gtrsim 10^6$ pc cm^{-6}, with the possibility of *maser* amplification. In fact, dense or compact H II regions have higher values of the emission measure.

Let us consider the simple case of optically thin recombination lines, with $\tau_L \ll 1$, where the maser effect may be neglected. Let r be the ratio between line-continuum excess intensity and continuum intensity. Remembering that in the region where $h\nu/kT \ll 1$ we may use the Rayleigh–Jeans distribution (3.18b), ratio r may be expressed in terms of brightness temperature

$$r = \frac{T_{\mathrm{bL}} - T_{\mathrm{bC}}}{T_{\mathrm{bC}}}. \tag{8.90}$$

The emitted intensity may be obtained directly from integral $\int j_\nu \mathrm{d}s$ [see (3.22)]. Using (8.87), the Rayleigh–Jeans distribution and the Kirchhoff law (3.24), we obtain

$$r \simeq \frac{b_m \tau_{\mathrm{L}}^*}{\tau_{\mathrm{C}}}, \tag{8.91}$$

where we assume $k_\nu = k_{\mathrm{C}} = k_{\mathrm{C}}^*$ for the continuum and $k_\nu^* = k_{\mathrm{C}} + k_{\mathrm{L}}^*$ in the line center. The line optical depth τ_{L}^* is given by (8.88). For the continuum, we can use the free–free absorption coefficient given by (2.8) with $n_{\mathrm{p}} \simeq n_{\mathrm{e}}$. The result may be written

$$\frac{rb}{c} \simeq \frac{3\sqrt{3\pi}}{2} \frac{b_m \, \Delta n \, M\,(\Delta n) Z^2}{g_{\mathrm{ef}}} \frac{h\nu}{kT}. \tag{8.92}$$

Quantity rb/c is equivalent to W_λ/λ, where W_λ is the equivalent width of the emission line. Therefore, (8.92) allows us to determine the electron temperature of H II regions with emission measures EM $\lesssim 10^6$ pc cm^{-6}. For values of emission measure EM ~ 10^5–10^7 pc cm^{-6} and n_{e} ~ 10^4 cm^{-3}, we obtain dimensions of the order of 0.001 to 1 pc. For instance, using (8.92) for the H76α line with $b_m \simeq 1$, $\Delta n \simeq 1, M(\Delta n) \simeq 0.19, Z \simeq 1, \nu \simeq 1.47 \times 10^{10}\,\mathrm{Hz}, r \simeq 1, b \simeq 1.3 \times 10^6\,\mathrm{cm\,s^{-1}}$, and $g_{\mathrm{ef}} \simeq 1.3$, we have $T \simeq 10^4$ K. More accurate values fall in the interval 6,000–15,000 K, in accordance with other methods, as we shall see in Sect. 8.7 for the forbidden line analysis.

An approximate expression for α-transitions ($\Delta n = 1$) with $\tau \ll 1$ may be written

$$T^{1.15} \simeq \frac{2.3 \times 10^4 \, \nu^{2.1}}{(T_{\mathrm{L}}/T_{\mathrm{C}})\Delta\nu_{\mathrm{h}}}, \tag{8.93}$$

where ν is the frequency in the line center in GHz, $\Delta\nu_{\mathrm{h}}$ is the FWHM in kHz, and $T_{\mathrm{L}}/T_{\mathrm{c}}$ is the ratio between line and continuum brightness temperatures. Applying this expression to the H76α line, we have $\nu = 14.7$ GHz, $\Delta\nu_{\mathrm{h}} \simeq 2b\nu\sqrt{(\ln 2)}/c \simeq 800$ kHz, and $b \simeq 10^6$ cm s^{-1} [see (3.49)]. With $T_{\mathrm{L}}/T_{\mathrm{c}} \simeq 0.2$, we have $T \simeq 10^4$ K, in accordance with the previous estimate.

Some electron temperature determinations in galactic H II regions from the H66α line ($\nu = 22.4$ GHz) indicate for RCW 38, $T_L/T_C \simeq 0.14$ and FWHM = 2,500 kHz and for RCW 57, $T_L/T_C \simeq 0.22$ and FWHM = 2,100 kHz. Using (8.93), we obtain $T_e \simeq 11{,}100$ K and $T_e \simeq 8{,}700$ K, respectively, which can be compared with results $T_e \simeq 10{,}600$ K and $T_e \simeq 8{,}500$ K, obtained with a more accurate expression.

8.6.3 Maser Emission in H II Regions

We saw in Chap. 3 that, for $h\nu/kT \ll 1$, the integrated absorption cross section may be written

$$\sigma = \sigma_u \frac{h\nu}{kT}\chi \tag{8.94}$$

(3.68), where the χ term is again the correction factor for stimulated emissions, taking into account the deviations relative to TE [see (8.84)]. When $\chi < 0$, σ and the absorption coefficient k_ν are negative so that an eventual radiation in the line may be amplified. Let us analyze the case where line intensity is sufficiently weak so as not to affect the relative populations in the involved levels, that is, we suppose a non-saturated maser (microwave amplification by stimulated emission of radiation) process, with χ more or less constant. Let us assume a spherical emitting region of radius r and a line of sight passing at distance d from the center of the sphere. In this case, the radiation path inside the sphere is simply

$$L = 2\sqrt{a^2 - d^2} = 2a\left(1 - \frac{d^2}{a^2}\right)^{1/2} \tag{8.95}$$

(note that $L = 2a$ for $d = 0$) so that the total optical depth $\tau_{\nu r}$ is

$$\tau_{\nu r} = \left(1 - \frac{d^2}{a^2}\right)^{1/2} 2a\, k_\nu = \left(1 - \frac{d^2}{a^2}\right)^{1/2} \tau_0 \mathrm{e}^{(\lambda\Delta\nu/b)^2}, \tag{8.96}$$

where τ_0 is the value of the optical depth in the center of the line following a trajectory that passes through the sphere's center, and we use the thermal Doppler profile (cf. Sect. 3.5). Let us consider the blackbody background radiation at 2.7 K as the incident radiation. In the limit $h\nu/kT \ll 1$, we can use the Rayleigh–Jeans distribution (3.18b) and the Kirchhoff law (3.24) in the transfer equation solution (3.23), obtaining

$$I_\nu = I_\nu(0)\mathrm{e}^{-\tau_\nu(r)} + B_\nu(T)\left[1 - \mathrm{e}^{-\tau_\nu(r)}\right] \tag{8.97}$$

or in terms of brightness temperature T_b:

$$T_b = T_{b0}\,e^{-\tau_\nu(r)} + T\left[1 - e^{-\tau_\nu(r)}\right] \simeq 2.7\,e^{-\tau_\nu(r)}, \tag{8.98}$$

where we neglect spontaneous emissions. Determinations of brightness temperature associated with OH maser emission at $\lambda \simeq 18$ cm and H_2O ($\lambda \simeq 1.35$ cm) in H II regions result in values $T_b \sim 10^{11}-10^{15}$ K, much higher than the electron temperature values in these regions and even higher than the effective temperatures of central stars, pointing to a strong deviation from TE. From (8.98), we see that the total optical depths must reach values of the order of $-\tau_{\nu r} \sim 24$–34, that is, $-\tau_{\nu r} \gg 1$. Term e^{-r} is therefore positive and constitutes a radiation *input*, instead of a loss. Population inversion in the energy levels must be caused by some kind of *pumping*, which may be radiative or collisional. The emerging radiation beam is concentrated both in frequency and in apparent dimension. The observed frequency interval corresponds to velocities of the order of 0.2 km s^{-1} or temperatures $T \sim mv^2/k \sim 100$ K. With turbulence velocities of the order of 1 km s^{-1}, the gas temperature would be at most of the order of 2,500 K. The observed angular diameters are small, of the order of fractions of arcseconds. In this case for distances of the order of 1 kpc, linear dimensions would be of the order of $10^{13}-10^{14}$ cm, which is a fraction of the H II region total dimension.

8.7 Physical Conditions in Ionized Nebulae

Physical conditions in photoionized nebulae may be determined from the analysis of emission lines observed in the optical and ultraviolet spectrum. This process generally involves two steps; first, a plasma diagnostics must be performed, that is, electron temperature T or T_e and electron density n_e are determined, and next, abundances of the principal chemical elements are determined. Alternatively, detailed photoionization models may be considered, which assume the central star to have a blackbody spectrum or adopt more accurate atmospheric models in conjunction with a photoionization code.

8.7.1 Plasma Diagnostics

Let us initially assume that the electron temperature has already been determined, for instance, from radio recombination lines, $T_e = 10{,}000$ K, so now we need to estimate electron density. In Chap. 5, we have seen a method to determine n_e using [S II] 6716/6731 lines. Let us now determine the electron density from O II 3726/3729 Å emission lines. The three lower levels of this ion are $^4S_{3/2}$ (level 1), $^2D_{5/2}$ (level 2), and $^2D_{3/2}$ (level 3). In Chap. 5 (Fig. 5.1; Table 5.1), some properties of these levels were presented. Levels 2 and 3 have energies close to each other, $E \simeq 3.3$ eV. More accurately, energies are given by $\lambda^{-1} = E/hc = 26{,}810.7\ \mathrm{cm}^{-1}$ and $26{,}830.2\ \mathrm{cm}^{-1}$, respectively, for levels 2 and 3, corresponding to the 3,729 Å and 3,726 Å lines. Statistical weights are $g_1 = 4$, $g_2 = 6$, and $g_3 = 4$. Probabilities A_{kj} and constants $\Omega(j,k)$ are given in Table 5.1, knowing that the rate between levels 2 and 3 is $A_{32} = 1.3 \times 10^{-7}\ \mathrm{s}^{-1}$. Collision rates γ_{kj} may be determined from (5.32) and rates γ_{jk} from (5.34), as we have seen in the example of Sect. 8.5.

Writing the statistical equilibrium equation for levels 3 and 2, and neglecting induced transitions, we have

$$n_3[n_e(\gamma_{32} + \gamma_{31}) + (A_{32} + A_{31})] = n_2\, n_e\, \gamma_{23} + n_1\, n_e\, \gamma_{13}, \tag{8.99}$$

$$n_2[n_e(\gamma_{21} + \gamma_{23}) + A_{21}] = n_1\, n_e\, \gamma_{12} + n_3(n_e\, \gamma_{32} + A_{32}). \tag{8.100}$$

From these relations, we may write

$$\frac{n_3}{n_2} = \frac{\gamma_{13}\left[(\gamma_{21} + \gamma_{23}) + \frac{A_{21}}{n_e}\right] + \gamma_{12}\gamma_{23}}{\gamma_{12}\left[(\gamma_{31} + \gamma_{32}) + \frac{A_{31}+A_{32}}{n_e}\right] + \gamma_{13}\left[\gamma_{32} + \frac{A_{32}}{n_e}\right]}. \tag{8.101}$$

The ratio between line intensities may be written

$$\frac{I_{21}}{I_{31}} = \frac{n_2 A_{21} h\nu_{21}}{n_3 A_{31} h\nu_{31}}, \tag{8.102}$$

so that we can write an expression for the I_{3729}/I_{3726} ratio as a function of electron density. Using numerical values from Table 5.1, we have $\gamma_{21} = 1.2 \times 10^{-8}\ \mathrm{cm}^3\ \mathrm{s}^{-1}$, $\gamma_{32} = 2.5 \times 10^{-8}\ \mathrm{cm}^3\ \mathrm{s}^{-1}$, $\gamma_{31} = 1.2 \times 10^{-8}\ \mathrm{cm}^3\ \mathrm{s}^{-1}$, $\gamma_{12} = 3.7 \times 10^{-10}\ \mathrm{cm}^3\ \mathrm{s}^{-1}$, $\gamma_{13} = 2.5 \times 10^{-10}\ \mathrm{cm}^3\ \mathrm{s}^{-1}$, and $\gamma^{23} = 1.7 \times 10^{-8}\ \mathrm{cm}^3\ \mathrm{s}^{-1}$. We can therefore write

$$R(\mathrm{O\,II}) = \frac{I_{3729}}{I_{3726}} \simeq 0.2\frac{a}{b}, \tag{8.103}$$

where

$$a = 3.7 \times 10^{-10}\left[3.7 \times 10^{-8} + \frac{1.8 \times 10^{-4}}{n_e}\right] + 2.5 \times 10^{-10}\left[2.5 \times 10^{-8} + \frac{1.3 \times 10^{-7}}{n_e}\right], \tag{8.104}$$

$$b = 2.5 \times 10^{-10}\left[2.9 \times 10^{-8} + \frac{3.6 \times 10^{-5}}{n_e}\right] + 6.3 \times 10^{-18}. \tag{8.105}$$

The intensity ratio is shown in Fig. 8.8 as a function of electron density. For instance, for planetary nebula NGC 6210, we obtain $I_{3729}/I_{3726} = 0.46$. From Fig. 8.8, we have $n_e \simeq 7 \times 10^3\ \mathrm{cm}^{-3}$ for $T_e \simeq 10^4$ K, which can be compared with the value $n_e = 7.7 \times 10^3\ \mathrm{cm}^{-3}$ obtained from more accurate methods.

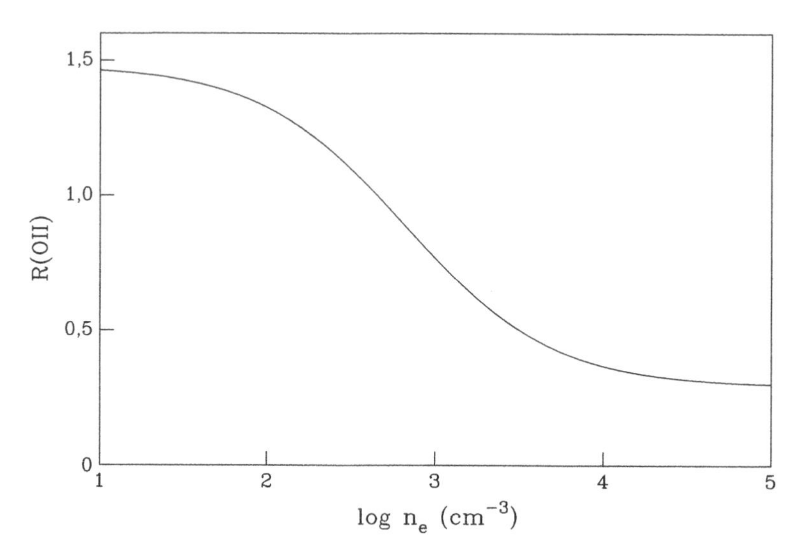

Fig. 8.8 Intensity ratio of the O II lines as a function of the electron density

Let us now consider the electron temperature T_e determination from the [O III] 4959,5007/4363 Å lines. Other lines can also be used, such as the [N II]6548,6584/5754 Å lines. For the O III first levels shown in Fig. 5.1, a similar procedure to the previous one leads us to the following result:

$$R(\mathrm{O\,III}) = \frac{I_{4959} + I_{5007}}{I_{4363}} = \frac{7.73\exp[(3.29\times 10^4)/T_e]}{1 + 4.5\times 10^{-4}\left(n_e/T_e^{1/2}\right)}. \tag{8.106}$$

In the lower densities limit ($n_e \ll 10^5$ cm^{-3} for $T_e \simeq 10^4$ K), the above ratio only depends on temperature:

$$R(\mathrm{O\,III}) \simeq 7.73\exp\left[\frac{3.29\times 10^4}{T_e}\right], \tag{8.107}$$

as shown in Fig. 8.9. As an example, for planetary nebula NGC 3132, we obtain R(O III) $= 392.2$, already corrected from interstellar extinction. Using (8.107), we get $T_e \simeq 8{,}380$ K, which can be compared with more accurate values for [O III] lines, $T_e \simeq 8{,}400$ K and for [N II] lines, $T_e = 9{,}590$ K. These results are consistent with electron density $n_e = 700$ cm^{-3} obtained from the [SII]6716/6730 Å lines.

8.7.2 *Abundances*

Once the plasma diagnostics is completed, we have determined, in principle, electron temperature and density. Interstellar extinction to the nebula (Chap. 9)

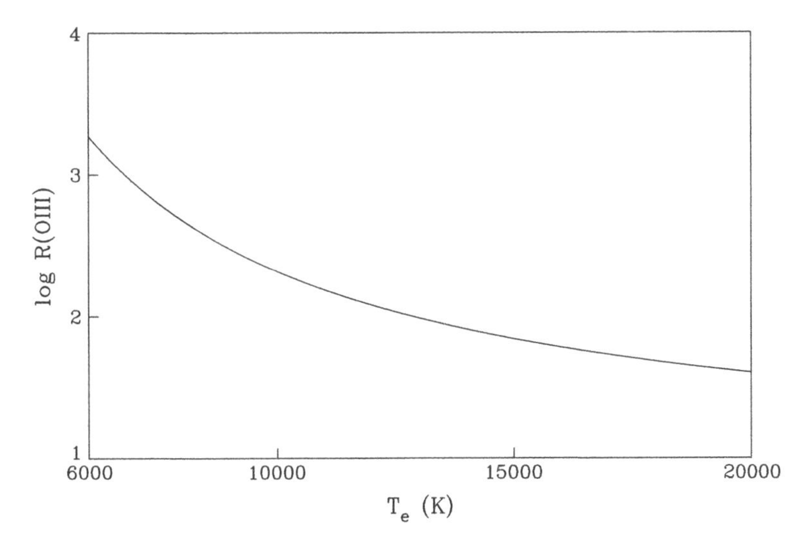

Fig. 8.9 Intensity ratio of the O III lines as a function of the electron temperature in the low density limit

must also be determined using, for instance, the Balmer decrement method or comparing emission at radio (unaffected by extinction) and optical wavelengths. The abundance of the main ions can be calculated by empirical methods or by means of a photoionization model, from emission lines observed intensities, using relations similar to (8.102). Abundances for He, O, N, etc., may be obtained by summing the contributions of several ions and correcting for stages that are not observed by means of ionization correction factors (ICF) determined theoretically.

The results show abundances generally similar to cosmic abundance, being H II regions and planetary nebulae mainly composed of H with about 10 % of He (in number of atoms). The most abundant heavy elements, O, N, S, etc., can reach values of about 10^{-4} of the H abundance. The analysis of the interstellar ionized nebulae abundances—in particular the oxygen case, whose determination in the stars is rather complex—has deep implications on the study of stellar nucleosynthesis and chemical evolution of the galaxies.

Exercises

8.1 Table 8.2 relates effective temperatures ($T_{\rm eff}$) with radius (R_*) of main sequence hot stars. Consider these data and estimate the number of ionizing photons emitted per second by the stars. Use blackbody fluxes and compare the results with the values given in the table, which were obtained from atmosphere models.

8.2 H free–free emission (bremsstrahlung) can contribute to the cooling function in H II regions. For an ion with density n_i and charge Z_i, the energy loss per cm^3 per second is given by

$$\Lambda_{ff} = \frac{2^5 \pi e^6 Z_i^2}{3\sqrt{3}\, h\, m_e\, c^3} \left[\frac{2\pi k T}{m_e}\right]^{1/2} n_i\, n_e\, g_{ff}.$$

(a) Estimate $\Lambda_{ff}/n_p n_e$ for an H II region of pure H with $T = 10^4$ K. Take the Gaunt factor to be $g_{ff} \simeq 1$. (b) Estimate $\Lambda_{ff}/n_p n_e$ for an H I region with H and He. Assume that both are ionized once and consider a normal abundance for He. (c) Compare the results with the equivalent values obtained for cooling due to collisional excitation.

8.3 Measurements of oxygen and sulfur line intensities in planetary nebula NGC 6302 give the following results, already corrected from interstellar extinction, on a scale where I(Hβ) = 100.0. [O III]: I(4959) = 361.4, I(5007) = 1,352.0, I (4363) = 17.0, [O II]: I(3729) = 51.1, I(3726) = 53.9, [SII]: I(6716) = 11.3, and I(6731) = 11.1. (a) Estimate the nebula temperature and density using the O II and O III lines. (b) How would the above results be affected with the sulfur lines inclusion?

8.4 A main sequence star with 2.0 $M_\odot$ reaches the giant branch where it maintains a mass loss rate of 10^{-6} $M_\odot$ per year during a 10^6-year period. At the top of the asymptotic giant branch (AGB), it ejects a planetary nebula with mass M_{pn}, whose central star evolves into a 0.7 $M_\odot$ white dwarf. (a) What is the mass of the planetary nebula? (b) Supposing the timescale of the planetary nebula to be 20,000 years, what is the mean mass loss rate necessary to form the nebula? Neglect mass loss during the main sequence phase.

8.5 The ionized mass of a planetary nebula may be written $M_i = (4/3)\pi\, R_i^{\,3}\, \mu\, n_e \epsilon m_H$, where R_i is the ionized radius; μ is the mean molecular weight; n_e is the electron density; ϵ is the filling factor, which takes into account the ionized gas distribution in the nebula; and m_H is the H atom mass. It can be shown that the electron density is proportional to $F^{1/2}\, \epsilon^{-1/2}\, R_i^{-3/2}\, d$, where F is the observed free–free flux at 5 GHz and d is its distance. (a) Show that the ionized mass may be written in the form

$$M_i = \text{constant} \times F^{1/2}\, \epsilon^{1/2}\, \theta^{3/2}\, d^{5/2},$$

where $\theta = R_i/d$ is the nebula angular radius. (b) Show that the distance to the nebula may be written as

$$d = K\, F^{-1/5}\, \epsilon^{-1/5}\, \theta^{-3/5}\, M_i^{2/5},$$

where K is a constant. This is the Shklovsky method for determining distances to planetary nebulae. (c) Constant $K \simeq 50$, if the flux at 5 GHz is in mJy, θ is in arcseconds, M_i is in solar masses, and d is in kpc. Determine the distance to the nebula NGC 7009 with the following data: $F \simeq 700$ mJy, $\epsilon \simeq 1$, $\theta \simeq 15''$, and $M_i \simeq 0.2\, M_\odot$ (1 mJy = 10^{-29} W m^{-2} Hz^{-1}).

Bibliography

Abraham, Z., Lépine, J.R.D., Braz, M.A.: H66-alpha radio recombination line observations of southern H II regions. Mon. Notices Roy. Astron. Soc. **193**, 737 (1980). Determination of electron temperature in H II regions from radio recombination lines. The results mentioned in Section 8.6 are from this reference

Aller, L.H.: Physics of Thermal Gaseous Nebulae. Kluwer, Dordrecht (1984). Basic text about physical processes in ionized nebulae, with a discussion on the heating processes, excitation conditions, and abundance determination

Bowers, R.L., Deeming, T.: Astrophysics II. Jones and Bartlett, Boston (1984). Referred to in Chapter 1. Includes a good discussion on ionized hydrogen regions

Costa, R.D.D., Chiappini, C., Maciel, W.J., Freitas, P.J.A.: New abundances of southern planetary nebulae. Astron. Astrophys. Suppl. **116**, 249 (1996). Plasma diagnostics and abundance determination in planetary nebulae in the Galaxy. The results mentioned in Section 8.7 are taken from this reference

Dyson, J., Williams, D.A.: The Physics of the Interstellar Medium. Institute of Physics Publishing, London (1997). Referred to in Chapter 1. Includes a quite accessible discussion on ionized H interstellar regions, their temperatures, and dynamical processes

Gurzadyan, G.S.: The Physics and Dynamics of Planetary Nebulae. Springer, Berlin (1997). Very complete monograph on planetary nebulae

Kwok, S.: Origin and Evolution of Planetary Nebulae. Cambridge University Press, Cambridge (2000). Recent book about the principal aspects of astrophysics of planetary nebulae, including their origins and evolution

Maciel, W.J., Pottasch, S.R.: Photoelectric heating of H II regions. Astron. Astrophys. **106**, 1 (1982). Study of photoelectric heating by grains in H II regions. The results from Section 8.5 and Figure 8.6 are based on this reference

Osterbrock, D.: Astrophysics of Gaseous Nebulae and Active Galactic Nuclei. University Science Books, Mill Valley (1989). Referred to in Chapter 1. Classical text on ionized gaseous nebulae, with an extension to active galactic nuclei. Excellent discussion on spectral analysis, abundance determination, and plasma diagnostics, including tables with atomic constants of the main ions observed in these nebulae. Table 8.3 and Figures 8.3 and 8.5 are based on this reference

Pottasch, S.R.: Planetary Nebulae. Reidel, Dordrecht (1984). Classical monograph about planetary nebulae, with a particularly interesting discussion on abundances

Spitzer, L.: Physical Processes in the Interstellar Medium. Wiley, New York (1978). Referred to in Chapter 1. Includes an advanced treatment of interstellar ionized nebulae, in particular its thermodynamic aspects, Strömgren radius determination, abundances, and dynamical evolution. Tables 8.2 and 8.4 and Figure 8.7 are based on this reference

Chapter 9
Interstellar Dust Grains

9.1 Introduction

Interstellar dust grains are, along with the gas, the two main components of the interstellar medium. Until now, we have basically studied the gaseous component, in particular, the neutral "cold" gas of interstellar clouds and the ionized "hot" gas of photoionized nebulae. We saw that these regions contain solid particles that induce observable effects, such as infrared emission in H II regions and photoelectric heating of interstellar gas, one of the principal effects of interstellar grains (Chaps. 7 and 8). In this chapter, we will study in more detail the physical characteristics of dust grains and some of its observable properties in the interstellar medium. These processes include interstellar extinction, starlight reddening, scattering of stellar radiation, interstellar polarization, and thermal emission of the grains. Another aspect, the formation of H_2 molecules on the surface of the grains, will be considered in Chap. 10.

9.2 Efficiency Factors

The most obvious effects of interstellar grains are starlight extinction and reddening of the most distant stars, that is, an alteration in magnitude (or flux) and color. Though considered a problem for the study of stellar physics, as we saw for instance in the study of the radiation field in Chap. 2, extinction may be used to infer information about the physical properties of the grains, such as sizes, chemical compositions, and characteristics of the radiation emission and absorption process. We shall first physically characterize an interstellar dust grain, and next, we will examine its effects on starlight.

W.J. Maciel, *Astrophysics of the Interstellar Medium*,
DOI 10.1007/978-1-4614-3767-3_9,

9.2.1 Definitions

When a light beam is intercepted by a solid grain, part of the radiation is absorbed by the grain and part of it is scattered. Let $\sigma_a(\nu)$ and $\sigma_s(\nu)$ be the cross sections of the absorption and scattering processes, respectively, functions of frequency ν. If σ_g is the geometric cross section of the grains, we can define the efficiency factors for absorption $Q_a(\nu)$ and for scattering $Q_s(\nu)$ by

$$Q_a(\nu) = \frac{\sigma_a(\nu)}{\sigma_g}, \tag{9.1}$$

$$Q_s(\nu) = \frac{\sigma_s(\nu)}{\sigma_g}. \tag{9.2}$$

The combined effect of absorption and scattering by the dust grains is the *interstellar extinction*, characterized by the extinction cross section

$$\sigma_e(\nu) = \sigma_a(\nu) + \sigma_s(\nu) \tag{9.3}$$

and by the extinction efficiency factor

$$Q_e(\nu) = Q_a(\nu) + Q_s(\nu). \tag{9.4}$$

We see that a relation similar to (9.1) and (9.2) exists for the extinction efficiency factor, that is, $Q_e(\nu) = \sigma_e(\nu)/\sigma_g$. Besides the three efficiency factors already mentioned, we may also define the efficiency factor for radiation pressure, as we shall see later on. Efficiency factors depend, besides frequency, on the nature and size of the grains.

9.2.2 Extinction Efficiency Factor

We saw in Chap. 3 that the radiative transfer equation solution in terms of specific intensity can be put in the form

$$I_\nu = I_\nu(0)e^{-\tau_{\nu r}} + \int_0^{\tau_{\nu r}} \frac{j_\nu}{k_\nu} e^{-\tau_\nu} d\tau_\nu. \tag{9.5}$$

When we observe stellar radiation attenuated by interstellar extinction, we use flux F_ν which is the observed quantity:

$$F_\nu = I_\nu \cos\theta \, d\omega, \tag{9.6}$$

where the integral must be calculated over the solid angle defined by the star. In this case, the contribution of scattered radiation for the integral of (9.5) is generally negligible, and we may write

$$F_\nu = F_\nu(0)\, e^{-\tau_{\nu r}}. \tag{9.7}$$

Let A_λ be the extinction in magnitudes, that is, the increase in magnitude of a star whose original flux is $F_\nu(0)$, being F_ν the observed flux. We have

$$A_\lambda = -2.5 \log \frac{F_\nu}{F_\nu(0)} = -2.5 \log e^{-\tau_{\nu r}} = 1.086\, \tau_{\nu r}. \tag{9.8}$$

Let n_d (cm^{-3}) and N_d (cm^{-2}) be the volumetric density and column density of the grains, assumed identical. In this case, for a homogeneous medium,

$$\tau_{\nu r} = \int k_\nu ds = \int n_\mathrm{d} \sigma_\mathrm{e}(\nu) ds = \sigma_\mathrm{e}(\nu) N_\mathrm{d}. \tag{9.9}$$

In terms of the extinction efficiency factor, the optical depth can be written as

$$\tau_{\nu r} = N_\mathrm{d} \sigma_\mathrm{g} Q_\mathrm{e}(\nu), \tag{9.10}$$

so that

$$A_\lambda = 1.086 N_\mathrm{d} \sigma_\mathrm{g} Q_\mathrm{e}(\nu). \tag{9.11}$$

This equation is valid for identical grains. For different kinds of grains, (9.11) must be transformed in a summation for all species.

9.2.3 Scattering Efficiency Factor

Let us now examine the scattered radiation by grains in an extended region, such as a nebula or an interstellar cloud. In this case, the radiation-specific intensity I_ν can be observed, and we may write the scattered radiation (erg cm^{-3} s^{-1} Hz^{-1} sr^{-1})

$$j_{\nu s}(\mathbf{k}) = n_\mathrm{d} Q_\mathrm{s}(\nu) \sigma_\mathrm{g} \int I_\nu(\mathbf{k}') F(\mathbf{k} - \mathbf{k}') d\omega', \tag{9.12}$$

where $\mathbf{k}'$, $\mathbf{k}$ are unit vectors of the incident and scattered photons directions, respectively; $d\omega'$ is the solid angle interval around $\mathbf{k}'$; and F($\mathbf{k} - \mathbf{k}'$) is the normalized phase function, so that $\int F d\omega = 1$. Product $I_\nu(\mathbf{k}') F(\mathbf{k} - \mathbf{k}') d\omega'$ gives us the scattered intensity inside an unitarian solid angle around direction $\mathbf{k}$.

Factor $Q_s(\nu)$ plays an important role in the determination of the grain contribution to the scattered radiation in nebulae, such as reflection nebulae. This factor can be related to the grain albedo $\gamma(\nu)$, that is, the fraction of scattered radiation relative to the total extinguished radiation:

$$\gamma(\nu) = \frac{\sigma_s(\nu)}{\sigma_e(\nu)} = \frac{Q_s(\nu)}{Q_e(\nu)}. \tag{9.13}$$

From (9.12), we see that the radiation scattered by the grains can give us information about the grain properties. A typical value for the optical albedo is $\gamma \simeq 0.6$. We can study the grain scattering from, for instance (1) diffuse galactic light, that is, galactic starlight scattered by grains situated in the galactic plane; (2) scattered light in the neighborhood of hot stars, such as H II regions; (3) reflection nebulae, where radiation from a moderately hot star is reflected by dust clouds, such as in the Pleiades; and (4) X-ray halos around X-ray point sources, produced by scattering by dust grains at grazing angles.

9.2.4 Absorption Efficiency Factor

Let us now find a relation between the absorption efficiency factor and the grain emissivity. After absorbing radiation (generally in the optical or ultraviolet), the grains reemit it in another region of the spectrum (generally in the infrared). A good approximation consists in estimating the grain emissivity $j_{\nu d}$ assuming that they are perfect radiators at temperature T_d. In this case, we may write

$$j_{\nu d} = n_d Q_a(\nu) \sigma_g B_\nu(T_d), \tag{9.14}$$

given in erg cm^{-3} s^{-1} Hz^{-1} sr^{-1}, where $B_\nu(T_d)$ is the Planck function at the grain temperature. As we shall see later on, the grain-infrared emission can be used to determine some properties of the regions and the grains.

9.2.5 Efficiency Factors and Mie Theory

To determine cross sections or efficiency factors, it is necessary to have complete knowledge of light diffraction theory by small particles. Relatively simple quantitative results can be obtained for a few cases, such as for homogeneous spherical grains with isotropic optical properties. We can then define a dimensionless parameter

$$x = \frac{2\pi a}{\lambda}, \tag{9.15}$$

where a is the radius of the grains. Let n be the refraction index of the material that constitutes the grain and k the grain absorption index. In general, the refraction index is complex and may be written as

$$m = n - ik, \tag{9.16}$$

that is, the imaginary part originates the absorption. In fact, when an electromagnetic wave with electric vector characterized by $E \propto e^{i(Ks-\omega t)}$ propagates in a direction s through a medium with refraction index m, m being complex, its velocity is $v = c/m = \omega/K$. Since m is complex, K is also complex or $K = \alpha + i\beta$, and the wave has the form $E \propto e^{i(\alpha s+i\beta s-\omega t)}$ or $E \propto e^{i(\alpha s-\omega t)}\ e^{-\beta s}$, that is, its amplitude decreases exponentially with the distance, according to an absorption coefficient given by the imaginary part of the refraction index or $k \propto \beta$.

Determining cross sections means solving the Maxwellian equations with boundary conditions suitable for a spherical surface. This problem was initially solved by Mie and Debye around 1908, and so the formulae used to obtain the cross section are generally determined by the so-called *Mie theory*. The computational problem is complex, and detailed solutions exist only for spheres, spheroids, and infinite homogeneous cylinders.

Even for spherical grains, the expressions used to calculate σ or Q are relatively complex and need to be written in the form $Q = Q(x,m)$. For spherical grains, the geometric section is $\sigma_g = \pi a^2$, so that efficiency factors may be written as $Q = \sigma/\pi a^2$.

Figure 9.1 shows the extinction efficiency factor for some species representative of interstellar dust grains. Curve A has $m = \infty$, corresponding to a completely reflecting sphere, that is, with an infinite dielectric constant; curve B has $m = 1.33$ corresponding to dielectric ice particles; curve C has $m = 1.33-0.09i$ corresponding to "dirty ice," that is, ice spheres with absorbing impurities; and curve D has $m = 1.27-1.37i$, characteristic of iron absorbing spheres.

For small values of x ($x \lesssim 3$) or large values of λ, Q_e behaves quite differently depending on the nature of the grain and on x. For $x \gg 3$, corresponding to small values of λ, all extinction efficiency factors tend to a value of the order of 2. Efficiency factors for absorption by spherical particles have a similar behavior but instead tend to a value close to 1. In this case, the energy taken away from the beam is approximately two times larger than the incident energy in the solid sphere. This can be explained by the *Babinet principle*, according to which the diffraction pattern due to an obstacle is identical to the one with an aperture having the same cross section and containing the same energy as the one that falls upon the obstacle. In other words, the diffracted radiation is lost by the incident beam.

We see in Fig. 9.1 that, in a general way, the efficiency factor behavior is quite complex, even for homogeneous spheres. Some special cases have simple solutions. For $x \ll 1$, we have:

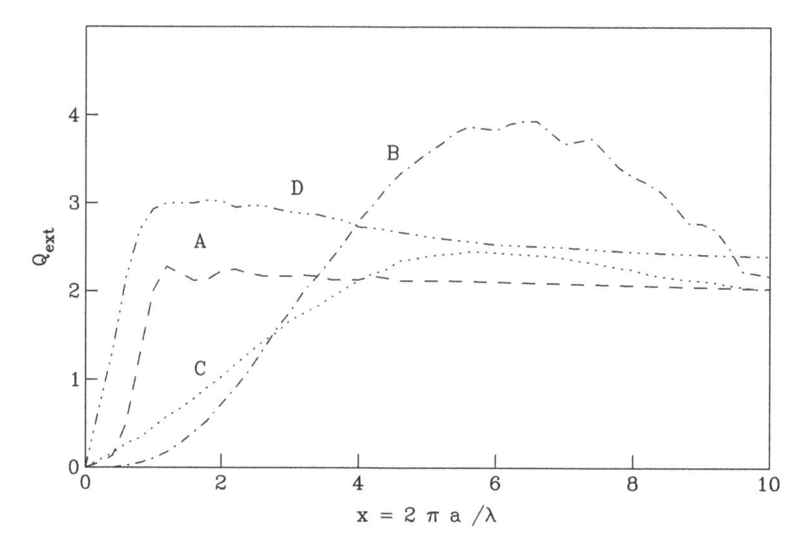

Fig. 9.1 Examples of the variation of the extinction efficiency factor as a function of the wavelength parameter x

(1) Pure dielectric spheres, $k = 0$, m real:

$$Q_a = 0, \tag{9.17}$$

$$Q_e = Q_s = \frac{8}{3} x^4 \left[\frac{m^2 - 1}{m^2 + 2} \right]^2 . \tag{9.18}$$

(2) Absorbing spheres, $k \neq 0$, m complex:

$$Q_s = \frac{8}{3} x^4 \left[\frac{m^2 - 1}{m^2 + 2} \right]^2 , \tag{9.19}$$

$$Q_e = -4x \, \mathrm{Im} \left[\frac{m^2 - 1}{m^2 + 2} \right] . \tag{9.20}$$

Besides efficiency factors, the Mie theory allows us to determine the scattering phase function for spherical grains. Letting θ be the angle between the incident and scattered beam directions, this function may be written as $F = F(\theta, a, \lambda, m)$. As an illustration, Fig. 9.2 shows the mean value of the phase parameter $\langle \cos\theta \rangle$ as a function of x for dielectric ice particles with $m = 1.33$. This mean value characterizes the scattered radiation direction. We see that, if $x \ll 1$, $\langle \cos\theta \rangle \simeq 0$ and $\langle \theta \rangle \simeq \pm\pi/2$. For $x \gg 1$, $\langle \cos\theta \rangle \simeq 1$ and $\langle \theta \rangle \simeq 0$, that is, the scattering process tends to remove less energy from the original beam direction. Isotropic scattering corresponds to $\langle \cos\theta \rangle \simeq 0$, and case $\langle \cos\theta \rangle \simeq 1$ is a forward-throwing scattering.

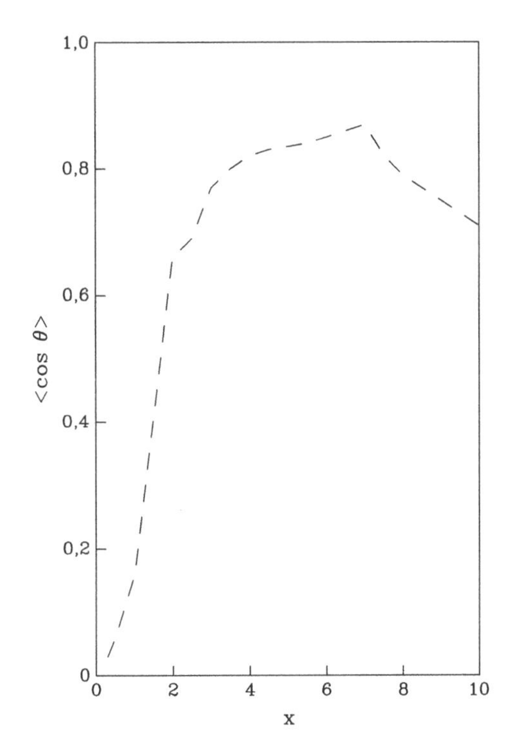

Fig. 9.2 Average values of the phase parameter as a function of the wavelength parameter x

Besides homogeneous spheres, there are results from the Mie theory for composite concentric spheres and infinite cylinders. For spheroid or irregular grains, there are numerical solutions.

9.2.6 *Efficiency Factor for Radiation Pressure*

Let us consider the interaction between a radiation beam with flux F_ν and a spherical grain with radius a and efficiency factors Q_a, Q_s, and Q_e. The energy absorbed by the grain is proportional to $F_\nu \pi a^2 Q_a$. In the same way, the energy scattered by the grain, which is returned to the original direction, is proportional to $F_\nu \pi a^2 Q_s \langle \cos\theta \rangle$, where θ is the angle between the scattering and original directions. The energy removed from the beam's original direction due to scattering is proportional to $F_\nu \pi a^2 Q_s (1 - \langle \cos\theta \rangle)$, so that the total energy removed from the original direction is proportional to

$$\begin{aligned} F_\nu \pi a^2 [Q_a + Q_s(1 - \langle \cos\theta \rangle)] &= F_\nu \pi a^2 (Q_e - Q_s \langle \cos\theta \rangle) \\ &= F_\nu \pi a^2 Q_p, \end{aligned}$$

where we introduce the efficiency factor for radiation pressure $Q_{\rm p}$:

$$Q_{\rm p} = Q_{\rm e} - Q_{\rm s}\langle\cos\theta\rangle. \tag{9.21}$$

As in the previous cases, we can define a cross section for radiation pressure by $\sigma_{\rm p} = \pi a^2 Q_{\rm p}$.

A photon with energy $h\nu$ transports a momentum $h\nu/c$ in its propagation direction. The momentum per unit time per unit surface per frequency interval transported by a radiation beam with flux F_ν is simply F_ν/c given in erg cm^{-3} Hz^{-1} = dyne cm^{-2} Hz^{-1}, which is the pressure exerted by radiation per frequency interval. The radiative force exerted on a grain is given by

$$F_{\rm r} = \frac{1}{c}\int \pi a^2 Q_{\rm p}(\nu) F_\nu d\nu, \tag{9.22}$$

where we integrate the flux over a certain frequency interval and point out the efficiency factor dependency on frequency. For instance, for a grain situated at a distance r from a star with luminosity L, if $Q_{\rm p}(\nu) \sim Q_{\rm p}$, we have

$$F_{\rm r} \simeq \frac{\pi a^2 Q_{\rm p} L}{4\pi r^2 c}. \tag{9.23}$$

Expression (9.23) illustrates one of the dynamical aspects involved in the study of interstellar dust grains, which can for instance be applied to the grain ejection process and consequent sweeping of the gas in circumstellar envelopes. Other dynamical processes include (1) Poynting–Robertson effect, that is, the velocity lost by a grain moving in a perpendicular direction to the direction of radiation propagation in circumstellar envelopes, and (2) the effect of electromagnetic forces on electrically charged grains, in particular, during grain escape from the neighborhood of stars with magnetic fields.

9.3 Interstellar Extinction

9.3.1 Color Excess

Besides general extinction A_λ measured in magnitudes, the presence of dust grains in interstellar clouds causes a reddening of starlight, that is, a change in color or a relative change in stellar magnitudes according to wavelength. The determination of the general extinction A_λ is difficult, but selective extinction, characterized by the variation of A_λ as a function of wavelength, can be more easily obtained. Let us consider two stars A and B with identical spectra, that is, of the same spectral type and luminosity class. If $r_{\rm A}$ and $r_{\rm B}$ are the distances to the stars in pc, relations between apparent magnitudes (m) and absolute magnitudes (M) of the two stars at two wavelengths λ_1 and λ_2 may be written as

$$m_{\mathrm{A},\lambda_1} = M_{\lambda_1} + 5\log r_{\mathrm{A}} - 5 + A_{\mathrm{A},\lambda_1}, \tag{9.24a}$$

$$m_{\mathrm{A},\lambda_2} = M_{\lambda_2} + 5\log r_{\mathrm{A}} - 5 + A_{\mathrm{A},\lambda_2}, \tag{9.24b}$$

$$m_{\mathrm{B},\lambda_1} = M_{\lambda_1} + 5\log r_{\mathrm{B}} - 5 + A_{\mathrm{B},\lambda_1}, \tag{9.24c}$$

$$m_{\mathrm{B},\lambda_2} = M_{\lambda_2} + 5\log r_{\mathrm{B}} - 5 + A_{\mathrm{B},\lambda_2}, \tag{9.24d}$$

so that

$$\Delta m_{\lambda_1} = m_{\mathrm{A},\lambda_1} - m_{\mathrm{B},\lambda_1} = 5\log\frac{r_{\mathrm{A}}}{r_{\mathrm{B}}} + (A_{\mathrm{A},\lambda_1} - A_{\mathrm{B},\lambda_1}). \tag{9.25a}$$

Similarly,

$$\Delta m_{\lambda_2} = m_{\mathrm{A},\lambda_2} - m_{\mathrm{B},\lambda_2} = 5\log\frac{r_{\mathrm{A}}}{r_{\mathrm{B}}} + (A_{\mathrm{A},\lambda_2} - A_{\mathrm{B},\lambda_2}). \tag{9.25b}$$

Calculating the difference,

$$\begin{aligned} \Delta m_{\lambda_1} - \Delta m_{\lambda_2} &= (A_{\mathrm{A},\lambda_1} - A_{\mathrm{B},\lambda_1}) - (A_{\mathrm{A},\lambda_2} - A_{\mathrm{B},\lambda_2}) \\ &= (A_{\mathrm{A},\lambda_1} - A_{\mathrm{A},\lambda_2}) - (A_{\mathrm{B},\lambda_1} - A_{\mathrm{B},\lambda_2}) \\ &= (A_{\lambda_1} - A_{\lambda_2})_{\mathrm{A}} - (A_{\lambda_1} - A_{\lambda_2})_{\mathrm{B}} \\ &= \Delta(A_{\lambda_1} - A_{\lambda_2}). \end{aligned} \tag{9.26}$$

Assuming that one of the stars is free of extinction or reddening, we have

$$\Delta m_{\lambda_1} - \Delta m_{\lambda_2} = A_{\lambda_1} - A_{\lambda_2} = E(\lambda_1, \lambda_2), \tag{9.27}$$

where $E(\lambda_1,\lambda_2)$ is the color excess for the star suffering from extinction at wavelengths λ_1 and λ_2. Considering Johnson UBV system, with $\lambda_1 = 4{,}350$ Å and $\lambda_2 = 5{,}550$ Å, we may write for the standard color excess

$$\Delta m_B - \Delta m_V = E_{B-V} = E(B - V). \tag{9.28}$$

We can use the color excess (1) by setting wavelengths λ_1 and λ_2, in such a way that the observation of many stars allows us to study color excess variation, that is, the spatial distribution of the grains, or (2) by setting wavelength λ_2 and varying λ_1, which allows us to obtain the selective extinction variation with wavelength.

9.3.2 Spatial Distribution of Dust Grains

Photoelectric measurements of a large number of stars allow us to estimate a color excess value per kpc, for any direction, inside 100 pc of the galactic center and at less than 1 kpc from the Sun:

$$\frac{\langle E_{B-V} \rangle}{L} \simeq 0.6 \text{ mag kpc}^{-1}, \tag{9.29}$$

where L is the mean distance to the source. Due to the interstellar medium irregularity, this equation cannot be used to calculate absorption at a certain distance but can instead be used to determine the total quantity of dust in the galactic disk. Using a mean value $k \simeq 4$ clouds per kpc, the mean extinction per cloud is

$$E_0 \simeq \frac{\langle E_{B-V} \rangle}{L} \frac{1}{k} \simeq 0.1 \text{ mag}. \tag{9.30}$$

For a more detailed treatment, we must consider different kinds of clouds.

We saw in Chap. 4 that there is a good correlation between H column density and color excess,

$$N_{\rm H} \simeq 6 \times 10^{21} E_{B-V} \text{ mag}^{-1} \text{ cm}^{-2}. \tag{9.31}$$

Using a mean value $E_0 \simeq E_{B-V} \simeq 0.1$ mag for a "standard" interstellar cloud, we obtain a typical density $N_{\rm H} \simeq 4 \times 10^{20}$ cm^{-2}, in accordance with value $N_{\rm H} \simeq 3 \times 10^{20}$ cm^{-2} determined from measurements of the 21 cm absorption line in extragalactic sources, as seen in Chap. 4. Combining (9.29) with (9.31), we obtain $n_{\rm H} \simeq N_{\rm H}/L \simeq 1$ cm^{-3} for the typical volumetric density of the interstellar cloud.

9.3.3 Extinction Curve

Studies of extinction variation with wavelength produce results such as the ones shown in Fig. 9.3. In this figure, the ordinate is $E(\lambda,V)/E_{B-V}$, so that for $\lambda = V = 5{,}500$ Å, $1/\lambda = 1.8$ μm^{-1}, $E(V,V)/E_{B-V} = 0$ and for $\lambda = B = 4{,}350$ Å, $1/\lambda = 2.3$ μm^{-1}, $E(B,V)/E_{B-V} = 1$. In the optical domain, the extinction is approximately proportional to $1/\lambda$, giving rise to the well-known starlight reddening.

In general, the extinction increases for shorter wavelengths, having a particular prominence, also referred to as the "hump" at $\lambda \simeq 2{,}175$ Å $\simeq 0.22$ μm or $\lambda^{-1} \simeq 4.6$ μm^{-1}. This feature is not well understood, being generally attributed to some form of carbon grains, particularly graphite. In the infrared domain, two peaks are

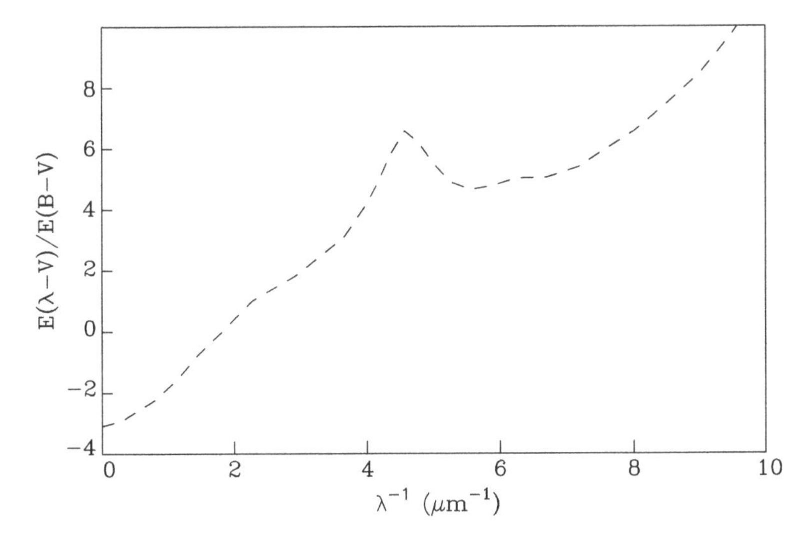

Fig. 9.3 An example of the interstellar extinction curve

generally observed (not shown in the figure) at $\lambda \simeq 3.1$ μm ($\lambda^{-1} \simeq 0.32$ μm^{-1}) and $\lambda \simeq 9.7$ μm ($\lambda^{-1} \simeq 0.10$ μm^{-1}), attributed to ice and silicate grains, respectively.

In principle, the mean extinction curve may be reproduced by computing curves corresponding to different grain species. This is a complex task because it involves identification and knowledge of the physical properties of many species. Besides that, the solution is generally not unique. The curve of Fig. 9.3 may be approximately reproduced by including graphite grains with radii of the order of 100 Å, responsible for ultraviolet extinction; SiC grains with radii of the order of 750 Å, affecting essentially the optical part of the spectrum; and silicates with radii of the order of 450 Å, important in the far ultraviolet.

A mean curve, good enough to calculate the reddening of stars, is given in Table 9.1. Columns show wavelength λ in μm, inverse wavelength λ^{-1} (μm^{-1}), and $E(\lambda,V)/E_{B-V}$ and A_λ/E_{B-V} ratios (see below). Some photometric bands are also indicated.

9.3.4 Total and Selective Extinction

Let us define ratio R_λ between the total absorption in magnitudes A_λ at wavelength λ and the standard selective extinction, measured by the color excess E_{B-V}:

$$R_\lambda = \frac{A_\lambda}{E_{B-V}}. \qquad (9.32)$$

Table 9.1 Data for the average extinction curve of Fig. 9.3

	λ (μm)	λ^{-1} (μm^{-1})	$E(\lambda, V)/E_{B-V}$	A_λ/E_{B-V}
	∞	0	−3.10	0.00
L	3.4	0.29	−2.94	0.16
K	2.2	0.45	−2.72	0.38
J	1.25	0.80	−2.23	0.87
I	0.90	1.11	−1.60	1.50
R	0.70	1.43	−0.78	2.32
V	0.55	1.82	0	3.10
B	0.44	2.27	1.00	4.10
	0.40	2.50	1.30	4.40
	0.344	2.91	1.80	4.90
	0.274	3.65	3.10	6.20
	0.250	4.00	4.19	7.29
	0.240	4.17	4.90	8.00
	0.230	4.35	5.77	8.87
	0.219	4.57	6.57	9.67
	0.210	4.76	6.23	9.33
	0.200	5.00	5.52	8.62
	0.190	5.26	4.90	8.00
	0.180	5.56	4.65	7.75
	0.170	5.88	4.77	7.87
	0.160	6.25	5.02	8.12
	0.149	6.71	5.05	8.15
	0.139	7.18	5.39	8.49
	0.125	8.00	6.55	9.65
	0.118	8.50	7.45	10.55
	0.111	9.00	8.45	11.55
	0.105	9.50	9.80	12.90
	0.100	10.00	11.30	14.40

For $\lambda = V = 5{,}500$ Å,

$$R_V = \frac{A_V}{E_{B-V}}. \tag{9.33}$$

We have then

$$\begin{cases} R_\lambda - R_V = \dfrac{A_\lambda - A_V}{E_{B-V}} = \dfrac{E(\lambda, V)}{E_{B-V}} \\ \dfrac{A_\lambda}{E_{B-V}} = R_V + \dfrac{E(\lambda, V)}{E_{B-V}}. \end{cases} \tag{9.34}$$

Ratio $R_\lambda = A_\lambda/E_{B-V}$ is given in Table 9.1 for the mean extinction curve. Considering long wavelengths, $\lambda \to \infty$ (or $1/\lambda \to 0$), we should expect that $A_\lambda = A_\infty \to 0$, that is, $R_\lambda = R_\infty \to 0$, or

$$R_V = -\frac{E(\infty, V)}{E_{B-V}.} \tag{9.35}$$

In Fig. 9.3, we see that $E(\infty,V)/E_{B-V} \to -3$, so that

$$\begin{cases} R_V \simeq 3 \\ A_V \simeq 3E_{B-V}. \end{cases} \tag{9.36}$$

Values of R_V have been determined from infrared extinction, resulting in $R_V \simeq 3.4$. Independently, $R_V \simeq 3.2$ is obtained from the determination of A_V for stars with known excess color. The value of R_V is not necessarily constant, and it can reach values of the order of 4–6 in the direction of some molecular clouds. However, for most directions, variation is small.

Let us use the obtained value for the total and selective extinction ratio (R_V) and determine interstellar grain-integrated properties. We saw in Fig. 9.1 that the extinction efficiency factor Q_e tends to 2.0 when $x \gg 1$, that is, in the ultraviolet part of the spectrum. Let us consider $\lambda = 1{,}000$ Å so that $E(1000, V)/E_{B-V} \simeq 10$ (see Fig. 9.3). Since $A_\lambda = A_V + E(\lambda,V)$ and $A_V = R_V\ E_{B-V} \simeq 3\ E_{B-V}$, we have $A_{1000} = 13\ E(B-V)$. Let us estimate the mean total area exposed by the grains per cm^3, $\langle n_d \sigma_g \rangle$. From (9.11) and (9.29),

$$\langle n_d \sigma_g \rangle \simeq \frac{\langle N_d \sigma_g \rangle}{L} \simeq \frac{A_{1000A}}{1.086\, Q_{e\,1000} L} \simeq \frac{13 E_{B-V}}{1.086 Q_{e\,1000} L} \simeq 1.2 \times 10^{-21}\ \text{cm}^{-1}, \tag{9.37}$$

where we use $Q_{e1000} = Q_e(1000\ \text{Å}) \simeq 2$. Introducing the total area exposed by the grains per hydrogen atom, we have

$$\Sigma_d = \frac{\langle n_d \sigma_g \rangle}{\langle n_H \rangle} \simeq 1.0 \times 10^{-21}\ \text{cm}^2, \tag{9.38}$$

where we use $n_H \simeq 1.2$ cm^{-3}.

9.3.5 Grain-to-Gas Ratio

For spherical grains, we saw that factor Q_e may be determined as a function of wavelength. Besides that, it is possible to show that

$$\int_0^\infty Q_e d\lambda = 4\pi^2 a \left(\frac{\epsilon_0 - 1}{\epsilon_0 + 2} \right) = 4\pi^2 a F_k, \tag{9.39}$$

where ϵ_0 is the grain dielectric constant and F_k is a dimensionless parameter. For $\lambda \to \infty$, $\epsilon_0 = m^2$ and F_k is defined. Let ρ_d be the grain mean density along the line of sight and let s_d be the density of the solid material that constitutes the grain. We have

$$\rho_d = n_d m_d = \frac{N_d}{L} m_d = \frac{N_d}{L} \frac{4}{3} \pi a^3 s_d. \tag{9.40}$$

Considering (9.39) and (9.11),

$$\begin{aligned} 4\pi^2 a F_k &= \int_0^\infty Q_e d\lambda = \int_0^\infty \frac{A_\lambda d\lambda}{1.086 N_d \sigma_g} \\ &= \frac{1}{1.086 N_d \sigma_g} \int_0^\infty A_\lambda d\lambda. \end{aligned} \tag{9.41}$$

Using (9.40),

$$\begin{aligned} \rho_d &= \frac{4}{3} \pi a^3 s_d \frac{1}{(1.086\, \sigma_g)(4\pi^2 a F_k)} \int \frac{A_\lambda d\lambda}{L} \\ &= \frac{1.0 \times 10^{23} s_d}{F_k} \int \frac{A_\lambda d\lambda}{L}, \end{aligned} \tag{9.42}$$

where L is in kpc and density ρ_d is in g cm^{-3}. Considering that $A_\lambda = A_V + E(\lambda,V)$ and the mean extinction curve given in Fig. 9.3, we can estimate the integral in (9.42) taking $A_v/E_{B-V} = R_V \simeq 3$. In this case,

$$\langle \rho_d \rangle \simeq 1.3 \times 10^{-27} s_d \left\langle \frac{A_V}{L} \right\rangle \frac{\epsilon_0 + 2}{\epsilon_0 - 1}, \tag{9.43}$$

where L is in kpc. We can numerically estimate density ρ_d considering $s_d \simeq 3$g cm^{-3} and $\varepsilon_0 \simeq 4$, which is suitable for the grains shown in Fig. 9.3:

$$\left\langle \frac{A_V}{L} \right\rangle \simeq \left\langle \frac{R_V E_{B-V}}{L} \right\rangle \simeq R_V \left\langle \frac{E_{B-V}}{L} \right\rangle \simeq 3 \times 0.61 \text{ mag kpc}^{-1},$$

or $\langle \rho_d \rangle \simeq 1.4 \times 10^{-26}$ g cm^{-3}, where we use (9.29). More detailed calculations for integral $\int A_\lambda d\lambda$ correct this value to $\langle \rho_d \rangle \simeq 1.8 \times 10^{-26}$ g cm^{-3}, for the same values of ε_0 and s_d. We see that the interstellar matter density in the form of grains is approximately $6 \times 10^{-24}/2 \times 10^{-26} \simeq 300$ times smaller than the Oort limit. Considering the above value for grain density and mean density $n_H \simeq 1.2$ cm^{-3} for the interstellar gas, we have that the grain-to-gas or dust-to-gas ratio in the interstellar medium is

$$\frac{\rho_d}{\rho_{gas}} \simeq \frac{\rho_d}{\rho_H + \rho_{He}} \simeq \frac{\rho_d}{n_H m_H + 4 n_{He} m_H}. \tag{9.44}$$

Assuming $n_{He}/n_H \simeq 0.1$, we have $\rho_d/\rho_{gas} \simeq \rho_d/(1.4\ n_H m_H) \simeq 1/160 \simeq 6 \times 10^{-3}$. In reality, these relations correspond to lower limits because there may be grains of large sizes that are not detected.

9.3.6 Diffuse Interstellar Bands

One of the persistent problems in the study of the interstellar medium, possibly associated with interstellar dust grains, is the presence of diffuse interstellar bands, an ensemble of about three hundred absorption features in the optical and infrared spectrum of reddened stars. Examples of these bands occur at $\lambda = 4{,}430$ Å, 5,780 Å, 6,177 Å, and 6,282 Å with widths between 1 and 30 Å, thus much broader than typical atomic interstellar line widths.

Diffuse interstellar bands present some correlation with interstellar extinction suggesting, since the 1930s, their association with interstellar dust grains responsible for interstellar extinction. However, more recent works show a large scattering, thus putting this correlation in some doubt, at least for a part of the diffuse bands.

Besides solid grains, other types of particles could be responsible for these bands, such as free molecules of CO_2, NH_4, CH_4^+, etc. More recently, long carbonated molecules have been proposed, such as fullerene (C_{60}) and PAHs (Sect. 9.5), in particular, naphthalene ($C_{10}H_8$), pyrene ($C_{16}H_{10}$), and coronene ($C_{24}H_{12}$). Pyrene was apparently successfully identified as being responsible for features at 9,577 Å and 9,632 Å. Most probably, we need more than one "carrier" to explain the origin of the bands, as is the case of other features attributed to interstellar grains.

9.4 Interstellar Polarization

Information about the nature and distribution of interstellar grains may be obtained from interstellar polarization measurements, independently discovered by Hiltner and Hall in 1949. Light from reddened stars generally presents some linear polarization, and so grains that are responsible for extinction are also probably responsible for the observed polarization. Light polarization means that the waves propagating along different directions of the radiation electric vector are absorbed in different ways creating anisotropy. The grains responsible for polarization must thus have some anisotropy, which sets a preference for nonspherical, elongated particles or else anisotropic grains, if spherical. For polarized radiation of total intensity I, being I_M and I_m, the intensities along the maximal and minimal propagation directions, respectively, we have

$$I = I_M + I_m, \tag{9.45}$$

$$P = \frac{I_M - I_m}{I}, \tag{9.46}$$

where P is the *degree of polarization*, typically of a few percent for the reddened stars in the galactic plane direction. Circular polarization is also observed for some stars, but shows even lower values, of the order of 10^{-4}. Polarization can also be measured in magnitudes, defined by $p = 2.5 \log I_M/I_m$. If $P \ll 1$, we have $p \simeq 2.17$ P. Extinction efficiency factors, such as the ones seen in Fig. 9.1, also depend on the observed particle orientation relative to the electric field direction of the incident radiation. In fact, the necessary anisotropy for the existence of polarization can be expressed in terms of the difference between the efficiency factors corresponding to the particle orientation, parallel or perpendicular to the electric vector.

Initially observed at optical wavelengths, the interstellar polarization database has recently been expanded by ultraviolet measurements (WUPPE project). In addition to the degree of polarization, information concerning the nature of interstellar grains may be obtained by studying polarization variation with extinction, its wavelength dependency, and its variations with galactic longitude.

9.4.1 Variation with Extinction

In a general way, low-extinction objects also have low polarization, whereas a high extinction can be associated with a degree of polarization within a relatively large interval, varying from $P = 0$ to P $\simeq 0.1\ E_{B-V}$, that is, $P(\%) \lesssim 10\ E_{B-V}$. Assuming that dust grains are responsible for extinction as well as for polarization, we may relate the degree of polarization P with the extinction efficiency factor. Using definition (9.46), assuming that $I \propto e^{-\tau} \propto e^{-N_d \sigma_\nu} \propto e^{-N_d \sigma_g Q_e}$, and applying I_M and I_m, we may expand the exponential term, if $P \ll 1$, to obtain

$$P \simeq \frac{1}{2} N_d \sigma_g (Q_{eM} - Q_{em}), \tag{9.47}$$

where Q_{eM} and Q_{em} are the maximum and minimum values of Q_e as the electric vector is rotated. Using (9.11), we obtain

$$\frac{P}{A_\lambda} \simeq 0.46 \frac{Q_{eM} - Q_{em}}{Q_e}. \tag{9.48}$$

For instance, with $A_V/E_{B-V} \simeq 3$, we have $P \simeq 0.1\ E_{B-V} \simeq 0.1\ A_V/3$ or $P/A_V \simeq 0.03$. Considering $m = 1.33$ at $\lambda = 5{,}500$ Å and $x = 2\pi a/\lambda \simeq 3$, theoretical calculations suggest $Q_e \simeq 2$ and $Q_{eM} - Q_{em} \simeq 0.35$, that is, $P/A_V \simeq 0.08$. Different materials produce different degrees of polarization, eventually approaching the limit value $P \simeq 0.1\ E_{B-V}$.

9.4.2 Variation with Wavelength

In the optical region of the spectrum (4,000–8,000 Å), polarization varies considerably with wavelength, reaching different maximum values $P(\lambda_M)$ for different wavelengths (λ_M). The polarization normalized to the maximum value as a function of wavelength relative to λ_M is essentially the same for all stars and may be adjusted by a simple expression, the *Serkowski law*, given by

$$\frac{P(\lambda)}{P(\lambda_M)} = \exp\left[-K \ln^2\left(\frac{\lambda}{\lambda_M}\right)\right]. \tag{9.49}$$

In this relation, we can consider $K \simeq 1.15$, though some recent works show some variation of this parameter with the maximum polarization wavelength, in the form $K \simeq \alpha + \beta\lambda_M$, where the wavelength is in μm, $0.01 \gtrsim \alpha \gtrsim -0.10$ and $1.9 \gtrsim \beta \gtrsim 1.6$. This relation gives good fits for optical and near-infrared regions but shows increasing deviations for $\lambda \gtrsim 3$ μm, where other fits are necessary, such as a power law.

Wavelength variation of maximum λ_M for different stars apparently reflects the presence of grains with different properties, in particular size distribution, in interstellar clouds. Objects with higher values of λ_M have higher extinction in the infrared, presenting selective extinction curves slightly different from the mean extinction curve, so that $R_V \propto \lambda_M$, with values for ratio R_V falling in the interval $2 \lesssim R_V \lesssim 4$. The maximum value of the degree of polarization $P(\lambda_M)$ depends on the grain columnar density and on their optical properties, which affect the polarization intensity and depend on the nature of the grains.

9.4.3 Galactic Magnetic Field

The observed polarization excludes perfectly spherical and isotropic grains, so the particles must probably be elongated. In this case, in order not to be destroyed, the polarization produced by a cloud of particles needs some type of alignment of the rotation axis of the grains. The classical model for this alignment is the so-called Davis and Greenstein mechanism, according to which the interstellar magnetic field plays a major role, so that the galactic distribution of the grains—for instance, of the polarization vectors—can also provide information about the galactic magnetic field. The galactic disk region is permeated by a general magnetic field of low intensity, much lower than the ones existing on Earth and in the Solar System, which are typically of the order of 1 Gauss. Grains, assumed elongated, when subjected to a magnetic field, tend to align their minor axis with the field direction. Due to frequent collisions with gas atoms, grains develop a rotation movement around the field lines. Radiation absorption by an ensemble of aligned grains in a cloud has a direction dependency, generating polarization. Thus, the observed polarization direction indicates the magnetic field direction responsible for the alignment.

We can estimate the rotation frequency ω of a grain with radius a, density s_g, and moment of inertia I in a cloud with temperature T. We have

$$\frac{1}{2} I \omega^2 \simeq \frac{3}{2} kT,$$

from which we obtain

$$\omega^2 \simeq \frac{3kT}{s_g \frac{4}{3}\pi a^5}. \tag{9.50}$$

With $a \simeq 1{,}000$ Å $= 10^{-5}$ cm, $s_g \sim 1$ g cm^{-3}, and $T \simeq 100$ K, we have $\omega \simeq 10^5$ Hz.

To understand Davis and Greenstein mechanism, let us assume the grains to be paramagnetic. In this case, the field will induce a magnetic moment on the grain. With the rotation movement of the grain, the magnetic moment orientation varies to the expense of the grain rotational energy. This effect depends on the rotation axis orientation relative to the field, being more pronounced when the rotation axis is perpendicular to the direction of the field, which tends to damp this movement, leaving the grain with the rotation axis preferentially aligned parallel to the magnetic field.

Application of the Davis and Greenstein mechanism to interstellar grains encounters some problems, in particular, due to the fact that the relaxation time is relatively high. Other processes have been proposed as possible alternatives, such as Gold mechanic alignment, Purcell superthermal paramagnetic alignment, and alignment by radiative torques.

The distribution of polarization vectors as a function of galactic coordinates (longitude l, latitude b) shows a clear association of the field with the galactic plane ($b = 0$), and stars with low latitudes generally have polarization vectors parallel to the plane. For the $l \simeq 90°$ and $l = 270°$ regions, the distribution is more complex because these directions coincide with the orbit of the Sun, more precisely with the Local Standard of Rest (LSR) relatively to the galactic center. For higher latitudes, the field presents a longitude dependency and also an association with neutral H gas radio emission.

Interstellar polarization is not the only means by which information regarding the galactic magnetic field can be obtained. Zeeman effect and Faraday rotation measurements allow us to obtain data concerning different phases of the interstellar medium, and radio synchrotron emission measurements can be applied to other galaxies. The Zeeman effect may be observed as a splitting of the H 21 cm line by the magnetic field in dense clouds. In molecular clouds, the OH molecule 18 cm line may also be measured. Faraday rotation is the rotation of the polarization plane that occurs when a polarized radio wave passes through a region containing ionized gas and a magnetic field. This rotation depends on the field intensity, on wavelength, and on the gas density so that the field intensity can be obtained from a density estimate. Synchrotron emission is also polarized, being produced by cosmic electrons rotating around the magnetic field of the galaxies. All these observations lead to an interstellar magnetic field with typical intensities of 1–5 μGauss, around a million times weaker than the geomagnetic field.

Some observations concerning the galactic magnetic field direction may be obtained from the study of the variation of the polarization direction, that is, the vibration plane orientation with galactic longitude. This variation, measured in terms of the Stokes parameters, depends on the characteristics of the grains present in the clouds for a certain direction and on the magnetic field. Results coincide in a general way with the ones obtained by other methods but suggest a more complex configuration for the galactic magnetic field direction, in particular due to the presence of fluctuations.

9.5 Physical Properties of the Grains

9.5.1 Sizes

Interstellar grain sizes may be determined, to a first approximation, from features of the extinction curve. Generally, it is necessary to introduce a size distribution in order to fit the curve for a certain direction, a process that is not necessarily univocal, because the chemical composition of the grains also affects the extinction curve due to the different optical properties of the different grain species. For instance, a mixture of graphite and silicates allows to fit the extinction curve reasonably well in the interval $0.1 \lesssim \lambda\ (\mu\mathrm{m}) \lesssim 1$. A model known as MRN (from Mathis, Rumpl, and Nordsieck), proposed toward the end of the 1970s, gives a good fit for the extinction curve of diffuse clouds with $R_V = 3.1$. In this case, we have a distribution of the type $n(a) \propto a^{-3.5}$, with $50 \lesssim a(\text{Å}) \lesssim 10{,}000$ for graphite and $250 \lesssim a(\text{Å}) \lesssim 12{,}500$ for silicates. In the near and intermediate infrared, emission suggests the presence of small size particles, of the order of 5–50 Å (nanoparticles) and of a nonequilibrium thermal process, whereas emission in the far infrared radiation points to larger grains ($a \gtrsim 100$ Å) with lower temperatures.

We may obtain a rough estimate of the size distribution of interstellar grains by means of a simple reasoning. Considering $n(a,t)$ the number of grains with radii between a and $a + da$ at time t, the equilibrium equation for the grain size distribution may be written in a simplified way:

$$\frac{dn(a)}{da}\frac{da}{dt} + P(a)n(a) = 0, \tag{9.51}$$

where $da/dt = \dot{a}$ is the grain-growing rate, assumed constant, and $P(a)$ is the probability per unit time of a grain with radius a being destroyed. Assuming that $P(a) \propto a^{\xi}$, we obtain

$$\frac{dn(a)}{n(a)} + \text{constant } a^{\xi}\, da \simeq 0, \tag{9.52}$$

whose solution is

$$n(a) \propto \mathrm{e}^{-a^{(\xi+1)}}. \tag{9.53}$$

For instance, for $\xi = 2$, the destruction probability is proportional to the area of the grain surface, and we obtain $n(a) \propto \mathrm{e}^{-a^3}$.

9.5.2 Temperature

In neutral H regions as well as in ionized H regions, dust grain temperatures may be determined assuming equilibrium between the energy absorbed by the grains per unit time and the energy emitted by them, in a similar way as the gas processes analyzed in Chap. 7. Grain energy input comes essentially from stellar or diffuse radiation absorption and from collisions with gas particles, whereas losses are due to emission, mostly in the infrared part of the spectrum.

The grain energy input per unit time per unit projected area of the grains (G) may be expressed in terms of radiative and collisional components:

$$G = G_{\mathrm{r}} + G_{\mathrm{c}}, \tag{9.54}$$

where radiative input (G_{r}) depends on the radiation field energy density U_{λ} (units: erg cm^{-3} Å^{-1}) and on the absorption efficiency factor, $Q_{\mathrm{a}}(\lambda)$:

$$G_{\mathrm{r}} = c \int_0^{\infty} Q_{\mathrm{a}}(\lambda) U_{\lambda} d\lambda, \tag{9.55}$$

with units erg cm^{-2} s^{-1}, for instance. Collisional input is more complicated, and term G_{c} is generally proportional to the product $n\sigma vE$, where n is the gas particle density, σ is the relative cross section, v is the mean velocity of the particles, and E is the mean energy input of the grains during collisions.

The energy lost by the grains per unit area per unit time, assuming they emit as a blackbody at temperature T_{d}, is given by

$$L_{\mathrm{r}}(T_{\mathrm{d}}) = 4\pi \int_0^{\infty} Q_{\mathrm{a}}(\lambda) B_{\lambda}(T_{\mathrm{d}}) \mathrm{d}\lambda, \tag{9.56}$$

depending explicitly on grain temperature. Equation (9.56) may be directly obtained from (9.55) with $U_\lambda = (4\pi/c)B_\lambda(T_d)$ [cf. (2.40)].

When the $G = L$ condition is fulfilled, it allows in principle to determine T_d and can be applied to neutral or ionized H regions. For neutral regions, term G_c is negligible relative to G_r, basically due to the observed values of the radiation field (Chap. 2) and to the relatively low mean energy input values of the grains from collisions with H atoms, lower or of the order of 1 eV. In fact, for $T \simeq 100$ K and $Q_a \simeq 0.1$, we have $G_r/G_c \simeq 2 \times 10^4/n_H$, that is, $G_r \gg G_c$ for typical diffuse clouds, where $n_H \simeq 0.1-10$ cm^{-3}. Only in very dense clouds, where $n_H \gtrsim 10^4$ cm^{-3}, collisions begin to have some importance and only then must we consider molecular H. In this case, we have

$$c\int_0^\infty Q_a(\lambda)U_\lambda d\lambda = 4\pi\int_0^\infty Q_a(\lambda)B_\lambda(T_d)d\lambda. \qquad (9.57)$$

This equation can be solved if we know the interstellar radiation field and the detailed variation of the absorption efficiency with wavelength, which implies some knowledge of the nature and optical properties of the grains. If Q_a is assumed λ-independent, the solution of (9.57) is trivial, obtaining

$$cQ_a\int U_\lambda d\lambda = 4\pi Q_a\int B_\lambda(T_d)d\lambda,$$

which may be integrated, so that

$$T_d = \left(\frac{cU}{4\sigma}\right)^{1/4} = \left(\frac{U}{a}\right)^{1/4}, \qquad (9.58)$$

where we use the Stefan–Boltzmann radiation constant, $\sigma = 5.67 \times 10^{-5}$ erg cm^{-2} s^{-1} K^{-4}, and the constant $a = 4\sigma/c = 7.57 \times 10^{-15}$ erg cm^{-3} K^{-4}. For a typical value of energy density $U_\lambda \sim 10^{-12}$ erg cm^{-3}, we obtain $T_d \simeq 3$ K, essentially the microwave background radiation temperature. More realistic values are obtained when introducing a λ dependency on the efficiency factor. For instance, adopting $Q_a \propto 1/\lambda$ for the optical and the infrared and representing the radiation field by a blackbody with temperature 10^4 K and with a dilution factor of 10^{-14}, we obtain $T_d \simeq 20$ K. For ice, graphite, and silicate grains with radii in the interval 50–1,000 Å, we obtain grain temperatures of 10–50 K, for typical conditions in diffuse interstellar clouds.

As we have seen in Sect. 8.5, in H II regions, photoelectric heating by grains can contribute significantly to gas heating, in particular, by absorption of Lyman-α photons and of stellar radiation field ionizing photons by the grains. Supposing an idealized situation where Lyman-α radiation photons are produced and absorbed at a constant rate, the grain energy input may be written as

$$G_{L\alpha} \simeq Q_a F_{L\alpha} h \nu_{L\alpha}, \tag{9.59}$$

where $F_{L\alpha}$ is the flux of Lyman-α photons that pass through the grain surface per unit projected area per unit time and $h\nu_{L\alpha}$ is the Lyman-α photon energy. In the steady state, the number of emissions per cubic centimeter per second is equal to the number of absorptions, that is,

$$\sigma_g n_d Q_a F_{L\alpha,} \simeq 0.70\, n_p n_e \alpha, \tag{9.60}$$

where we take again the mean value $f \simeq 0.7$ suitable for $T \sim 10^4$ K for the $n \geq 2$ level recombination fraction that produces Lyman-α photons and $\alpha \simeq \alpha^{(2)}$ [see (8.91)]. Using (9.38) in (9.60) with $n_e \simeq n_p \simeq n_H$, we have

$$F_{L\alpha} \simeq \frac{0.70 n_p n_e \alpha}{\sigma_g n_d Q_a} \simeq 2.1 \times 10^8 \frac{n_H}{Q_a}, \tag{9.61}$$

where the last relation is in units of photons per square centimeter per second. Here, we assume $\alpha \simeq 3 \times 10^{-13}$ cm^3 s^{-1}, adequate for $T \simeq 10^4$ K. Replacing (9.61) in (9.59),

$$G_{L\alpha} \simeq 2.1 \times 10^8 n_H h \nu_{L\alpha} (\text{erg cm}^{-2}\, \text{s}^{-1}). \tag{9.62}$$

In typical H II regions, ratio $G_c/G_{L\alpha} \ll 1$ and thus term G_c can be neglected on a first approximation, the same being true for H I regions. Thus, grain temperatures in H II regions are essentially defined by the equilibrium between radiative input $G_{L\alpha}$ and energy loss by infrared emission, as we saw in (9.57). These temperatures are generally higher than the temperatures found in H I regions, typically reaching $T_d \simeq 20$–60 K for ice and graphite grains with sizes 100–1,000 Å. Grain temperature in H II regions vary with the distance to the central star, increasing for smaller distances until eventually reaching values that lead to the evaporation of grains near the central star. For instance, calculations taking into account absorption of high- and low-energy photons from the diffuse stellar radiation field suggest temperatures of the order of 100–200 K for smaller distances or of the order of 0.01 r_s, where r_s is the Strömgren radius (see Sect. 8.5).

9.5.3 Electric Charge

Several physical processes involving dust grains in interstellar regions can lead to a residual electric charge, in particular collision with electrons and protons and photoelectric emission. In an equilibrium situation, a mean electric charge may be determined, with the definition of an electric potential for a given grain distribution. Similarly to the case for the temperature, this charge may be determined considering equilibrium between processes of capture and loss of ions and electrons.

In the simplest situation, where photoelectric ejection is negligible, the grain charge is defined by the equilibrium between collisions with electrons and positively charged ions. Let us consider a grain with charge $-Z_d e$ and radius a that can collide with ions of mass m_i, moving at speed u_i when they are far away from the grain and at speed v when they are at a distance p_0 ("impact parameter") where the ion may be captured by the grain through electrostatic attraction. For distances, $p < p_0$ capture occurs, whose cross section is

$$\sigma_i = \pi p_0^2. \tag{9.63}$$

From the conservation of angular momentum, we have

$$u_i p_0 = a\, v, \tag{9.64}$$

and from energy conservation,

$$\frac{1}{2} m_i u_i^2 = \frac{1}{2} m_i v^2 - \frac{Z_d e^2}{a}. \tag{9.65}$$

From (9.63), (9.64), and (9.65), the ion capture cross section becomes

$$\sigma_i = \pi a^2 \left[1 + \frac{2 Z_d e^2}{m_i u_i^2 a} \right]. \tag{9.66}$$

In the same way, for the capture of electrons with mass m_e and velocity u_e, when away from the grain, we have

$$\sigma_e = \pi\, a^2 \left[1 - \frac{2 Z_d e^2}{m_e u_e^2 a} \right]. \tag{9.67}$$

In an equilibrium situation, the total number of collisions of ions and electrons with grains per second must balance, taking into account the velocity distribution of the particles. We can roughly assume that the particles have mean velocities $\bar{u}_e$ and $\bar{u}_i$ and the equilibrium condition may be written as

$$n_e \sigma_e \bar{u}_e \simeq n_i \sigma_i \bar{u}_i. \tag{9.68}$$

In a non-ionized cloud, $n_e \simeq n_i$, and so (9.66), (9.67), and (9.68) enable us to write

$$Z_d \simeq \frac{3a\, kT}{2e^2} \frac{\bar{u}_e - \bar{u}_i}{\bar{u}_e + \bar{u}_i} \simeq \frac{3a\, kT}{2e^2}, \tag{9.69}$$

where we consider $m_e u_e^2 \simeq m_i u_i^2 \simeq 3\,kT$. From (9.69), we can estimate the grain charge for a typical H I interstellar cloud with $T \simeq 100$ K and grains with radii $a \simeq 1{,}000$ Å. This leads to the result $Z_d \simeq 1$, whereas for an H ionized region with $T \simeq 10^4$ K, we will have $Z_d \simeq 90$. As we have seen in Sects. 7.7 and 8.5, photoelectric ejection is important in neutral (H I) and ionized (H II) interstellar regions, but expression (9.69) can be applied to collisionally ionized regions where $T \sim 10^6$ K, so that $Z_d \simeq 10^4$ for spherical grains with radii $a \simeq 1{,}000$ Å.

The inclusion of photoelectric emission tends to decrease the number of electrons of the grain, making the grain more positively charged, especially if the emission rate exceeds the electron capture rate. In this case, the grains' charge depends on the photoelectric process characteristics that, as we have seen in Sect. 7.7, are not well known, in particular the photoelectric yield y. This parameter can have very low values in the optical spectrum, $y \sim 10^{-4}$, but can reach $y \sim 0.1$ in the ultraviolet ($\lambda \sim 1{,}000$ Å), as already mentioned. In this case, in typical H I or H II regions, the photoelectric effect dominates and charges near zero or even positive can be obtained, depending on the efficiency of the absorption of diffuse Lyman-α radiation or ionizing photons from the central star followed by photoelectric emission. Values of $Z_d \simeq 120$ are obtained for grains with $a \simeq 1{,}000$ Å in a region with $T \simeq 8{,}000$ K, whereas near zero or positive charges are obtained in the neighborhood of stars with effective temperatures in the interval $5{,}000 \lesssim T_{ef}(\mathrm{K}) \lesssim 20{,}000$.

9.5.4 Chemical Composition

A first idea about the chemical composition of interstellar grains may be obtained from the study of typical interstellar abundances and mainly from the depletion factors observed for different chemical elements. In Sect. 4.7, we have seen that elements such as Fe, Ca, Al, Si, and Mg present high depletion factors, that is, they are extremely underabundant in the interstellar medium relative to "cosmic" abundance measured in meteorites and in the Sun. This underabundance may be explained if we admit that these elements are not in the form of interstellar gas but condense instead to form solid grains.

This aspect is made clearer by inspection of Fig. 9.4, where we show the abundance relative to cosmic abundance or depletion factor f_d defined in (4.65) for the direction of the star ζ Oph as a function of *condensation temperature* of the different species. This temperature is defined as the temperature needed for half of the atoms of an element to condensate in thermodynamic equilibrium, so that for temperatures lower than the condensation temperature, solid particles of this element material are stable.

In the figure, we see that the elements presenting a higher depletion are indeed the ones that have higher condensation temperatures. Stars with different degrees of reddening present a similar depletion pattern, though not identical, thus indicating chemical composition variety among the interstellar grains.

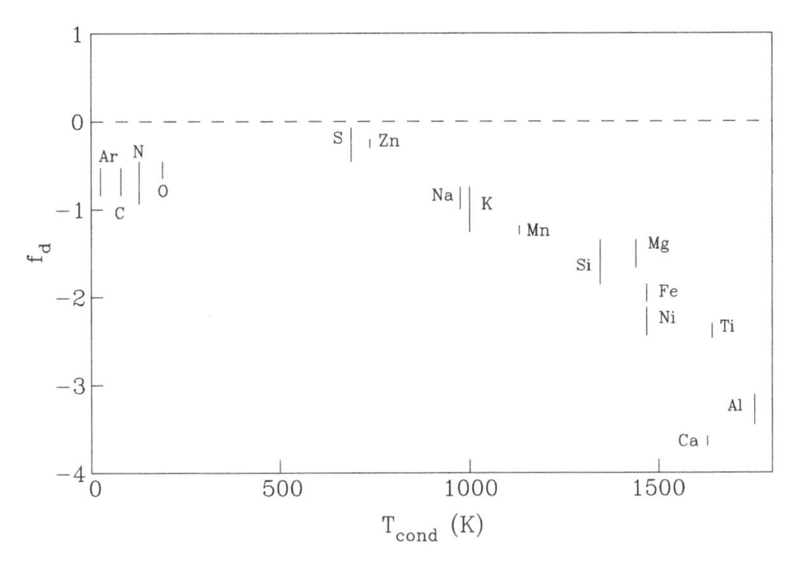

Fig. 9.4 The interstellar depletion factors in the direction of Zeta Oph as a function of the condensation temperature for different types of grains

More information about grain composition may be obtained from the identification of features in the extinction curve and in the observed thermal emission spectrum. The most common species are:

- Carbon, in the form of graphite or amorphous, with small sizes, $a \simeq 50$ Å, showing at ultraviolet absorption bands at $\lambda \simeq 2{,}175$ Å or $\lambda^{-1} \simeq 4.6\ \mu\text{m}^{-1}$ and expected in stars with ratio abundances O/C < 1. Graphite is an orderly and stable form of carbon, with a resonance near $\lambda = 2{,}175$ Å, precisely where a "bump" in the extinction curve is observed.
- Silicates, with larger sizes than carbon grains, showing emissions at 9.7 μm and 18 μm, in protostellar and circumstellar envelopes and absorption at the same wavelengths in the direction of the galactic center and background stars. Some examples are $MgSiO_3$ and Mg_2SiO_4 (magnesium silicate, olivine) and Fe_2SiO_4 (iron silicate). Besides pure metallic grains, grains presenting, for instance, metallic centers and organic mantles of H_2O or CO are also considered.
- SiC, silicon carbide, showing in infrared emission for $\lambda \gtrsim 10$ μm in carbon-star envelopes.
- Oxides, such as SiO, FeO, and Al_2O_3. MgO has infrared emission features similar to oxygen-type star envelopes, where O/C > 1.
- Ices, pure or "dirty," that is, containing impurities, showing in absorption bands at 3 μm in molecular clouds. They may contain compounds of H_2O (water), CO (carbon monoxide), NH_3 (ammonia), CH_3OH (methanol), etc.
- PAH or polycyclic aromatic hydrocarbons, suggested by more recent works, capable of producing infrared emission bands from 3.3 to 11.3 μm, and sufficiently strong for PAH to be considered important among interstellar species.

Strongest bands are at 3.3, 6.2, 7.7, 8.6, and 11.3 μm, corresponding to vibrations in the C–H and C–O connections of these compounds. PAHs are essentially molecules that have several benzene rings, where six carbon atoms form a ring where H and other atoms can connect. For instance, naphthalene ($C_{10}H_8$) has two of these rings, whereas anthracene ($C_{14}H_{14}$) has three. These molecules are very efficient in absorbing optical radiation emitted from stars and redistributing it in the infrared according to their vibration, torque, and rotation modes. They have some structural similarity with graphite grains, in associations of C and H hexagonal rings. Some examples are pyrene and coronene, containing 16 and 24 carbon atoms, respectively. In fact, these "grains" have some similarity with the so-called Platt particles, proposed around 40 years ago.

Biological grains were also proposed, basically composed of carbon and hydrogen. These are organic refractory grains, involving CH_2 and CO connections, that would be responsible for near-infrared absorption bands from 3 to 7 μm. Other models consider grains as coagulation products of fractal structures.

9.6 Energy Emission by Grains

Taking into account the relatively low temperatures of interstellar grains, it is not surprising to find their emission primarily in the infrared part of the spectrum, from the near infrared, approximately 1–10 μm, covering the intermediate infrared, 10–30 μm, to the far infrared, for wavelengths longer than 30 μm. Generally, hotter grains associated with ionized hydrogen regions are responsible for emission at shorter wavelengths, from 3 to 10 μm, whereas colder grains situated in neutral or molecular H regions produce emission at 100 μm or even longer.

From Wien's law, the maximum emission wavelength of a grain with temperature T_d is

$$\lambda T_d \simeq 0.29\,\mathrm{cm\,K}. \tag{9.70}$$

For $T_d \simeq 30$ K, we have $\lambda \simeq 100$ μm. Spectral observations in the far infrared were much improved by satellites like IRAS (Infrared Astronomical Satellite), COBE (Cosmic Background Explorer), and more recently, ISO (Infrared Satellite Observatory), operating in the wavelength interval 2.5–240 μm.

The detailed study of grain-infrared emission is extremely complex because grains can survive in different physical conditions, such as cold star external layers, circumstellar regions associated with H II regions and planetary nebulae, diffuse interstellar clouds, and molecular clouds. Grain emission can also play an essential role in the star-formation process, since the gravitational energy excess of collapsing clouds is removed from the system via infrared radiation of the grains. Besides that, there is evidence of a great variety of chemical composition and sizes, with different optical properties, that reach different equilibrium temperatures and thus have different emissivities.

9.6.1 Infrared Bands

Interstellar grains are probably responsible for a set of unidentified diffuse infrared bands observed between 3.3 μm and 11.3 μm; the stronger ones occurring at 3.3, 6.2, 7.7, 8.6, and 11.3 μm. These wavelengths are quite near of what would be expected by vibration of the connections C–H and C–C in aromatic compounds, such as benzene or flat molecules like PAHs.

Emission in infrared bands is observed throughout the whole Galaxy, being responsible for 10–20 % of the total radiation emitted by dust. It is also observed in nebulae such as H II regions and planetary nebulae, particularly in carbon regions, where O/C > 1. Apparently, there is a connection between interstellar emission bands and diffuse bands already mentioned in Sect. 9.2, though this connection is still not very clear. Some of the absorption bands were observed in emission, suggesting that at least some of the absorption band families are common to emission bands.

9.6.2 Continuum Emission

Continuum emission of the grains occurs by means of thermal radiation or fluorescence. Fluorescence can be observed in reflection nebulae in the red part of the visible spectrum, at wavelengths not longer than $\lambda \simeq 7{,}000$ Å, sometimes associated with H_2 emission, like in NGC 2023. According to the usual interpretation, this emission would originate from an intense ultraviolet flux falling on hydrogenated amorphous carbon particles or PAHs. If the ultraviolet flux is sufficiently strong, hydrogen molecules and solid particle components may be dissociated, thus restraining emission.

Thermal continuum emission is responsible for the major part of the emission observed in the infrared and points to the temperatures of interstellar grains as we have already seen. Measurements of interstellar grain-infrared radiation observed for instance by satellites IRAS, COBE, and ISO register emission from $\lambda \simeq 1$ μm to $\lambda \simeq 1{,}000$ μm with typical peaks at 140 μm and shorter wavelengths, for example, around $\lambda \simeq 10$ μm. In the near infrared, with $\lambda \lesssim 12$ μm, there is an overlap with the infrared bands at 3.3, 7.7, 11.3 μm, etc. This emission is also observed in reflection nebulae, planetary nebulae, H II regions, as well as in circumstellar dust, being attributed to grains of small size ($a \lesssim 50$ Å) with temperatures near the upper limit of $T_d \simeq 200$ K, containing carbon compounds and PAHs. In the far infrared, $\lambda > 30$ μm, "classical grains" are more important, with larger sizes ($a \gtrsim 100$ Å) and colder ($T_d \simeq 20$ K). For even longer wavelengths, with frequencies 10–60 GHz typical of microwaves, we also observe excess interstellar emission relative to infrared thermal emission, possibly produced by grains of small size that are rapidly rotating and by metallic incrusted grains.

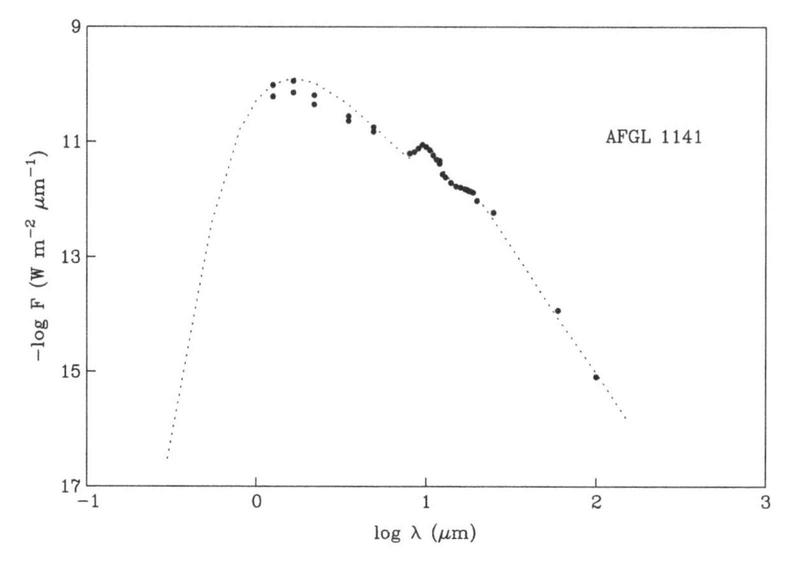

Fig. 9.5 Infrared emission for AFGL 1141 (Silvia Lorenz, Federal University of Rio de Janeiro)

Hotter grain emission, at shorter wavelengths, is more easily detected, especially circumstellar grains associated with bright stars, as seen in Fig. 9.5 for the oxygen-rich red giant AFGL 1141. The figure includes photometric measurements in J, H, K, L, and M bands; observational data from the IRAS satellite; and a theoretical model considering a circumstellar envelope containing silicate grains. Below $\lambda \simeq 7$ μm, the emission is typical of a blackbody with $T \simeq 2{,}000$ K, but for longer wavelengths, there is clearly a radiation excess, attributed to the circumstellar envelope grains. The fitted model has an effective temperature $T_{ef} = 1{,}900$ K, and the envelope has an internal radius $R_1 = 4R_*$ and an external radius $R_2 = 1{,}000R_*$. The observed peak at $\lambda \simeq 9.7$ μm corresponds to emission by silicate grains with sizes of 2,600 Å.

9.7 Formation of Interstellar Grains

Studies of grain formation in interstellar clouds encounter great difficulties in producing enough grains to explain the observed properties, such as extinction, scattering, polarization, and infrared emission.

Basically, these models are affected by the low densities of interstellar clouds, except in denser regions of molecular clouds. Because of that, preferred regions for interstellar grain sources are atmospheres and envelopes of cold giants and supergiants, where densities are considerably higher than in interstellar clouds (Table 1.1), but kinetic temperature is compatible with condensation temperatures of the main candidates of interstellar grain components.

These stars can have different chemical compositions, in particular, ratios O/C > 1 or O/C < 1, which basically define the type of grains that can be produced. Therefore, oxygen-type stars tend to produce silicate grains, whereas carbon stars contribute with graphite and SiC grains. In some cases, we believe that stars only produce condensation nuclei and that the grain external layers are added during and after the ejection process. Infrared observations of these stars show the presence of a spectral feature at 11.5 μm, attributed to SiC in carbon stars. Oxygen-type stars of the Mira type normally present the silicates feature at 9.7 μm. Parts of these grains were probably produced in the external cold layers of the stars, and we cannot exclude the possibility of the presence of remnant grains from the protostellar cloud.

Once the grains—condensation nuclei—formed in the external layers of cold stars, they are pushed away from the star by the radiation pressure. These grains interact with the gas of the stellar envelope and may in fact be responsible for the mass loss of these stars.

Our own Solar System is a rich region of solid particles, the "interplanetary grains." These particles may be observed in the form of meteorites, interplanetary dust particles, and cometary grains. Meteorites originate in asteroids and contain several carbon compounds, SiC, and silicates. Interplanetary dust particles originate in comets or asteroids, containing silicates, carbon, etc. Cometary grains have similar compounds, besides ice containing C, H, O, and N. Several types of pre-solar solid grains were identified in meteorites, such as carbon grains (diamond, graphite), SiC, and corundum (Al_2O_3). These grains present isotopic anomalies that suggest that they were not formed in the solar nebula but are instead part of an interstellar grain population that existed prior to the nebula formation.

Supernovae can also be considered as regions of grain formation, due to the abundance of heavy elements and their probable condensation during the process of mass ejection.

The difficulty of forming grains in interstellar clouds may be assessed by the long time it takes for the grains to reach the observed sizes. In fact, assuming that the grain radius increases with addition of type i atoms, with density n_i, mass m_i, and mean thermal velocity v_i, the grain radius a will be a function of time, given approximately by

$$a \simeq \epsilon \frac{n_i m_i v_i t}{4 s_g}, \tag{9.71}$$

where s_g is again the internal density of the grain and ε is the sticking coefficient. Taking typical values, like $s_g \simeq 1\,\mathrm{g\,cm^{-3}}$, $a \simeq 1{,}000\,\text{Å} \simeq 10^{-5}\,\mathrm{cm}$, $n_H \simeq 10\,\mathrm{cm^{-3}}$, $n_i \simeq 10^{-4}\, n_H \simeq 10^{-3}\,\mathrm{cm^{-3}}$, $m_i \simeq 20\, m_H \simeq 3.3 \times 10^{-23}\,\mathrm{g}$, $T \simeq 100\,\mathrm{K}$, and $v_i \simeq \sqrt{(kT/m_i)} \simeq 2 \times 10^4\,\mathrm{cm\,s^{-1}}$, we obtain for the timescale $t \simeq 4 s_g a / n_i m_i v_i \varepsilon \simeq 2 \times 10^9/\varepsilon$ years. If the sticking coefficient is $\varepsilon \lesssim 1$, then the time needed for grain growing is too long, longer than the age of the main interstellar space components and even longer than the ages of the majority of the stars. Naturally, if growing starts from already formed condensation nuclei, this time will decrease, but even so, it remains very long relative to typical interstellar timescales. For instance, an H II region is associated with hot young stars, with typical ages of 10^6 years. A planetary nebula with radius of the order of 0.2 pc and expansion velocity 20 km s^{-1} has a dynamical age of the order of 10^4 years.

In the more external regions of cold giant stars, densities are high, whereas temperatures remain in the interval 1,000–2,000 K, and so grain formation timescales are shorter. For some molecules, partial pressure can exceed vapor pressure of the corresponding solid. In this case, in even more external regions, where temperatures are even lower, the gas becomes highly saturated, and as a consequence, there will be solid particle nucleation.

According to the classical theory of homogeneous nucleation, in an equilibrium gas with n_0 atoms per unit volume, a globule will grow by aggregation of j particles until it reaches density n_j given by

$$n_j = n_0 \exp\left(-\Delta G_j / kT\right), \tag{9.72}$$

where ΔG_j is the globule formation free energy, a thermodynamic quantity that depends on the solid global properties, such as superficial energy and molecular volume. Function ΔG_j varies with j in a complex way, and n_j increases for sufficiently high values of j, when the globule becomes stable. Applying this theory to the main components of interstellar grains leads to satisfactory results, though the equilibrium hypothesis is frequently questioned, in particular, because gas in atmospheres and cold star envelopes is expanding, and nucleation timescales are also affected by dynamical timescales.

Once formed, grains tend to escape the star due to radiation pressure, according to (9.23). Several studies show that acceleration due to stellar radiation may dominate gravitational acceleration, resulting in a grain "drift velocity" relative to the gas, thus effectively escaping from the star. The final velocities reached can be higher than the final velocities of the gas, which are of the order of 5–20 km s^{-1} in giant and supergiant red stars. The presence of a magnetic field in the external layers of the stars can further complicate this process. The structure of these fields is not well known, thus being generally approximated by dipolar fields or fields similar to the solar magnetic field. Charged grains suffer a direct influence of these fields that alter their dynamical trajectories and escape times from the stellar envelopes.

Finally, during the propagation of the grains in stellar envelopes and interstellar clouds, several processes may occur, which can reduce their size or even destroy them. The main ones are (1) evaporation, especially in the neighborhood of hot stars, like the one occurring in H II regions and planetary nebulae; (2) sputtering, where the grain is hit by gas atoms at sufficiently high velocities to cause ripping of atoms from its surface, a process that may occur in shocks produced by supernovae, where velocities are higher than 50 km s^{-1}; and (3) shattering, due to collisions between the grains themselves.

Exercises

9.1 Diffuse galactic light is basically produced from stellar radiation scattered by interstellar grains. Consider a cloud with spherical grains with radius $a = 1{,}000$ Å and grain density n_d, in an interstellar cloud where $n_H = 10\ \text{cm}^{-3}$. (a) Use the grain-to-gas ratio determined in this chapter and estimate the grain density n_d. (b) Estimate the grain absorption coefficient defined per unit volume (cm^{-1}), $k_d \simeq \sigma_g\, n_d$, where σ_g is the geometric section of the grains. (c) Estimate the absorption coefficient corresponding to the gas atoms for Rayleigh scattering, where $\sigma_R \sim 10^{-24}\ \text{cm}^2$. Which process will dominate?

9.2 Show that the grain-to-gas ratio in the interstellar medium may be approximately written as

$$\frac{\rho_d}{\rho_H} \simeq \frac{(4/3)R_V a\, s_g}{1.086\, Q_e (N_H/E_{B-V}) m_H},$$

where ρ_d and ρ_H are grain and gas densities, respectively; Q_e is the extinction efficiency factor; N_H is the gas column density; $E_{B\text{-}V}$ is the color excess; and R_V is the ratio between general and selective extinction. Grains are assumed spherical with radius a and internal density s_g. (b) Estimate the ρ_d/ρ_H ratio using typical values for R_V, N_H, and $E_{B\text{-}V}$. Use $Q_e \simeq 1$, $s_g \simeq 3\ \text{g cm}^{-3}$, and typical sizes for silicate grains.

9.3 (a) From the definition of the degree of polarization P, show that polarization in magnitude is given by $p \simeq 2.17\text{P}$ for $P \ll 1$. (b) The maximum interstellar polarization in the direction of a star is 6.1 %, occurring for $\lambda = 5{,}400$ Å. Polarization measurements in this direction in the blue part of the spectrum give $P = 5.5$ % and $P = 5.1$ % for $\lambda = 4{,}000$ Å and $\lambda = 3{,}700$ Å, respectively. Apply Serkowski law and determine the mean value of constant K for this star.

9.4 A hot star has an interstellar reddening of $E_{B\text{-}V} = 0.3$. The equivalent width of the interstellar Na I D line ($\lambda = 5{,}890$ Å, $f = 0.65$) in the star direction is $W_\lambda = 700$ mÅ. (a) What is the H column density in the star direction? (b) Use the curve of growth given in Chap. 4 and estimate the Na interstellar abundance relative to H, that is, $\log(N_{Na}/N_H) + 12$. (c) What is the Na depletion factor, assuming a cosmic abundance of $\epsilon_{Na} = 6.3$?

9.5 Infrared object IRC+10216 has a diameter of 0.4 arcsec, corresponding to a dust layer. (a) Supposing that the object is at a 200 pc distance, what is the diameter of the dust layer in cm and in astronomical units (AU)? (b) The total luminosity of the object is 12,000 times higher than the one of the Sun. What would be its radius (in cm and in $R_\odot$), assuming an effective temperature of 2,000 K?

Bibliography

Dyson, J., Williams, D.A.: The Physics of the Interstellar Medium. Institute of Physics Publishing, London (1997). Referred to in Chapter 1. Includes a good discussion on interstellar grains and their physical properties, such as temperature and electric charge

Evans, A.: The Dusty Universe. Wiley, New York (1994). A very complete and up to date introduction to interstellar, circumstellar, and extragalactic dust grains

Greenberg, J.M.: In: McDonnell, J.A.M. (ed.) Cosmic Dust. Wiley, New York (1978). Excellent review article on interstellar dust grains. See also Setti, G.G. & Fazzio, G.G. (eds.). Infrared Astronomy. Dordrecht, Reidel, 1978 and the recent compilation of d'Hendrecourt, L.; Joblin, C. & Jones, A. (eds.). Solid Interstellar Matter: the ISO Revolution. Paris, EDP, 1999

Herbig, G.H.: The diffuse interstellar bands. Annu. Rev. Astron. Astrophys. **33**, 19 (1995). Recent review article about diffuse interstellar bands, including a catalogue of observed bands in the direction of HD 183143

Mie, G.: Beiträge zur Optik trüber Medien, speziell kolloidaler Metallösungen. Ann. Phys. **25**, 377 (1908). Classical work on the theory of the interaction of radiation with grains. See also Debye, P. Ann. Phys. vol. 30, p. 59, 1909

Osterbrock, D.: Astrophysics of Gaseous Nebulae and Active Galactic Nuclei. University Science Books, Mill Valley (1989). Referred to in Chapter 1. Includes a discussion on grains and interstellar extinction

Puget, J.L., Leger, A.: A new component of the interstellar matter - Small grains and large aromatic molecules. Annu. Rev. Astron. Astrophys. **27**, 161 (1989). Review article about polycyclic aromatic hydrocarbons (PAHs) and its applications to the interstellar medium physics

Savage, B.D., Mathis, J.S.: Observed properties of interstellar dust. Annu. Rev. Astron. Astrophys. **17**, 73 (1979). Detailed study of the interstellar extinction curve. Figure 9.3 and Table 9.1 are based on this reference. See also Mathis, J.S. Ann. Rev. Astron. & Astrophys. vol. 28, p.37, 1990; Rep. Prog. Phys. vol. 56, p.605, 1993 e Mathis, J.S.; Rumpl, W. & Nordsieck, K.H. Astrophys. J. vol. 217, p.425, 1977 (MRN)

Scheffler, H., Elsässer, H.: Physics of the Galaxy and Interstellar Matter. Springer, Berlin (1988). Referred to in Chapter 1. This book contains a good discussion on grains and their physical properties in connection with the structure of our Galaxy

Spitzer, L.: Physical Processes in the Interstellar Medium. Wiley, New York (1978). Referred to in Chapter 1. Includes an excellent discussion on interstellar grains, their optical properties, physical properties, and composition. Figures 9.1 and 9.4 are based on this reference

van de Hulst, H.C.: Light Scattering by Small Particles. Wiley, New York (1957). Basic text on light scattering theory by solid particles, with direct application to interstellar grains. See also Bohren, C.F. & Huffman, D.R. Absorption and Scattering of Light by Small Particles. New York, Wiley, 1983

Whittet, D.C.B.: Dust in the Galactic Environment. Institute of Physics Publishing, London (1992). A good general introduction to interstellar dust grains

Wickramasinghe, N.C.: Interstellar Grains. Chapman & Hall, London (1967). A complete discussion on interstellar grains. See also Hoyle, F. & Wickramasinghe, N.C. The theory of cosmic grains, Dordrecht, Kluwer, 1991; Wickramasinghe, N.C.; Kahn, F.D. & Mezger, P.G. (eds.). Interstellar Matter}, Genebra, Saas-Fee, 1972, and Wickramasinghe, N.C. & Nandy, K.. Rep. Prog. Phys. vol. 35, p.157, 1972. Figure 9.2 is based on this last reference

Wynn-Williams, G.: The Fullness of Space. Cambridge University Press, Cambridge (1992). Referred to in Chapter 1. Includes a chapter on interstellar grains, their physical properties, composition, and on the distribution of polarization vectors in our Galaxy

Chapter 10
Interstellar Molecules

10.1 Introduction

As we have seen in Chap. 1, the presence of molecules such as CN, CH, and CH^+ in the interstellar medium was already known by the end of the 1930s. Meanwhile, detection of more abundant species has only occurred forty years later. The H_2 molecule, the most abundant in interstellar regions—and in the universe—has only been identified in 1970, by rocket-based observations in Lyman ultraviolet absorption bands in the direction of star ξ Per. Other abundant species, such as the hydroxyl radical (OH), water vapor (H_2O), ammonia (NH_3), formaldehyde (H_2CO), and carbon monoxide (CO), were detected some years before by millimeter and centimeter observations.

Nowadays, around 130 molecules are known, having 2–13 atoms (Table 10.1), observed in as distinct objects as planetary atmospheres, comets, stellar atmospheres and envelopes, Herbig–Haro objects, star-forming regions, H II regions, planetary nebulae, interstellar clouds, supernovae ejecta and remnants, and active galaxies.

In the interstellar medium, molecules are mainly concentrated in dense clouds or molecular clouds. They are generally organic molecules and present a wide variety, including hydrides, simple oxides, sulfites, acetylene derivatives, aldehydes, alcohols, ethers, cyclic molecules, and radicals. The major part is electrically neutral, but some ions are also observed. In fact, the total number of detected molecules is even greater than indicated in Table 10.1, if we consider all different isotopes. For instance, CO has been detected in all possible combinations of ^{12}C and ^{13}C with ^{16}O, ^{17}O, and ^{18}O. Besides that, some molecules are detected in their linear (l) or cyclic (c) geometric form, such as l-C_3H and c-C_3H.

Molecular astrophysics, or astrochemistry, is nowadays one of the most dynamical research fields in astrophysics, involving (1) observations of molecular transitions in the interstellar medium, stars, etc.; (2) development of laboratory

W.J. Maciel, *Astrophysics of the Interstellar Medium*,
DOI 10.1007/978-1-4614-3767-3_10,

Table 10.1 Observed molecules in the interstellar medium

Atoms	Molecular species
2	H_2, OH, SO, SH, SO^+, SiO, SiS, SiC, SiN, HCl, NaCl, KCl, AlCl, AlF, NH, NO, NS, HF, CH, CH^+, CN, CO, CO^+, C_2, CS, CP, PN, PO
3	H_2O, H_2S, HNO, HCO, HCO^+, H_3^+, N_2H^+, NH_2, N_2O, OCS, C_2H, HCS^+, CO_2, C_2O, C_2S, C_3, MgCN, MgNC, NaCN, HCN, HNC, KCN, CH_2, SO_2, SiH_2, SiC_2, HOC^+
4	NH_3, H_3O, H_2CO, H_2CS, HNCO, HNCS, C_3N, HCO_2^+, C_3H, C_3O, C_3S, C_2H_2, CNH_2^+, HC_2N, H_2CN, H_3O^+, SiC_3
5	SiH_4, CH_4, HCOOH, HC_3N, CH_2NH, NH_2CN, H_2C_2O, C_4H, CH_2CN, C_5, SiC_4, C_3H_2, HC_2NC, HC_3N, H_2COH^+
6	CH_3OH, NH_2CHO, CH_3CN, CH_3NC, CH_3SH, C_5H, HC_2CHO, CH_2CH_2, H_2C_4, HC_3NH^+, C_5N, C_6^-, C_5S
7	CH_3CHO, CH_3NH_2, CH_3C_2H, CH_2CHCN, HC_5N, C_6H, C_7^-, CH_2OCH_2
8	CH_3CO_2H, CH_3C_3N, C_7H, H_2C_6, C_8^-
9	CH_3CH_2OH, CH_3OCH_3, CH_3CH_2CN, CH_3C_4H, HC_7N, C_8H, C_9^-
10	CH_3COCH_3, CH_3C_5N
11	HC_9N
13	$HC_{11}N$

projects, such as the study of the properties of molecular species, chemical reactions, and spectroscopy; (3) theoretical studies of the kinetics of reactions of astrophysical importance, molecular spectroscopy, etc.; (4) study of excitation conditions in several astronomical environments; and (5) general effects on interstellar medium and galactic evolution.

Some physical processes involving interstellar molecules were considered in previous chapters. For instance, we saw in Chap. 4 that observations of line intensities in H_2 Lyman bands allow the determination of column densities of this molecule in the direction of hot stars and of the absorbing region temperature, with typical results of $N(H_2) \sim 10^{20}$ cm^{-2} and $T \sim 80$ K, respectively. In Chap. 5, we mentioned the H_2 ($J = 2$) and CO ($J = 1$) de-excitation process by collisions with H atoms and H_2 molecules, respectively. These processes occur at typical rates of 3×10^{-12} cm^3 s^{-1} for $T \sim 100$ K. In Chap. 7, we saw that the collisional excitation of H_2 molecules by H atoms followed by radiative transitions may contribute to the cooling process in interstellar clouds. In this case, H_2 molecules can collide between themselves or with H atoms, getting excited until reaching an upper rotational level. The excited molecule, after a certain time, will decay to the ground level while emitting radiation, which constitutes an energy loss by the interstellar cloud. In Chap. 8, we have analyzed H_2O and OH maser emission in H II regions. Another effect of molecules is that their infrared emission may be used as a tracer for interstellar regions excited both by radiation and by collisions. Besides that, molecules play a fundamental role in the fractional ionization control in dense clouds, which has important consequences for gravitational collapse and star formation (Chap. 11).

10.2 Molecular Structure

Molecular structure is considerably more complex than atomic structure. Besides electron transitions, similar to the ones of the atoms and imprinting marks in the ultraviolet, molecules also present vibrational transitions in the infrared and rotational transitions at microwave or millimeter wavelengths. These transitions may be schematically represented in energy level diagrams, as in the case of atoms, which can be seen in Fig. 10.1 for the H_2 molecule. This diagram shows the potential energy of the system as a function of internuclear distance, reflecting the stronger attraction between atoms forming the molecule as the internuclear distance decreases, until a limit where repulsion between protons starts to dominate.

The curve shows the electron ground state for a typical diatomic molecule, H_2, represented by X, or more accurately $X^1\Sigma_g^+$. The horizontal line segments represent a vibrational level v and some rotational levels J. The upper electron levels are slightly shifted to the right, with a smaller depth relative to the electron ground level, reflecting the fact that upper-level electrons are less efficient in maintaining molecule connection, which are generally easier dissociated when in that states. Transitions between different vibrational and electron states can be represented by vertical lines in Fig. 10.1, that is, the internuclear distance is not altered during the transition. This characteristic is maintained in a quantum treatment, known as the Franck–Condon principle.

Vibrational transitions correspond to changes in the state of oscillations between the atoms that form the molecule, slightly changing the internuclear distance, for diatomic molecules. As a first approximation, we can assume a simple model of a harmonic oscillator for a diatomic molecule so that the vibrational energy of the atoms that constitute the molecule may be written

$$E_v = C\left(v + \frac{1}{2}\right), \tag{10.1}$$

where C is a constant and $v = 0, 1, 2, \ldots$ is the vibrational quantum number. This is an approximate expression, predicting vibrational levels equally spaced, whereas in

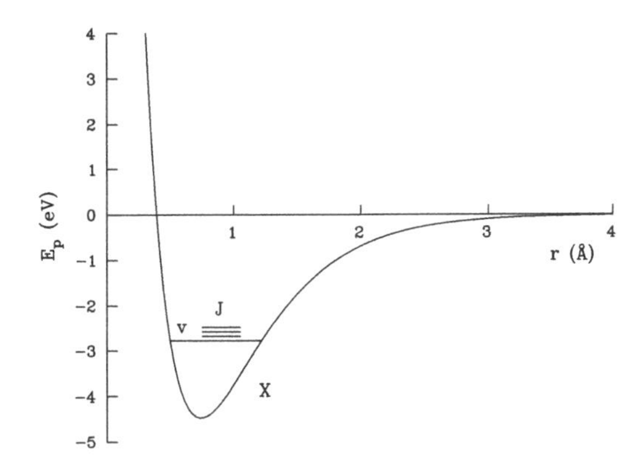

Fig. 10.1 Molecular energy levels

real molecules the separation tends to decrease as the quantum number v increases. In the case of an anharmonic oscillator, (10.1) is replaced by a series with terms $(v + 1/2)$, $(v + 1/2)^2$, etc., being necessary to introduce new constants for each additional term.

For instance, for CO, the transition of vibrational state $v = 1$ to state $v = 0$ in the electron ground state occurs at $\lambda = 4.6$ μm, so the energy difference between the two levels is $\Delta E_v = E_1 - E_0 = C = hc/\lambda \simeq 0.27$ eV. Transitions for polyatomic molecules may be described as a series of vibrations, each one with a different constant.

As the vibrational quantum number increases, vibration amplitude also increases and the value where the chemical connection breaks up and the molecule dissociates can be attained. This energy is the *dissociation energy*, D. For the H_2 molecule, $D = 4.48$ eV and for CO, $D = 11.1$ eV. The dissociation process may occur when a molecule in the electron ground state level X absorbs a photon with enough energy to reach an upper electron state A, where it has higher vibrational energy than the atoms that compose it. This process occurs with molecules such as OH, H_2O, and NH_3. For the most abundant molecule, H_2, the process is not very important because the available photons with energy $h\nu < 13.6$ eV are unable to excite the molecule to the continuum of the first excited state level. In this case the absorbed photons can produce excitation to some vibrational level of the first electron state level above the ground level. In most cases, the molecule decays to ground level X, reemitting the photon. In the other cases, de-excitation may occur at internuclear distances slightly larger than the equilibrium separation, in a classical description. Thus, the molecule ends up in the ground level vibrational continuum, with dissociation happening with the emission of a photon with lower energy than the original photon. Naturally, the energy difference between the two photons must be equal to the molecule dissociation energy plus the kinetic energy of the produced atoms.

Rotational transitions correspond to different energy states of a rotating molecule. For linear diatomic or polyatomic molecules with rigid rotation, the rotational energy can have the values

$$E_J = BJ(J+1), \tag{10.2}$$

where B is a constant and $J = 0, 1, 2, 3, \ldots$ is the rotational quantum number. In molecular spectroscopy, this constant is sometimes used in the form of *term values*, that is, an energy divided by hc with units cm^{-1}. In this case, the constant is $B' = B/hc$. Constant B is related to the molecule moment of inertia, I:

$$B = \frac{h^2}{8\pi^2 I} \tag{10.3}$$

and the moment of inertia is given by

$$I = \mu r^2, \tag{10.4}$$

where μ is the molecule reduced mass and r is its equilibrium internuclear separation. For these transitions, the selection rule $\Delta J = \pm 1$ is applied. Similarly to case (10.1), it is necessary to include terms of superior order in (10.2) for a more rigorous treatment of the rotational levels.

As an example, a transition between the first two rotational levels $J = 1 \rightarrow 0$ of the ground vibrational level of the electron ground state for CO corresponds to $\lambda = 2.6$ mm ($\nu = 115$ GHz), that is, the energy difference of rotational levels is $\Delta E_J = 2B = hc/\lambda \simeq 4.77 \times 10^{-4}$ eV or $B = 2.39 \times 10^{-4}$ eV $= 3.83 \times 10^{-16}$ erg. In this case, from (10.3) we have $I = 1.45 \times 10^{-39}$ g cm^2, and so the internuclear separation is $r = 1.12 \times 10^{-8}$ cm $= 1.1$ Å, with $\mu = 6.86\ m_{\mathrm{H}} = 1.15 \times 10^{-23}$ g.

For nonlinear polyatomic molecules, rotation is more complex. If the molecule has a symmetry axis, the rotational energy will include movements around and perpendicular to that axis. The rotational energy is then characterized by quantum numbers J (angular momentum) and K (J projection over the symmetry axis):

$$E = BJ(J+1) + (A - B)K^2, \qquad (10.5)$$

where constant A is defined as in (10.3) and the quantum number $K = -J, -J+1, -J+2, \ldots J$ with selection rules $\Delta J = 0 \pm 1$ and $\Delta K = 0$. Asymmetric polyatomic molecules have energy rotational levels even more complex and cannot be written as simple formulae. Besides that, several mechanisms produce degenerescency and multiplicity of energy levels, thus increasing even more the number of possible transitions, many of which are confirmed by observations. Naturally, the three transition types may occur simultaneously, creating a very rich molecular spectrum, especially in the case of polyatomic molecules.

10.3 H_2 in Interstellar Clouds

H_2 is the most abundant molecule in the interstellar space, constituting a part of the diffuse clouds and practically all mass of the dense clouds. We saw in Chap. 4 that the number of H nuclei—including atomic and molecular H—in a column with an area of 1 cm^2 is correlated with color excess in magnitude E_{B-V}. In fact, the quantity of H_2 observed in a certain direction is proportional to the quantity of dust in that direction, thus suggesting an association between solid grains and molecular gas. In other words, H_2 molecules are essentially situated in denser interstellar clouds, where the presence of solid grains prevents the propagation of ultraviolet radiation, which tends to dissociate them. These clouds are dense enough to support the approximation of two H atoms, thus fostering molecule formation.

We shall see that, besides acting as a shield against ultraviolet radiation from the interstellar radiation field, grains can also play an important role in the formation process of the H_2 molecule itself.

10.3.1 H–H Collision Timescale

Let us consider an H atom in an interstellar cloud with density n_H and temperature T so that the atom mean thermal velocity is $v \simeq \surd(kT/m_H)$. For a reaction between two atoms to occur, they must approach each other to less than a distance d, that is, the collision cross section is approximately equal to πd^2. The reaction timescale or collision time may then be written as

$$t_c \sim \frac{1}{\pi d^2 n_H v_H}. \tag{10.6}$$

Assuming $d \simeq 3$ Å, we have for a diffuse interstellar cloud, where $T \simeq 100$ K and $n_H \simeq 10$ cm^{-3}, a timescale $t_c \sim 10^8$ s. In a dense cloud, with $T \simeq 100$ K and $n_H \simeq 10^4$ cm^{-3}, we have $t_c \sim 10^5$ s (or a collision rate $1/t_c \sim 10^{-5}$ s^{-1}). These values are very high, meaning that the corresponding reactions are very slow. Comparing, for instance, with the atmosphere of a cold giant star, where $T \simeq 10^3$ K and $n_H \simeq 10^{15}$ cm^{-3}, we obtain $t_c \sim 10^{-6}$ s. In the Earth's atmosphere, in a reaction involving N or O, we have $T \simeq 300$ K and $n \simeq 10^{19}$ cm^{-3} (see Table 1.1), so that $t_c \sim 10^{-9}$ s.

10.3.2 H_2 Formation on the Surface of the Grains

When two H atoms collide, they may interact according to an attractive potential energy curve, such as the one of Fig. 10.1. Meanwhile, if the energy excess is not removed, the atoms will only suffer one collision and then will move apart without forming a molecule. Molecule formation may occur, for instance, if a third body removes part of the collision energy. In laboratory experiments, atoms recombination for molecule formation is facilitated by the presence of surfaces that act as a third body, thus stabilizing the formed molecule. Something similar occurs in interstellar clouds, where H atoms can combine on grain surfaces, which act as catalysts. Formation of H_2 in the surface of the grains involves the collision of two H atoms with the grain, followed by their migration through the solid surface, and finally molecule formation. H atoms that collide with the grain have a certain probability of suffering *adsorption*, adhering to the grain. For molecule formation to occur, the atoms must remain on the grain for a certain time, the *residence time*.

We can estimate the ratio or rate of H_2 formation from the collision rate between H atoms and grains. Let us consider spherical grains with radius a and numerical density n_d, so that the available area of the grains per unit volume is $\pi a^2 n_d$. Assuming H atoms with mean velocity v_H and density n_H, the atom flux is $n_H v_H$ and the H_2 molecule formation ratio (molecules per cm^3 per second) is

$$\frac{dn_{H_2}}{dt} \sim \frac{1}{2}\pi a^2 n_d\, n_H\, v_H, \tag{10.7}$$

where factor 1/2 takes into account the possible velocity orientations of the H atoms. In this expression, we assume that all H atoms that suffer collisions with the grain will adhere to it and effectively lead to the formation of a molecule. A stricter calculation involves the estimate of adsorption and molecule formation probabilities. Using mean values, $a \simeq 1{,}000$ Å $= 10^{-5}$ cm, $n_d/n_H \simeq 10^{-11}$, $T \simeq 100$ K, and $v_H \simeq \surd(kT/m_H) \simeq 10^5$ cm s^{-1}, we have, in orders of magnitude,

$$\frac{dn_{H_2}}{dt} \sim 10^{-16}\, n_H^2\, \mathrm{cm}^{-3}\, \mathrm{s}^{-1}. \tag{10.8}$$

For $n_H \simeq n_{H_2} \simeq 10^2$ cm^{-3}, we have $dn_{H_2}/dt \simeq 10^{-12}$ cm^{-3} s^{-1}.

Molecule destruction occurs mainly by absorption of radiation on the interstellar cloud edges, which acts as a shield for the molecules located in the more internal regions of the cloud (Sect. 10.5). For low values of column density, $N_{H_2} \lesssim 10^{14}$ cm^{-2}, the destruction rate is approximately 10^{-10} s^{-1}, but for higher densities, $N_{H_2} \sim 10^{20}$ cm^{-2}, this rate drops off to about 10^{-14} s^{-1}. Therefore, in a cloud interior with $n_{H_2} \sim 10^2$ cm^{-3}, the H_2 formation process in grains is efficient. For this to happen, it is necessary that one of the H atoms which collides with the grain to stay on the surface long enough to allow the collision and approximation of the second atom. This means that the collision rate of H atoms with the grain, τ_c, must be higher than the evaporation rate τ_e, or $\tau_c > \tau_e$. Collision rate is typically $\tau_c \sim \pi a^2 n_H v_H$, with values of the order of $\tau_c \sim 10^{-3}$ s^{-1}, for the above data and $T \simeq 100$ K. Evaporation rate estimates based on a classical description lead to rates higher than this value for temperatures observed for interstellar grains, but more suitable quantum descriptions produce results compatible with the process efficiency.

10.3.3 Detection of the Interstellar H_2 Molecule

The first detections of the H_2 molecule were made in the ultraviolet, from interstellar absorption lines in the direction of hot stars such as ξ Per and ζ Oph. Naturally, this technique is applied to relatively transparent interstellar clouds that allow the stellar radiation to pass through. The study of denser and opaque regions must be made from microwave and millimeter transitions.

The H_2 molecule is homopolar and symmetric, without an electric dipole moment. Rotational–vibrational transitions that give rise to microwave and millimeter lines are forbidden. Electric quadrupole transitions of low probability may occur, with $\Delta J = \pm 2$, with a minimum energy $\Delta E/k \simeq 510$ K, or $\Delta E \simeq 7.04 \times 10^{-14}$ erg $= 4.39 \times 10^{-2}$ eV, or $\lambda \simeq 28$ μm, corresponding to transition $J = 2 \rightarrow 0$. In the infrared, the lines are very weak and are not observed in absorption in the direction of strong infrared sources situated behind the molecular cloud. Under certain special conditions, however, emission lines at wavelengths $\lambda \simeq 2$ μm can be observed in molecular clouds. These lines indicate a high excitation temperature, much higher than the cloud temperature, which must be produced in shock waves in the clouds' interior or by a fluorescence mechanism. Therefore, in typical dense molecular clouds, where $T \lesssim 100$ K, millimeter and radio transitions of the H_2 molecule are not observed, and so the study of the molecule must be made indirectly, using other molecules. For instance, CO, the second most abundant molecule in the interstellar medium, has lines in this region of the spectrum.

10.3.4 CO and H_2 in the Interstellar Medium

CO is the second most abundant molecular species in molecular clouds, reaching values of 10^{-4} of the H_2 abundance. Collisions between these two molecules produce CO rotational level excitation, followed by emission of photons at millimeter wavelengths. This molecule detection allows us to determine its abundance relative to H_2 in the interstellar medium and also the abundance of the H_2 molecule itself, through a calibration. In fact, the CO 2.6 mm line in the Galaxy's molecular gas plays the same role as the atomic H 21 cm line in the mapping of diffuse interstellar gas (Chap. 4).

CO emission observations show a brightness temperature profile $T_b(v)$ as a function of radial velocity of the cloud gas (or of frequency, by Doppler effect). CO emission intensity may be written as

$$I_{CO} = \int T_b(v)\, dv, \tag{10.9}$$

where the integral is done over the whole radial velocity interval where the emission is detected. CO column density may, in principle, be obtained from (10.9), using a similar relation as for (4.20). Meanwhile, we are generally interested in determining the H_2 column density that may be written as

$$N_{H_2} = X_{CO} I_{CO}, \tag{10.10}$$

where we assume that the H_2 column density is proportional to the CO emission intensity. Measuring T_b in K and velocity in km s^{-1}, intensity I_{CO} has units K km s^{-1}. If N_{H2} is in cm^{-2}, the proportionality constant X_{CO} has units cm^{-2} K^{-1} (km s^{-1})$^{-1}$.

The necessary calibration to determine X_{CO} can be made in diffuse clouds, using H_2 ultraviolet measurements and the relation between H nuclei (including atomic and molecular H) and interstellar extinction (4.43). In denser molecular clouds, H_2 is not observed, but the grain-to-gas ratio may be used to obtain an independent calibration (Chap. 9). Naturally, the process is affected by the non-homogeneity of the interstellar medium, the lack of knowledge of the grain properties, and the uncertainties in the CO emission measurements themselves. Mean values obtained for dark clouds and giant molecular clouds (Sect. 10.7) are of the order of $X_{CO} \sim 1$–5×10^{20} cm^{-2} K^{-1} (km s^{-1})$^{-1}$ and can reach about 2×10^{21} cm^{-2} K^{-1} (km s^{-1})$^{-1}$ in the molecular cloud central regions. Calibrations can also be made from ^{13}CO emission observations, from γ-rays produced by cosmic rays interacting with interstellar gas (Chap. 2) and from the correlation between the CO emission and the cloud mass. A mean value is $X_{CO} \sim 4 \times 10^{20}$ cm^{-2} K^{-1} (km s^{-1})$^{-1}$. For instance, considering a region near the galactic plane, where the brightness temperature is $T_b \simeq 2$ K, approximately constant in the radial velocity interval $\Delta v \sim 20$ km s^{-1}, we have $I_{CO} \simeq 40$ km s^{-1} and $N_{H2} \simeq 1.6 \times 10^{22}$~$10^{22}$ cm^{-2}. If $N_{CO}/N_{H2} \sim 10^{-4}$, we obtain $N_{CO} \sim 10^{18}$ cm^{-2}. Assuming that the region has a size $L \sim 1$ pc, the H_2 volumetric density is $n_{H2} \sim N_{H2}/L \sim 10^4$ cm^{-3}, a typical value for a dense interstellar region (Table 1.1).

10.4 Molecular Reactions in Gaseous Phase

The molecule formation and destruction processes in the interstellar medium present four basic differences relative to known processes in the laboratory (1) Atomic and molecular species in interstellar clouds are mainly in the gas phase, excepting solid grains considered in Chap. 9. Therefore, the main processes leading to formation and fixation of a stable population of interstellar molecules are basically gaseous phase processes. (2) The low densities of interstellar clouds, even the densest ones (see Table 1.1), practically prevent reactions involving three or more molecules to occur. In contrast, however, very reactive radicals and molecules in laboratory, such as OH and N_2H^+, can be studied in a very detailed manner in interstellar clouds, due to their slower destruction and their higher stability in the clouds. Besides that, some species have only been identified in the interstellar medium, like HC_7N and HC_9N. (3) Interstellar cloud temperatures are very low, so the kinetic energy of atoms and molecules is also low and generally only exothermal reactions are important. Some exceptions occur when dynamical processes eject energy into the gas, which is higher than the mean thermal energy of the atoms in the clouds. For instance, in a typical interstellar cloud with $T \simeq 100$ K, the speed of sound (isothermal) is $c_s \simeq \sqrt{(kT/m_H)} \simeq 1$ km s^{-1}. Clouds have random motions with higher velocities than this value, $v_n \simeq 10$ km s^{-1}, so the collision of two clouds can produce shock waves with high enough energy to increase the local temperature by a factor of the order of a hundred. (4) In the interstellar medium, the oxygen abundance is much lower than the H abundance (for instance, from Table 4.3 we have $n_O/n_H \sim 10^{-3}-10^{-4}$). Therefore, oxidation reactions have much less importance in interstellar clouds than in laboratory conditions, where the oxygen high reactivity plays a major role in chemical processes. For instance, we saw that CO is the second most abundant molecule in the interstellar medium and is used to determine the abundance of the most important species, H_2. This is possible thanks to the high stability of CO in interstellar clouds and even in colder stellar atmospheres. In earthlike conditions, however, this molecule easily transforms into CO_2.

10.4.1 Ion-Molecule Reactions

Ion–molecule reactions are probably the most important reactions of the gaseous phase of the interstellar medium. Electric charge of one of the reagents acts as an attraction force, helping the reaction. Ions are always present inside molecular clouds due to ultraviolet radiation or propagation of energetic particles like cosmic rays, as we have seen in Chaps. 2 and 6, respectively.

Examples of main processes are:

$$C + h\nu \rightarrow C^+ + e^- \tag{10.11}$$

$$H_2 + \text{c.r.} \rightarrow H_2^+ + e^- + \text{c.r.} \tag{10.12}$$

In interstellar clouds, ion–molecule reactions frequently involve H_2. For instance, the $H^+{}_2$ ion produced by reaction (10.12) may trigger the following reactions in dense clouds:

$$H_2^+ + H_2 \rightarrow H_3^+ + H \tag{10.13}$$

$$H_3^+ + O \rightarrow OH^+ + H_2 \tag{10.14}$$

$$OH^+ + H_2 \rightarrow H_2O^+ + H \tag{10.15}$$

$$H_2O^+ + H_2 \rightarrow H_3O^+ + H \tag{10.16}$$

$$H_3O^+ + e^- \rightarrow H_2O + H \tag{10.17}$$

with H_2O production in dissociative recombination (10.17). This reaction may be replaced by OH formation,

$$H_3O^+ + e^- \rightarrow OH + H_2. \tag{10.18}$$

H_2O and OH formation may also result from reaction

$$O^+ + H_2 \rightarrow OH^+ + H, \tag{10.19}$$

followed by reactions (10.15)–(10.18).

Carbon chemistry in dense interstellar clouds is also based on ion–molecule reactions, especially the formation of CO. From the H_3^+ ion formed in (10.13) and in a similar way as in reactions (10.14) to (10.16), we have

$$H_3^+ + C \rightarrow CH^+ + H_2 \quad (10.20)$$

$$CH^+ + H_2 \rightarrow CH_2^+ + H \quad (10.21)$$

$$CH_2^+ + H_2 \rightarrow CH_3^+ + H \quad (10.22)$$

CO formation is thus accomplished by reactions

$$CH_3^+ + O \rightarrow HCO^+ + H_2 \quad (10.23)$$

$$HCO^+ + e^- \rightarrow CO + H \quad (10.24)$$

In diffuse clouds, radiation penetrates and produces reactive ions such as C^+, which associates radiatively with H_2:

$$C^+ + H_2 \rightarrow CH_2^+ + h\nu \quad (10.25)$$

The product (CH_2^+) may undergo reactions like (10.22)–(10.24) or form CH or CH_2 by reactions

$$CH_3^+ + e^- \rightarrow CH + H_2 \quad (10.26)$$

$$CH_3^+ + e^- \rightarrow CH_2 + H \quad (10.27)$$

In general, ion–molecule reactions are involved in the formation of the most abundant molecules that are observed in the interstellar medium, and also play an important role in long-chain molecule synthesis, such as cyanopolyacetylenes HC_9N and $HC_{11}N$. In fact, we believe that ion–molecule reactions involving NH_3 or NH_2 lead to the production of HCN, from which more complex molecules are formed by carbon atom pair increments.

10.4.2 Neutral–Neutral Reactions

Though less efficient than ion–molecule reactions, neutral atom–atom or atom–molecule collisions can lead to the formation of a new molecule. In this case, the interaction is weaker and there may exist an energy barrier that can hinder the reaction. The reaction may occur if the incident particle has enough energy or if it can cross the potential barrier by tunneling. Some examples are:

$$O + OH \rightarrow O_2 + H \quad (10.28)$$

$$N + NO \rightarrow N_2 + O \quad (10.29)$$

For the first reaction, the barrier is negligible but the second one is slow for $T \simeq 10$ K and faster for $T \simeq 300$ K, the barrier being important.

H_2O molecules may be produced in interstellar clouds from neutral O—more abundant than O^+ ions involved in reaction (10.19)—by the reaction

$$O + H_2 \rightarrow H_2O + h\nu \quad (10.30)$$

This reaction is exothermal, but has an activation energy barrier corresponding to a temperature of about 100 K. This temperature can be reached by means of shock waves, as mentioned before. Endothermic reactions, such as the reaction involving C and H_2 for hydrocarbon production or N and H_2 for ammonia production (NH_3), may also occur because of the same reason.

10.4.3 Radiative Association

Reactions of the type

$$A + B \rightarrow AB + h\nu \quad (10.31)$$

are the so-called *radiative association* reactions and can occur in the interstellar medium, where the excess energy of atoms A and B is transported by radiation, leaving them with not enough energy to separate. This type of reaction does not frequently occur between neutral atoms, but may be important for ion–atom or ion–molecule interactions, like (10.25) or

$$C^+ + H \rightarrow CH^+ + h\nu. \quad (10.32)$$

10.4.4 Radiative Recombination

In radiative recombination an ion recombines with an electron emitting a photon. For instance, in addition to reaction (10.19), an O^+ ion may be destroyed by

$$O^+ + e^- \rightarrow O + h\nu \quad (10.33)$$

10.4.5 *Dissociative Recombination*

This is a reaction where a molecular ion recombines with an electron, forming two neutral products, like in

$$OH_2^+ + e^- \rightarrow OH + H \quad (10.34)$$

$$OH_3^+ + e^- \rightarrow OH + H_2 \quad (10.35)$$

Generally these are faster recombinations than radiative recombinations. Other examples are reactions (10.17), (10.26), and (10.27).

10.4.6 *Charge Exchange Reactions*

In diffuse clouds, radiation is not sufficiently energetic to ionize O, but the O^+ ion may be formed by a charge exchange reaction of the type

$$O + H^+ \rightarrow O^+ + H \quad (10.36)$$

H^+ ions are produced by H ionization by cosmic rays (Chap. 6). The produced O^+ ion can then be used in reactions initiated with (10.19) up to H_2O or OH formation.

10.5 Molecule Destruction

The reactions of the above processes can also destroy some already formed molecules, as can be seen from the necessary reagents. The main processes of interstellar molecule destruction are *photodissociation* and *collisional dissociation.*

10.5.1 *Photodissociation*

Photodissociation is possibly the principal means of molecule destruction in diffuse interstellar clouds, like for instance, in reaction

$$OH + h\nu \rightarrow O + H \quad (10.37)$$

Molecules in interstellar clouds not protected by dust grains can be destroyed by ultraviolet radiation in relatively short timescales, of the order of a few hundred years. The necessary timescale for photodissociation can be estimated considering that the ultraviolet flux of the interstellar radiation field is $f_\lambda \simeq F_\lambda/h\nu \sim 10^5$

photons cm^{-2} s^{-1} Å^{-1} (see Chap. 2). The photodissociation cross section is typically of the order of $\sigma_{ph} \sim 10^{-17}$ cm^2 in a bandwidth $\Delta\lambda \sim 100$ Å. Therefore, the timescale of molecule life for photodissociation is of the order of

$$t_{ph} \sim \frac{1}{f_\lambda \sigma_{ph} \Delta\lambda} \simeq 300 \, \text{years}. \tag{10.38}$$

More accurately, we may introduce the photodissociation rate

$$\beta_{ph} = c \int n_\lambda \, \sigma_{ph} \, d\lambda, \tag{10.39}$$

where n_λ is the density of photons of the radiation field (photons per cm^3 per Å) and the integral is done between limits $\lambda = 912$ Å and a threshold wavelength that depends on the molecule under consideration. In (10.39), we assume that all absorbed photons lead to dissociation. Since the photon density is related to the energy density of the radiation field U_λ (erg cm^{-3} Å^{-1}) by

$$U_\lambda = h \, \nu \, n_\lambda = \frac{h \, c \, n_\lambda}{\lambda}, \tag{10.40}$$

we have

$$\beta_{ph} = \frac{1}{h} \int U_\lambda \lambda \sigma_{ph} d\lambda. \tag{10.41}$$

Note the similarity between relation (10.41) and the photoionization rate (6.15). The photodissociation timescale t_{ph} is simply

$$t_{ph} = \frac{1}{\beta_{ph}} = \frac{h}{\int U_\lambda \, \lambda \sigma_{ph} \, d\lambda}, \tag{10.42}$$

so that, besides the radiation field, it is necessary to know in detail the variation of the photodissociation cross section with wavelength.

Timescales for photodissociation become much longer as the interstellar cloud extinction increases, which tends to block the photons of the radiation field. Typically, t_{ph} increases by one to two orders of magnitude for each additional optical extinction magnitude absorbed by the clouds.

10.5.2 Collisional Dissociation

Inside dense clouds, where ultraviolet photons are unable to penetrate due to absorption by grains in the peripheral regions, the dominant process for molecular dissociation is collisional. In fact, the higher density of these regions favors gas phase reactions, with the possible dissociation of molecular species. Besides that, we saw that dynamical processes such as shock waves, providing enough energy to break through the activation energy barrier, can favor molecule formation. Meanwhile, if the shock is strong enough, with velocities of some tens of km s^{-1}, collisions will be able to dissociate molecules such as H_2, producing H atoms, which in turn, can collisionally dissociate other molecules like NH or NH_3. For instance, the atoms of a cloud with $v_{ic} \sim 50$ km s^{-1} have a mean kinetic energy $E_k \sim (1/2) m_H v_{ic} \sim 13$ eV, higher than the dissociation energy of some molecules like H_2 and CO.

10.6 Reaction Kinetics

The main molecular processes occur in dense interstellar clouds and the major part of the molecules of Table 10.1 is observed in these regions. In general, molecular abundances are not those obtained when chemical equilibrium is assumed, mainly due to the fact that many of the chemical reactions depend on gas temperature. Therefore, interstellar chemistry is basically controlled by reaction kinetics, and so reactions should be considered on an individual basis.

10.6.1 Reaction Rates

Let us consider a reaction of the type

$$A + B \rightarrow M + X. \tag{10.43}$$

If σ is the reaction cross section, we can define the *reaction constant* or *reaction coefficient* k by relation

$$k = \langle \sigma v \rangle, \tag{10.44}$$

where v is the relative velocity of the reagents. Assuming, for instance, a Maxwellian distribution for the velocities of the reagent atoms, the reaction constant may depend on gas temperature.

The reaction rate r measures the variation rate of the reagents and products involved in the reaction over time, that is,

$$r = -\frac{dn_A}{dt} = -\frac{dn_B}{dt} = \frac{dn_M}{dt} = \frac{dn_X}{dt}. \tag{10.45}$$

Rate r is a positive quantity and is in general proportional to the reaction constant and to the reagent densities, and can be written as

$$r = kn_A n_B, \tag{10.46}$$

where n_A and n_B are numerical densities of reagents A and B. Measuring σ in cm^2 and v in cm s^{-1}, constant k has units cm^3 s^{-1}, and measuring densities in cm^{-3}, rate r has units of cm^{-3} s^{-1}.

Reaction constants may strongly depend on temperature, in particular when there is an activation energy barrier, as we saw. For instance, in reaction (10.19), the reaction constant is $k \sim 10^{-9}$ cm^3 s^{-1} per molecule, corresponding to a cross section $\sigma \sim 10^{-14}$ cm^2 in interstellar clouds. Rates of this order are valid for many ion–molecule reactions. In neutral–neutral reaction

$$\mathrm{O} + \mathrm{CH} \rightarrow \mathrm{CO} + \mathrm{H} \tag{10.47}$$

we obtain $k \sim 10^{-11}$ cm^3 s^{-1} and $\sigma \sim 10^{-15}$ cm^2 for temperatures $T \sim 300$ K, which is a lower value than the one obtained for the ion–molecule reaction.

Reaction constant k dependency on temperature is often written using the classical *Arrhenius equation*,

$$k(T) = A\mathrm{e}^{-E_a/kT}, \tag{10.48}$$

where A is the *Arrhenius factor* or *pre-exponential factor* and E_a is the *activation energy*. From this relation we have

$$\ln k = \ln A - \frac{E_a}{kT}, \tag{10.49}$$

that is, the graph of ln k as a function of T^{-1} is a straight line where the slope is E_a/k and the y-intercept is ln Å. For instance, reaction (10.30) may lead to OH production via reaction

$$\mathrm{O} + \mathrm{H_2} \rightarrow \mathrm{OH} + \mathrm{H} \tag{10.50}$$

Laboratory measurements of the reaction constant at different temperatures are shown in Fig. 10.2.

The points are experimental results and the line is a fit obtained using the Arrhenius equation. The obtained values are ln $A \simeq -24.71$ or $A \simeq 1.86 \times 10^{-11}$ cm^3 s^{-1} and $E_a/k \simeq 4{,}620$ K or $E_a \simeq 0.40$ eV. Naturally, extrapolations for high temperatures ($1/T \rightarrow 0$) or, more frequently, low temperatures (dashed line) in interstellar clouds are dangerous because there may be an A and E_a dependency on temperature. Considering a cloud with $T = 100$ K, from (10.48) we obtain $k \simeq 1.6 \times 10^{-31}$ cm^3 s^{-1} (ln $k \sim -71$), a much lower value than the ones obtained at room temperature. In fact, reaction (10.50) may be important for H_2O synthesis in high-temperature regions ($T \sim 10^3$ K), followed by reaction OH + H_2 $\rightarrow$ H_2O + H.

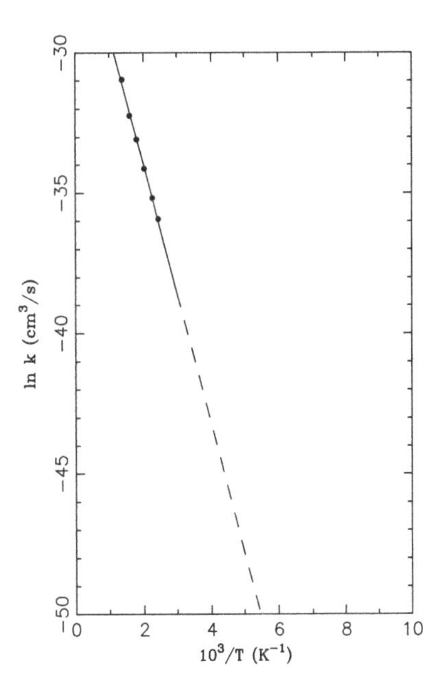

Fig. 10.2 Experimental data for the chemical reaction (10.50)

Table 10.2 Average values of the reaction constants for different reactions

Type of reaction	k (cm^3 s^{-1})
Ion–molecule reactions	10^{-9}
Neutral–neutral reactions	10^{-11}
Radiative association—diatomic	10^{-17}
Radiative association—polyatomic	10^{-9}
Radiative recombination	10^{-12}
Dissociative recombination	10^{-6}
Charge exchange reactions	10^{-9}

As another example, reaction (10.47)

$$O + CH \rightarrow CO + H \tag{10.51}$$

has a constant that is practically independent of temperature, $k \simeq 2.2 \times 10^{-11}$ cm^3 s^{-1}. We can compare the O atom removal rate given by (10.50) with the O atom removal rate given by (10.51) in an interstellar cloud with $T \simeq 100$ K, $n_{H_2} \sim 10^4$ cm^{-3}, $n_O/n_{H_2} \sim 10^{-3}$, and $n_{CH}/n_{H_2} \sim 10^{-8}$, that is, $r \sim -dn_O/dt \sim 10^{-26}$ cm^{-3} s^{-1}, and $r \sim 10^{-14}$ cm^{-3} s^{-1} for reactions (10.50) and (10.51), respectively. Therefore, though the abundance is higher for H_2 than for CH, the second reaction is more efficient for O removal. In fact, reaction (10.50) is only important for temperatures higher than the one considered above.

In general, reaction constants vary highly from one reaction to another and also often with temperature. Table 10.2 shows some mean values for the principal types of reactions in the gaseous phase seen in Sect. 10.4 and must be considered as a first approximation.

Neutral–neutral reaction constants are typically of the order of 10^{-11} cm^3 s^{-1} at room temperature, like in (10.51). Ion–molecule reactions generally have constants that are one or two orders of magnitude higher, $k \sim 10^{-9}$ cm^3 s^{-1}, like in reaction (10.21). In addition temperature dependency is weaker. The highest reaction constants of ion–molecule reactions are essentially due to a larger interaction cross section. In this case, as the molecule approaches an ion, its electrons undergo attraction by the ion, generating a nonuniform charge distribution that creates the molecular *polarizability*. Due to stronger attraction by the ion, the negative part of the molecule changes its initial trajectory, resulting in a collision between the molecule and the ion. A simple theory that considers the electrostatic forces responsible for this attraction is Langevin's theory, according to which $k_L \propto q(\alpha/\mu)^{1/2}$, where q is the ion charge, α is the molecule polarizability, and μ is the reduced mass of the ion–molecule system. Measuring q in electron charges, α in cm^3 and μ in grams, we have

$$k_L \simeq 2.67 \times 10^{-9}\, q(e) \left(\frac{\alpha}{\mu}\right)^{1/2}, \tag{10.52}$$

where k_L is in cm^3 s^{-1}. For instance, in reaction (10.19)

$$O^+ + H_2 \rightarrow OH^+ + H \tag{10.53}$$

we have $\mu \simeq (32/18) m_H \simeq 2.97 \times 10^{-24}$ g and $q = 1$. Taking $\alpha_{H2} \simeq 8.2 \times 10^{-25}$ cm^3, we find $k_L \simeq 1.4 \times 10^{-9}$ cm^3 s^{-1}. More precise formulations exist in the literature, such as the mean dipolar orientation theory, best suited to reactions involving polar molecules.

10.6.2 Abundance Calculation

The calculation of molecular abundances in interstellar clouds requires knowledge of the reactions that affect the considered molecule. Let us assume that molecule M is exclusively formed by reaction (10.43), whose reaction constant is k_1. Let us also assume that this molecule is destroyed by photodissociation with rate β_M (s^{-1}) and by reaction

$$M + C \rightarrow Y + Z, \tag{10.54}$$

whose constant reaction is k_2. In this case, variation with time of the abundance of M, that is, dn_M/dt is given by

$$\begin{aligned}\frac{dn_M}{dt} &= k_1\, n_A\, n_B - n_M\, \beta_M - k_2\, n_M\, n_C \\ &= k_1\, n_A\, n_B - n_M[\beta_M + k_2\, n_C].\end{aligned} \tag{10.55}$$

In time-dependent processes, such as in the study of molecular abundances in comets, this equation must be directly integrated. Assuming steady state, the molecule formation and destruction rates must be equal, that is, $dn_M/dt = 0$, and the density of molecule M is given by

$$n_M = \frac{k_1\, n_A\, n_B}{\beta_M + k_2\, n_C}, \tag{10.56}$$

that is, n_M is a function of n_A, n_B, and n_C, as well as k_1, k_2, and β_M. We need to know these quantities and write equations such as (10.43) and (10.54) for A, B, and C species. Therefore, to study the chemistry of interstellar clouds, an ensemble of all kinds of reactions mentioned here must be considered, along with their reaction constants for suitable temperatures. The problem is extremely complex, because hundreds of reactions are possible, even considering a limited number of chemical elements, and rates are not always well determined. The solution of the system of equations can, in principle, lead to the determination of the abundance of molecules involved in the reaction network.

10.7 Molecular Abundances

10.7.1 Diffuse Clouds

Some results suitable for diffuse clouds, such as the cloud in the direction of star ζ Oph (Chap. 4), are shown in Table 10.3. The cloud physical conditions are approximately $T \simeq 20$–60 K and $n_H \simeq 10^2 - 10^3$ cm^{-3}. The data of Table 10.3 may be written in abundance absolute values, adopting an H mean column density of $N_H \simeq 5.2 \times 10^{20}$ cm^{-2}.

Table 10.3 Average molecular abundances in diffuse interstellar clouds

Atom–molecule	Abundance
H	1.00
H_2	0.79
O	5.6×10^{-4}
C^+	1.9×10^{-4}
N	9.5×10^{-5}
C	6.2×10^{-6}
CO	3.8×10^{-6}
HD	4.0×10^{-7}
OH	9.2×10^{-8}
CH^+	5.5×10^{-8}
CH	4.8×10^{-8}
C_2	4.8×10^{-8}
CN	5.0×10^{-9}
CS	3.0×10^{-9}
NH	$<1.4 \times 10^{-8}$
H_2O	$<4.2 \times 10^{-9}$

10.7.2 Dense Molecular Clouds

As we have seen, in dense molecular clouds, especially in dark clouds, the main constituent is molecular hydrogen, H_2. In these clouds, interstellar grains play a double role (1) acting as a protective layer for molecular species inside the cloud, by absorbing the ultraviolet radiation of the interstellar field, and (2) acting as H_2 molecule formation sites. In these conditions, molecular abundances of the main species of Table 10.1 are relatively high, as can be seen in Table 10.4, which is representative of molecular regions in the Orion Nebula (Orion KL, from Kleinmann Low) and in the dark cloud TMC-1 (Taurus Molecular Cloud) in the Taurus constellation.

To give an idea of abundance absolute values, the H_2 column densities of these objects are $N_{H2} \sim 10^{23}$ cm^{-2} (Orion KL) and $N_{H2} \sim 10^{22}$ cm^{-2} (TMC-1). For comparison, numbers between parenthesis show the abundances of some of these molecules in the circumstellar envelope of infrared object IRC+10216. Denser molecular clouds ($n_{H2} \sim 10^5$–10^6 cm^{-3}), such as Orion KL, have temperatures $T \sim 30$–100 K, dimensions $L \sim 1$–5 pc, and masses $M \sim 10^2$–10^3 $M_{\odot}$, and are associated with star-forming regions. Dark clouds, such as TMC-1, do not present signs of star formation; have lower densities and temperatures, $n_{H2} \sim 10^3$–10^4 cm^{-3}, $T \sim 10$ K; and higher dimensions and masses, $L \sim 1$–10 pc, $M \sim 10^2$–10^5 $M_{\odot}$. In general, abundances relative to H_2 in these clouds and in dense molecular clouds are similar, and Table 10.4 shows mean values for these objects. Note that molecular clouds and dark clouds are non-homogeneous objects, where physical properties such as density and temperature show considerable variations throughout the cloud.

Giant molecular clouds (GMCs) are essentially molecular cloud complexes with intense molecular emission and signs of star formation. They have dimensions $L \sim 20$–80 pc and masses $M \sim 10^4$–10^6 $M_{\odot}$, being, along with some globular clusters, the most massive objects in the Galaxy. Examples of GMC are SgrB2 in the central region of the Galaxy and the molecular cloud in Orion, at a distance of 500 pc from the Sun. The most studied GMC is the one in Orion, which has a very complex internal structure, containing dense filaments, gas and dust layers, and clusters of matter mixed with low density bubbles and cavities. These structures have scales from a few thousandth of a parsec to a few parsecs. The system is in constant dynamical evolution, under the action of supernovae, intense stellar winds, radiative processes, and magnetic fields, with timescales $\lesssim 10^7$ years, shorter than the age of the cloud, which is of the order of 10^8 years. Associated with hot stars, GMC are much affected by ultraviolet radiation, creating photodissociation regions or photon dominated regions.

Giant molecular clouds are particularly important in the study of galactic structure, especially from the analysis of CO observations, with the aim to map the spatial and kinematic distribution of molecular clouds, and determine their dimensions and masses. One of the most interesting results of these studies is the indication that dense molecular gas is concentrated in the central region of the Galaxy, in a "ring" situated between 3 and 7 kpc from the galactic center, approximately. The major part of the H_2 mass is located in this ring and coincides with the region where star formation is more intense. Besides that, CO emission surveys

Table 10.4 Average molecular abundances in dense interstellar clouds

Molecule	Abundance	
H_2	1.0	
CO	10^{-4}	(10^{-3})
H_2O	10^{-6}?	(10^{-6})
CH_3OH	10^{-6}	
OH	10^{-7}	
H_2CO	10^{-7}	
NH_3	10^{-7}	
CN	10^{-8}	(10^{-6})
C_2H	10^{-8}	(10^{-6})
HCN	10^{-8}	(10^{-5})
CH_3OCH_3	10^{-8}	
CH_3CHO	10^{-8}	
OCS	10^{-8}	
CH_3C_2H	10^{-8}	
SO	10^{-8}	
SO_2	10^{-8}	
CS	10^{-8}	(10^{-6})
HCO^+	10^{-8}	
H_2S	10^{-8}	
H_2CS	10^{-9}	
CH_3CN	10^{-9}	(10^{-8})
HC_3N	10^{-9}	(10^{-6})
HNC	10^{-9}	(10^{-7})
HCS^+	10^{-9}	
HC_5N	10^{-9}	(10^{-5})
HC_7N	10^{-9}	(10^{-6})
HC_9N	10^{-9}	(10^{-6})
C_3N	10^{-9}	(10^{-6})
CH_3C_3N	10^{-9}	
CH_2CHCN	10^{-10}	(10^{-8})
HNCO	10^{-10}	
PN	10^{-10}	
$HC_{11}N$	10^{-10}	

show that this molecule can be used as one of the tracers of spiral structure in the Milky Way and in other galaxies, like M51, thus confirming the association of spiral arms with regions of intense star formation.

Exercises

10.1 A rough expression for the potential energy of a bound state of a diatomic molecule in a certain electron configuration is Morse potential, given by

$$E_{\mathrm{p}}(r) = D\left[1 - \mathrm{e}^{-a(r-r_0)}\right]^2,$$

where D, a, and r_0 are constants defined for each molecule. (a) Draw a graph of $E_p(r)$ as a function of r, taking $D = 4.48$ eV, $a = 2$, and $r_0 = 0.74$ Å. (b) Show that function $E_p(r)$ has a minimum for $r = r_0$ that is interpreted as the equilibrium separation. (c) Show that $E_p(r) \rightarrow D$ for $r \rightarrow \infty$. What happens for $r \rightarrow 0$?

10.2 The equilibrium internuclear separation of molecule CS is 1.535 Å. What is the wavelength of the rotational transition corresponding to $J = 1 \rightarrow 0$?

10.3 An H_2 molecule in its ground state level absorbs a photon with wavelength $\lambda = 1{,}000$ Å. After excitation, the molecule dissociates emitting a photon with wavelength $\lambda = 1{,}700$ Å. Assuming that the molecule dissociation energy is $D = 4.48$ eV, what is the mean kinetic energy of each H atom?

10.4 Assume that molecular abundances may be calculated from chemical equilibrium. In this case, equilibrium between the most abundant molecules, H_2 and CO, and molecules CH_4 and H_2O may be written as

$$CO + 3H_2 \rightarrow CH_4 + H_2O$$

Defining the equilibrium constant K, we may write

$$n_{CH_4}\, n_{H_2O} = K\, n_{CO}\, n_{H_2}^3.$$

The constant may be obtained from the reaction enthalpy variation, being $K \simeq 10^{-2}\,\mathrm{cm}^6$ for $T \simeq 200$ K. Interstellar clouds have $T \lesssim 100$ K, so that this value must be considered as a lower limit. Consider a cloud with $n_{H_2} \sim 10^4\,\mathrm{cm}^{-3}$ and $n_{CO}/n_{H_2} \sim 10^{-6}$. Assuming that $n_{CH_4} \sim n_{H_2O}$, what is the methane equilibrium abundance? Compare the result with H_2O abundance. What conclusion can you draw regarding the cloud's chemical equilibrium?

10.5 Show that, in orders of magnitude, (10.42) may be written in the form of (10.38).

Bibliography

Duley, W.W., Williams, D.A.: Interstellar Chemistry. Academic Press, London (1984). Excellent monograph, considering all principal aspects involved in the study of interstellar molecules. Table 10.2 is based on this reference

Dyson, J., Williams, D.A.: The Physics of the Interstellar Medium. Institute of Physics Publishing, London (1997). Referred to in Chapter 1. Discusses interstellar molecules, their role as interstellar cloud coolers, and H_2 formation on the surface of the grains

Hartquist, T.W., Williams, D.A.: The Chemically Controlled Cosmos. Cambridge University Press, Cambridge (1995). A good elementary introduction to molecular astrophysics. See also the more advanced text by the same authors, The Molecular Astrophysics of Stars and Galaxies. Oxford, Oxford University Press, 1998 and Hartquist, T.W. (ed.). Molecular Astrophysics. Cambridge, Cambridge University Press, 1990. Tables 10.3 and 10.4 are partially based on this last reference

Herzberg, G.: Molecular Spectra and Molecular Structure I. Spectra of Diatomic Molecules. Van Nostrand Reinhold, New York (1950). Classical and advanced book on molecular spectroscopy. See also Herzberg, G. The Spectra and Structures of Simple Free Radicals. Ithaca, Cornell University Press, 1971

Hollenbach, D.J., Thronson, H.A. (eds.): Interstellar Processes. Reidel, Dordrecht (1987). Referred to in Chapter 6. Excellent ensemble of advanced review articles about several aspects of interstellar medium astrophysics, including interstellar molecules

Scheffler, H., Elsässer, H.: Physics of the Galaxy and Interstellar Matter. Springer, Berlin (1988). Referred to in Chapter 1. Includes a general discussion on interstellar molecules, molecular clouds, and maser emission

Spitzer, L.: Physical Processes in the Interstellar Medium. Wiley, New York (1978). Referred to in Chapter 1. Advanced discussion on some of the aspects relative to interstellar molecules, such as their role in cloud cooling and maser emission

Wynn-Williams, G.: The Fullness of Space. Cambridge University Press, Cambridge (1992). Referred to in Chapter 1. Discusses interstellar molecules, observations, and their formation on the surface of the grains

Chapter 11
Dynamics and Equilibrium in the Interstellar Medium

11.1 Introduction

The occurrence of dynamical processes is an essential characteristic of the interstellar medium, defining the observed stable structures, like diffuse interstellar clouds and, ultimately, the equilibrium between several components of the space between the stars. For the major part of the processes analyzed up to now, we have sought to determine different population distributions from steady-state equations, as shown in Chaps. 5, 6, and 10. On the other hand, in dynamical processes, it is necessary to obtain the differential equations that govern time variations of the principal physical quantities, which involve hydrodynamic equation systems, frequently coupled with radiative processes.

The main sources of interstellar matter are mass loss processes in red supergiants, planetary nebulae, stellar winds, supernovae, novae, and matter infall of extragalactic origin, whereas interstellar material depletion occurs mainly due to star formation. These processes are extremely complex and a detailed treatment is beyond the scope of this text. In fact, they lie between the main areas of research and many answers are still being sought. In this chapter, we will present a general view taking into account some of the principal dynamical processes that affect the physics of the interstellar medium, obtaining some very rough estimates of the involved quantities.

11.2 Dynamical Processes

Among the main dynamical processes occurring in the interstellar medium, we can highlight (1) supernovae interaction with the interstellar medium, where the source of energy and momentum transmitted to the gas is the explosive ejection of a significant part of the mass of the star in its last evolutionary state; (2) H II regions and planetary nebulae expansion, where a hot and photoionized gas expands

W.J. Maciel, *Astrophysics of the Interstellar Medium*,

DOI 10.1007/978-1-4614-3767-3_11,

throughout the interstellar medium under the action of radiation and stellar winds; (3) dynamical processes of energy and momentum transfer by means of intense winds produced in the external layers of hot or cold stars, caused by a great diversity of physical processes; (4) interstellar cloud collisions, with shock wave propagation through the interstellar gas; and (5) dense interstellar cloud collapse and fragmentation processes, under the action of the cloud's own gravity and interstellar magnetic fields. Let us consider some of these processes, in a simplified way.

11.2.1 Supernovae and the Interstellar Medium

A star with mass $M_* \gtrsim 10\, M_\odot$ will end its evolutionary phase in a supernova explosion. The explosive ejection of a considerable fraction of the star mass constitutes an important source of energy and momentum for the interstellar gas.

For instance, a star with $M_* \simeq 20\, M_\odot$ can produce an ejecta with velocity of the order of one hundredth the speed of light, so that the kinetic energy of the ejected material is $E_{sn} \sim M_*\, v^2 \sim 10^{51}$ erg. A part of this energy is effectively transmitted to the interstellar gas, increasing its temperature and pressure and making it expand. Naturally, the involved velocities are much higher than the speed of sound in the medium, and thus, the explosion generates shock waves that propagate throughout the gas.

The process of supernovae interaction with the interstellar medium is extremely complex, both during the initial stages of the ejecta formation and in the following stages of gas expansion due to the shock wave, in the form of supernova remnants. Let us consider an ideal situation where explosion is instantaneous. In the early stages, the shock wave expands spherically with radius r and velocity $v(r)$ in an interstellar region that was initially at rest with density ρ. We may neglect radiative losses in the early phases, and the energy explosion E_{sn} must be equal to the total energy (kinetic and thermal) in the region immediately behind the shock. In a first approximation, for an adiabatic shock, the total energy per unit mass is of the order of $(1/2)v(r)^2$, so that

$$E_{sn} \simeq \frac{2}{3}\pi r^3 \rho v(r)^2. \tag{11.1}$$

Assuming $v(r) = dr/dt$, this equation can be easily integrated, to give

$$r \simeq \left(\frac{2}{5}\right)^{-2/5} \left[\frac{3\,E_{sn}}{2\,\pi\,\rho}\right]^{1/5} t^{2/5}, \tag{11.2}$$

where we assume $r \to 0$ in time $t \to 0$. From (11.2) we obtain the expansion velocity

$$v(r) \simeq \left(\frac{2}{5}\right)^{3/5} \left[\frac{3\,E_{sn}}{2\,\pi\,\rho}\right]^{1/5} t^{-3/5}. \tag{11.3}$$

From (11.2) to (11.3), we see that the radius of the formed bubble increases with time, whereas the shock velocity decreases. Therefore, the temperature in the post-shock region, proportional to $v(r)$, also decreases, inducing radiative cooling by collisional excitation of spectral lines. Cooling may be intense, forming a layer of colder material behind the shock wave, pushed by the pressure of hot gas heated during the early stages. In this phase, generally called snowplow, we may assume that the layer is expanding in such a way that the momentum is conserved. Let us assume that the shell forms in time t_0 in a transition region where $r = r_0$ and $v(r) = v(r_0) = v_0$. From momentum conservation we then have

$$\frac{4}{3}\pi r^3 \rho v(r) \simeq \frac{4}{3}\pi r_0^3 \rho v_0. \tag{11.4}$$

This expression may be integrated and the following relations are obtained:

$$r \simeq r_0 \left[1 + \frac{4\,v_0}{r_0}(t - t_0)\right]^{1/4}, \tag{11.5}$$

$$v(r) \simeq v_0 \left[1 + \frac{4\,v_0}{r_0}(t - t_0)\right]^{-3/4}. \tag{11.6}$$

For long enough times, $t \to \infty$ or $t \gg r_0/v_0$, we have $r \propto t^{1/4}$ and $v(r) \propto t^{-3/4}$. Comparing (11.5) and (11.6) with (11.2) and (11.3), we notice that in the momentum conservation stage, the radius slowly increases, whereas velocity rapidly decreases.

We can obtain some numerical estimates taking $E_{sn} \sim 10^{51}$ erg and $n \sim \rho/m_H \sim 1$ cm^{-3}. The velocity of the transition region v_0 may be estimated from a more detailed study of the gas cooling process after the shock wave has passed, being typically of the order of 250 km s^{-1}. Therefore, from (11.2) to (11.3), we have $t_0 \simeq 39{,}000$ years and $r_0 \simeq 25$ pc. The quantity of interstellar gas swept by the wave during this time is $M \sim (4/3)\pi r_0^3 \rho \simeq 1{,}600\, M_\odot$, a value exceeding many times the mass of the ejected material.

We can now estimate the efficiency of the conversion of the supernovae explosive energy into interstellar gas kinetic energy. During the snowplow phase, the fraction ϵ of the energy E_{sn} converted into kinetic energy is

$$\epsilon \sim \left[\frac{4}{3}\pi r^3 \rho\right]\left[\frac{1}{2}v(r)^2\right]\frac{1}{E_{sn}} \sim \frac{2\pi\rho}{3E_{sn}} r^3 v(r)^2. \tag{11.7}$$

Using (11.5) and (11.6) we obtain

$$\epsilon \sim \frac{2\pi\rho\, r_0^3\, v_0^2}{3E_{sn}}\left[1 + \frac{4\,v_0}{r_0}(t - t_0)\right]^{-3/4}. \tag{11.8}$$

Equation (11.8) can even be simpler, assuming $t \gg t_0$. The observed velocities of supernovae remnants vary from around 6,000 km s^{-1} for Cas A, a relatively recent event, some hundreds years old, to 300 km s^{-1} for the Cygnus Loop, with an age of the order of 20,000 years. Therefore, the timescale needed for the gas to reach cloud velocity, $v_{ic} \sim$ 10–50 km s^{-1}, must be even longer. Considering $t - t_0 \simeq t \sim 5 \times 10^{13}$ s $\sim 1.6 \times 10^6$ years, the characteristic time to reach velocities of the order of 10 km s^{-1} and using the numerical values given before, we find $\varepsilon \sim 4$ % for the conversion efficiency of explosive energy into gas kinetic energy. Part of the remaining energy is transformed into gas thermal energy (internal), used in ionization and dissociation processes or emitted in the form of radiation.

11.2.2 Expansion of H II Regions

We have seen in Chap. 8 that H II regions have electron temperatures $T_{HII} \sim 10^4$ K and densities $n_{HII} \sim 10^2$ cm^{-3}, whereas in the surrounding H I region, we have $T_{HI} \sim 10^2$ K and $n_{HI} \sim 10$ cm^{-3}. Since the pressure is given by $P \simeq nkT$, we have $P_{HII} \sim 10^{-10}$ dyne cm^{-2} and $P_{HI} \sim 10^{-13}$ dyne cm^{-2}, respectively, or $P_{HII}/P_{HI} \sim 10^3$ and $T_{HII}/T_{HI} \sim 10^2$. Therefore, an H II region contains a hot gas with relatively high pressure immersed in a low-pressure cold gas, so that the hot gas tends to expand, characterizing a dynamical process.

For the case of a stationary nebula, we saw in Chap. 8 that the transition region between ionized and neutral gas is small relative to the dimensions of the H II region. Considering the expansion, we have an ionization front that propagates through the neutral gas. Let us analyze this process in a very simplified way, assuming that the ionization front thickness is small relative to the dimensions of the nebula (plane-parallel geometry) and that the star is instantaneously "turned on", that is, we do not consider evolutionary timescales. Let us consider an ionization front moving with velocity $v(r) = dr/dt$ in a spherical region with radius r, initially at rest, with original density $\rho = n\, m_H$. If f_λ is the ionizing photon flux (photons cm^{-2} s^{-1}) at r, we have approximately

$$\frac{dr}{dt} \simeq \frac{f_\lambda}{n}, \tag{11.9}$$

that is, we assume that all the photons ionize the H I region. The flux f_λ may be obtained, considering that the number of ionizing photons emitted per second by the central star (Q_*) is equal to the number of photons that reach the ionization front per second ($=4\pi r^2 f_\lambda$) plus the ionization rate inside radius r. From Chap. 8, we know that this second component is given by the recombination rate for levels above the ground level, so that for the *on the spot* approximation

$$Q_* \simeq 4\pi r^2 f_\lambda + \frac{4}{3}\pi r^3 n^2 \alpha, \tag{11.10}$$

where we assume that $n_p n_e \sim n_e^2 \sim n^2$ and represent the recombination coefficient for the levels above the ground level $\alpha^{(2)}$ simply by α. Replacing (11.10) in (11.19), we have

$$\frac{dr}{dt} \simeq \frac{Q_*}{4\pi r^2 n} - \frac{1}{3} r\, n\, \alpha. \tag{11.11}$$

For a given central star and a given H I cloud, parameters Q_*, n, and α are well determined and as r increases, the ionization front velocity decreases. Considering the definitions of Strömgren radius r_s and recombination time t_r [see (8.19) and (8.81)], we have

$$Q_* = \frac{4}{3}\pi r_s^3 n^2 \alpha, \tag{11.12}$$

$$t_r = \frac{1}{n\alpha}. \tag{11.13}$$

Equation (11.11) becomes

$$\frac{d(r/r_s)}{d(t/t_r)} \simeq \frac{1}{3}\left[\frac{1-(r/r_s)^3}{(r/r_s)^2}\right], \tag{11.14}$$

whose solution may be written as

$$\frac{r}{r_s} \simeq \left[1 - e^{-t/t_r}\right]^{1/3}. \tag{11.15}$$

We saw that for $t \to 0$ or $t \ll t_r$, we have $r \to 0$ or $r \ll r_s$; for $t \geq t_r$, we have $r \to r_s$. Using typical values $n \sim 10^2\,\text{cm}^{-3}$, $Q_* \sim 10^{49}\,\text{s}^{-1}$, and $\alpha \sim 2 \times 10^{-13}\,\text{cm}^3\,\text{s}^{-1}$ (typical of a temperature of the order of 10^4 K), we have $t_r \sim 5 \times 10^{10}\,\text{s} = 1.6 \times 10^3$ years and $r_s \sim 3.4$ pc. The mean velocity in this region is $r_s/t_r \sim 2{,}100\,\text{km s}^{-1}$, continuously decreasing according to (11.14). The speed of sound in the ionized gas is of the order of 10 km s^{-1} and the initial movement is supersonic, except for $t \gg t_r$ or $r \gg r_s$. The effect of pressure increase by ionization propagates at the speed of sound, so that the ionized gas density is essentially the same as the one of the neutral gas. The ionization front expansion generates a shock wave that requires a more rigorous treatment than the one that can be made with simple numerical estimates.

This description can be applied to the initial expansion stages, when the gas occupies a region of the same order of magnitude as the Strömgren radius in timescales of the order of the recombination time. For the final stages ($t \gtrsim 10^6$ years), the total region swept by the expansion must be much larger, until it reaches conditions typical of H I regions. We can roughly estimate the final extension r_f of this region, considering that the gas affected by ionization has a final temperature T_f and reaches a final density n_f. The region must be in pressure equilibrium with the interstellar H I region, whose density and temperature are n and T, respectively, that is, $2n_fT_f \sim nT$, because ionized gas has about two times more free particles than neutral gas. The timescale for pressure equilibrium depends on the initial density, but we may suppose that equilibrium is reestablished before the central star has evolved significantly. Let us also assume that all ionizing photons emitted by the star are absorbed in the region with radius r_f according to rate Q^*. Using these approximations, we obtain $r_f \sim (2T_f/T)^{3/2} r_s$. For $T_f \sim 10^4$ K and $T \sim 10^2$ K, we obtain $r_f \sim 34\, r_s$. The total swept mass is $M_f \propto r_f^3 n_f$, thus of the order of two hundred times the mass inside the Strömgren radius.

11.2.3 Stellar Winds and the Interstellar Medium

Most stars show evidence for the presence of stellar winds, at least during part of their evolution. Several wind acceleration mechanisms were identified, such as the effect of stellar radiation pressure on ions and dust grains in stellar envelopes, energy dissipation by subphotospherical waves, pulsation, and pressure gradient effect. For instance, the solar wind has velocities of the order of 350 km s^{-1} near the Earth's orbit, corresponding to a continuous rate of mass loss by the Sun $dM/dt \sim 10^{-14}\, M_\odot$ per year. Cold giant stars have slower winds, with terminal velocities $v_t \sim$ 10–20 km s^{-1}, whereas hot OB stars or planetary nebula central stars have very fast winds, with terminal velocities higher than 1,000 km s^{-1}.

The mass loss rate of giant and supergiant stars, both cold and hot, may reach much higher values than the solar rate, $dM/dt \sim 10^{-6}$–$10^{-5}\, M_\odot$ per year. These stars can be an additional energy and momentum source for the interstellar medium, in particular in the case of hot stars with strong winds. Energy ejection in the interstellar medium by winds is not as high as for the case of supernovae, naturally, but may significantly affect limited dimension regions in the star's vicinity. For instance, considering a hot star, typical wind with terminal velocity $v_t \simeq$ 2,000 km s^{-1} and a mass loss rate $dM/dt \sim 10^{-6}\, M_\odot$ per year, the rate of mechanical energy ejection in the interstellar gas is

$$\frac{dE_v}{dt} \sim \frac{1}{2}\frac{dM}{dt} v_t^2 \sim 10^{36}\,\mathrm{erg\,s^{-1}}. \tag{11.16}$$

The stellar wind impact on the interstellar gas creates layers with different physical properties around the star. In the layer nearest to the star, the wind is moving at terminal velocity, supersonic relative to the interstellar gas velocity. In the most external layer, the interstellar gas is unaffected by the wind. In the regions in between a complex interaction occurs, between wind-generated shock waves and gas that is initially at rest. The wind acts like a piston, transferring energy and momentum to the gas and slowing itself down in the process. In the wind's referential, gas advances toward it, producing a shock and forming a "snowplow" layer, like in the previous case. For the intermediate regions, the temperatures reach very high values, of the order of 10^6–10^7 K, higher than the wind temperature, even for very hot stars. Radiative cooling in these regions tends to decrease the temperatures in the post-shock phase until finally reaching the low temperatures typical of molecular clouds.

A similar analysis to the one made for supernovae shows that the shock is effective in an internal region of radius r, given approximately by

$$r \simeq 2.5 \times 10^{-19} \left[\frac{1}{\rho} \frac{dE_v}{dt} \right]^{1/5} t^{3/5}, \quad (11.17)$$

where r is in parsecs, dE_v/dt is in erg s^{-1}, ρ is in g cm^{-3}, and t is in seconds. Similarly, the velocity of the shock at r is

$$v(r) = \frac{dr}{dt} \simeq 4.6 \times 10^{-6} \left[\frac{1}{\rho} \frac{dE_v}{dt} \right]^{1/5} t^{-2/5}, \quad (11.18)$$

where $v(r)$ is in km s^{-1}. The efficiency of the conversion of the wind mechanical energy into gas kinetic energy is higher than for the previous case, $\varepsilon \sim 20\%$. We may obtain some numerical estimates considering $n \sim 10^2$ cm^{-3} or $\rho \sim 10^{-22}$ g cm^{-3} and $dE_v/dt \sim 10^{36}$ erg s^{-1}. The timescale can also be estimated, supposing that the gas in front of the shock is ionized and that the ultraviolet photons emitted per second by the star are totally absorbed by the expanding gas layer. For hot OB stars, this scale is of the order of 70,000 years. Using these values in (11.17) and (11.18), we obtain $r \sim 2.5$ pc and $v(r) \sim 20$ km s^{-1}. We note that the affected region is considerably smaller than for the case of supernovae explosions. This process and the expansion of H II regions associated with very hot stars have several effects on the gas. In fact, the action of stellar winds is a good example, illustrating the transition between circumstellar and interstellar gas. Another important example is the formation of planetary nebulae. The model generally assumed correct is the *interaction winds model*, where a slow and cold wind coming from a red giant star is overtaken by a rapid hot wind coming from a newborn central star. The interaction between these two winds produces a layer similar to the "snowplow" that propagates on timescales of the order of 10^4 years. This is the typical lifetime of a planetary nebula with radius $R \sim 0.3$ pc and an expansion velocity $v_{\rm exp} \sim 20$ km s^{-1}.

11.2.4 Diffuse Cloud Collisions

We saw in Chap. 10 that a diffuse interstellar cloud has spatial velocities $v_{\rm ic} \sim 10$ km s^{-1}. Adopting $T \sim 100$ K and densities $n \sim 10$ cm^{-3} or $\rho \sim 10^{-23}$ g cm^{-3}, the cloud pressure is $P \sim nkT \sim 10^{-13}$ dyne cm^{-2} and the speed of sound is $c_s \sim \sqrt{(P/\rho)}$

~ 1 km s^{-1}. Therefore, collisions between clouds are supersonic, and there may be considerable kinetic energy dissipation into heat generated by shock waves. The steady-state energy dissipation rate by the clouds must balance the kinetic energy ejection rate, basically coming from supernovae. Let us estimate these rates and show that, indeed, they balance each other.

Let us consider spherical clouds with radius $R \sim 5$ pc or volume $V_{\rm ic} \sim 1.5 \times 10^{58}$ cm^3. These clouds are distributed throughout the galactic disk with a total of N clouds in a volume $V_{\rm d} \sim \pi R_{\rm d}^2 h \simeq 6 \times 10^{66}$ cm^3, where $R_{\rm d} \simeq 15$ kpc and $h \simeq 300$ pc (see Chap. 1). The cloud collision cross section is given approximately by $\sigma \sim \pi R^2 \sim \pi (V_{\rm ic}/4)^{2/3} \sim V_{\rm ic}^{2/3}$. We saw in Chap. 4 that there is around $\nu \sim 5$ clouds per kpc in a certain direction, so that the total number of clouds in the disk is

$$N \sim \frac{\nu V_{\rm d}}{V_{\rm ic}^{2/3}}, \tag{11.19}$$

that is, $N \sim 1.6 \times 10^7$ for the given values. The mean free path for collisions between clouds is

$$\lambda \sim \frac{V_{\rm d}}{N V_{\rm ic}^{2/3}}, \tag{11.20}$$

that is, $\lambda \sim 200$ pc. The timescale between collisions is

$$t_{\rm c} \sim \frac{\lambda}{v_{\rm ic}}, \tag{11.21}$$

or $t_{\rm c} \sim 2 \times 10^7$ years. The collision rate is therefore $N/t_{\rm c}$ given in s^{-1}. The dissipated energy in each collision is $E_{\rm c} \sim n m_{\rm H} V_{\rm ic} v_{\rm ic}^2$, so that the total energy dissipated per unit volume per second is

$$\mathcal{L}_{\rm ic} \sim \frac{N E_{\rm c}}{t_{\rm c} V_{\rm d}} \sim \frac{N^2 V_{\rm ic}^{5/3}\, n\, m_{\rm H}\, v_{\rm ic}^3}{V_{\rm d}^2}. \tag{11.22}$$

For these values we obtain $\mathcal{L}_{\rm ic} \sim 10^{-27}$ erg cm^{-3} s^{-1}.

The estimated rate of supernovae is approximately of one event every $t_{\rm sn} \sim 30$ years. The total energy freed by a supernova is $E_{\rm sn} \sim 10^{51}$ erg, from which a fraction $\varepsilon \sim 4\%$ is effectively transformed into kinetic energy of the diffuse clouds. Assuming that supernovae are distributed throughout the galactic disk, the cloud kinetic energy ejection rate is approximately

$$\mathcal{G}_{\rm sn} \sim \frac{E_{\rm sn}\, \epsilon}{t_{\rm sn} V_{\rm d}}. \tag{11.23}$$

We see that $\mathcal{G}_{sn} \gtrsim 10^{-27}$ erg cm^{-3} s^{-1}, of the same order of magnitude of the energy dissipated by the clouds. Therefore, the energy ejected by supernovae is enough to maintain the balance of diffuse clouds, a result that is confirmed by more accurate calculations than the above estimates.

11.3 Interstellar Medium Equilibrium

As seen in previous chapters, the five main interstellar regions are (I) diffuse clouds, (II) dense molecular clouds, (III) intercloud medium, (IV) ionized hydrogen regions, and (V) coronal gas. Table 11.1 shows the mean values of particle density n (cm^{-3}), temperature T (K), and pressure P (dyne cm^{-2}) in these regions. We saw that "low-pressure regions" I, III, and V have similar pressures, occupying the whole interstellar space, whereas "high-pressure regions" II and IV are incrusted in the general interstellar space.

The similarity between pressures reveals another characteristic of the interstellar medium, that the mean energy density of its constituents is similar. In fact, the energy density of those regions, including the radiation field, cosmic rays, and magnetic field, is of the order of 0.5–1.0 eV cm^{-3}.

Low-pressure regions can interchange material. For instance, when parts of the intercloud medium suffer shocks due to supernovae, they may convert into coronal gas. These regions are approximately at pressure equilibrium, and, in fact, pressure gradients are removed on a timescale given by the time needed for a perturbation to traverse the region at the speed of sound. For instance, in a diffuse cloud (I) with $R \sim 5$ pc, $P \sim 10^{-13}$ dyne cm^{-2}, and $\rho \sim 10^{-23}$ g cm^{-3}, we obtain a timescale $t_p \sim R/c_s \sim 5 \times 10^6$ years. However, we saw in Sect. 11.2 that diffuse clouds collide with a timescale between collisions $t_c \sim 2 \times 10^7$ years. The timescale to reach equilibrium t_{eq} must, in principle, be shorter than this value. Since $t_{eq} \sim t_p$, the timescale for a perturbation moving at the speed of sound to cross the considered region, we have $t_{eq} < t_c$. In time t_p sound waves travel about 40 pc in the intercloud medium and 130 pc in the coronal gas, much larger distances than the mean dimensions and separations between clouds, so there is a strong tendency for the existence of pressure equilibrium in these regions.

Table 11.1 Average particle densities, temperatures an pressures in the interstellar medium

Region	n (cm^{-3})	T (K)	P (dyne cm^{-2})
Diffuse clouds	10	100	10^{-13}
Dense molecular clouds	10^4	<100	10^{-11}
Intercloud medium	10^{-1}	10^4	10^{-13}
Ionized H regions	10^2	10^4	10^{-10}
Coronal gas	10^{-2}	10^5	10^{-13}

11.4 Gravitational Collapse and Star Formation

As already mentioned in Chap. 10, giant molecular clouds are true birthplaces of stars, where young and bright objects are immersed in gas and dust clouds. On the other hand, the star formation theory from cloud collapse and fragmentation still has many unknown aspects, making star formation the least known stage in the study of stellar evolution.

11.4.1 Jeans Mass

Let us consider a dense interstellar cloud, spherical and homogeneous, with mass M, radius R, and density ρ. In this case $M \propto R^3\rho$ and the potential energy per unit mass is $E_p \simeq G\,M/R \propto R^2\rho$. The kinetic energy of a cloud particle per unit mass is $E_k \simeq c_s^2$, where $c_s \simeq \sqrt{(P/\rho)}$ is again the speed of sound in the gas. For the contraction and formation of a star to happen, it is necessary that $E_p \simeq E_k$, a condition occurring for dimensions R_J, such as $R_J \propto c_s/\sqrt{\rho}$, where R_J is the so-called Jeans radius. The minimum mass needed for contraction to take place is the *Jeans mass*, $M_J \propto R_J^3\rho \propto c_s^3/\rho^{1/2}$ or $M_J \propto T^{3/2}\,\mu^{-3/2}\,\rho^{-1/2}$, where we consider $c_s^2 \propto kT/\mu$, μ being the mean molecular weight of the gas. In a more accurate calculation, the last proportionality can be written as

$$M_J \simeq 1.4 \times 10^{-10} \frac{T^{3/2}}{\mu^{3/2}\rho^{1/2}}, \qquad (11.24)$$

where M_J is in $M_\odot$, T is in K, and ρ is in g cm^{-3}. When the Jeans critical mass is reached, the cloud collapses in a timescale of the order of the free-fall time t_{ff}, that is, the time needed for a cloud to collapse under the gravitational force. This time can be estimated by $t_{ff}^2 \sim 2R_J/g$, where $g \simeq GM_J/R_J^2$ or $t_{ff} \propto 1/\sqrt{\rho}$. Explicitly, we have $t_{ff} \simeq \sqrt{(3/(4\pi G\rho))}$, which can be written as

$$t_{ff} \simeq 6 \times 10^{-5} \frac{1}{\sqrt{\rho}}, \qquad (11.25)$$

where ρ is in g cm^{-3} and t_{ff} is in years. In an interstellar cloud with $T \simeq 100$ K, $\mu \simeq 1$, and $n \simeq 1$ cm^{-3}, we obtain $M_J \sim 10^5\ M_\odot$ and $t_{ff} \sim 10^8$ years, that is, masses are of the order of the masses of globular clusters and giant molecular clouds. As the collapse progresses, density increases and both M_J and t_{ff} decrease. For instance, considering a region with $T \simeq 50$ K, $\mu \simeq 1$, and $n \simeq 10^6$ cm^{-3}, we obtain $M_J \sim 40\ M_\odot$ and $t_{ff} \sim 10^4$ years. In this case, the region that is collapsing can, in fact, become a single star.

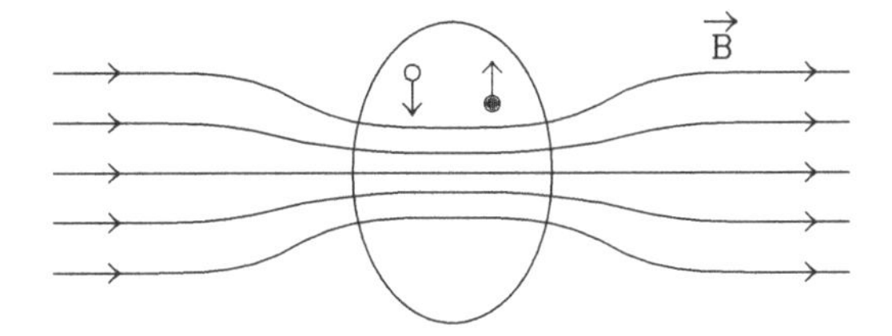

Fig. 11.1 Ambipolar diffusion in the process of star formation

11.4.2 Ambipolar Diffusion

Diffuse interstellar clouds are in equilibrium due to the pressure of the more diluted and hot gas that surrounds them. In denser, optically thick clouds, the cloud's own gravity plays an important role, leading to cloud collapse and probably to protostar formation. If the cloud is immersed in a magnetic field (Chap. 9), this field may counterbalance gravity, increasing the necessary Jeans mass for collapse to occur. The result is a quasi-static structure, where the collapse is hampered in the directions perpendicular to the field lines. Initially, the collapse takes place along field lines, extending to the perpendicular directions only when the density of the cloud is very high.

In the formed structure, the magnetic force in the region where collapse is impeded, is parallel and has the opposite direction of the gravitational force (11.1). The magnetic field affects charged particles that tend to move away from the central region of the cloud (closed circle in Fig. 11.1), whereas neutral atoms, attracted only by gravity, follow an opposite trajectory (open circle in Fig. 11.1), constituting ambipolar diffusion. This diffusion originates friction between charged and neutral particles, decreasing their speed. The result is a delay in the gravitational collapse as well as frictional heating of the cloud. The timescale of ambipolar diffusion depends on the gas fractional ionization, being of the order of $t_d \sim 4 \times 10^5$ years for a degree of ionization $x \sim 10^{-8}$ and $t_d \sim 4 \times 10^6$ years for $x \sim 10^{-7}$. After this period, collapse progresses up to the formation of the protostar.

11.4.3 Collapse and Fragmentation

The complexity of the star formation problem may be assessed by a simple comparison between sizes and masses of molecular clouds and stars. GMC have sizes larger or of the order of 5 pc and masses above $10^4\ M_\odot$, whereas a star has a radius of the order of 10^{11} cm and masses of the order of the Sun, that is, the mean densities of the stars are 20 orders of magnitude higher than the ones of the clouds.

The basic idea of star formation, proposed in the 1950s, considers that the successive fragmentation and collapse stages are efficient enough to reduce the masses of molecular clouds to stellar sizes. This process is called hierarchical fragmentation and depends on the cloud physical conditions, such as its mass and size, on the presence of magnetic fields, rotation, non-homogeneities, radiation emission, and several heating and cooling processes of the interstellar gas. Emission of radiation plays an essential role because the temperature of the central regions increases during collapse and part of the energy excess must be lost in the form of radiation in order for the collapse to proceed. Cloud fragmentation makes this process more efficient because the smaller parts are capable of collapsing in shorter timescales. The process proceeds up to the formation of a very dense and opaque structure, in which the temperature is high enough to trigger thermonuclear reactions. The energy produced is ultimately responsible for the pressure gradient that balances the gravitational attraction of the gas mass, forming an object in hydrostatic equilibrium. From an observational point of view, several young stellar objects (YSO) are identified, generally associated with interstellar gas and dust. Besides young OB stars, other examples are T Tauri stars and Herbig–Haro objects, with the presence of jets and other dynamical structures.

CO emission observations reveal the existence of clumps inside giant molecular clouds, with masses below 100 $M_\odot$, identified as new stars in formation. The gravitational collapse and fragmentation process leads to the formation of stellar objects within mass interval 100 $M_\odot \gtrsim M \gtrsim 0.01\ M_\odot$, approximately. These stars complete their evolutionary trajectories within timescales between millions and several billion years and then return part or all of their constituent matter to the place where it all came from, the interstellar medium.

Exercises

11.1 Prove relations (11.2), (11.3), (11.5), and (11.6).

11.2 Show that (11.15) is the solution of (11.14).

11.3 Consider the interaction between a stellar wind and the interstellar medium. From (11.17) to (11.18) show that $r \simeq (5/3)v(r)t$ (units cgs).

11.4 What temperature should a gas cloud with radius $R = 0.5$ pc and density $n = 10^4\ \mathrm{cm}^{-3}$ have to prevent gravitational collapse?

11.5 Consider a spherical cloud with radius R, mass M, and density ρ, so that $M = (4/3)\pi R^3\rho$. (a) Show that, in orders of magnitude, the free-fall time of the cloud is $t_{ff} \simeq \sqrt{(3/(4\pi G\rho))}$. (b) Show that, more accurately, $t_{ff} \simeq \sqrt{(3\pi/(32G\rho))}$. (c) What is the free-fall time of a cloud with $R = 30$ pc and $M = 500\ M_\odot$?

Bibliography

Dyson, J., Williams, D.A.: The Physics of the Interstellar Medium. Institute of Physics Publishing, London (1997). Referred to in Chapter 1. Excellent discussion of the basic equations of dynamical processes, shock waves, and some applications to the interstellar medium. Section 11.2 is partially based on this reference

Lamers, H.J.G.L.M., Cassinelli, J.P.: Introduction to Stellar Winds. Cambridge University Press, Cambridge (1999). Complete and advanced discussion on stellar winds, acceleration mechanisms, observational evidence, and winds interaction with the interstellar medium

Scheffler, H., Elsässer, H.: Physics of the Galaxy and Interstellar Matter. Springer, Berlin (1988). Referred to in Chapter 1. Includes a detailed analysis of interstellar dynamical processes

Spitzer, L.: Physical Processes in the Interstellar Medium. Wiley, New York (1978). Referred to in Chapter 1. Considers some of the main dynamical processes, interstellar medium equilibrium, and star formation

Erratum to:

Astrophysics of the Interstellar Medium

Walter J. Maciel

W.J. Maciel, *Astrophysics of the Interstellar Medium*,
DOI 10.1007/978-1-4614-3767-3, © Springer Science+Business Media New York 2013

10.1007/978-1-4614-3767-3_12

The publisher regrets the following information was mistakenly omitted on the copyright page during the production process:

Translation from the Portuguese language edition: *Astrofísica do Meio Interestelar* by Walter J. Maciel © Editora da Universidade de São Paulo (Edusp). All rights reserved.

The online version of the original book can be found at
http://dx.doi.org/10.1007/978-1-4614-3767-3

The following corrections given by the author were mistakenly omitted during the production process:

PAGE 18, SECTION 2.2.1, SECOND PARAGRAPH

Plasma emissivity, defined as the total power emitted per unit volume per unit solid angle per unit frequency interval between ν e $\nu + d\nu$, is given by

$$\epsilon_\nu = \frac{n_e}{4\pi} \int P(v, \nu)\, f(v) dv, \tag{2.2}$$

and it is generally measured in erg $\text{cm}^{-3}\ \text{s}^{-1}\ \text{sr}^{-1}\ \text{Hz}^{-1}$, where $f(v)$ is the distribution function of electron velocities and $P(\nu,v)$ is the total power emitted per unit frequency interval during the collision between an electron with velocity v and an ion with density n_i. If $f(v)$ is given by the Maxwellian distribution, the emissivity is...

PAGE 57, AFTER EQUATION (4.13)

where we neglect the radiation intensity falling on the region opposite to the observer and $\tau_{\nu r}$ is again the total optical depth of the emitting region...

PAGE 80, AFTER EQUATION (5.15)

From (5.6), we see that coefficient γ_{jk} gives the collision probability per unit time per field particle so that $n_c\, \gamma_{jk}$ gives the number of excitations per second, and the product $n_t\, n_c\, \gamma_{jk}$ gives the number of excitations per cubic centimeter per second, where n_t is the test particle density...

PAGE 126, EQUATION (7.18)

$$\Gamma_{ei} = n_e\, n_i \sum_j \left[\langle \sigma_{cj}\, v \rangle\, \bar{E}_2 - \langle \sigma_{cj}\, v\, E_1 \rangle \right], \tag{7.18}$$

Constants and Units

Boltzmann constant:

$$k = 1.3807 \times 10^{-16}\ \text{erg/K} = 1.3807 \times 10^{-23}\ \text{J/K}$$

Electron charge:

$$e = 4.8032 \times 10^{-10}\ \text{cm}^{3/2}\ \text{g}^{1/2}\ \text{s}^{-1} = 1.6022 \times 10^{-19}\ \text{C}$$

Electron mass:

$$m_e = 9.1094 \times 10^{-28}\ \text{g} = 9.1094 \times 10^{-31}\ \text{kg}$$

Gravitational constant:

$$G = 6.6726 \times 10^{-8}\ \text{cm}^3\ \text{g}^{-1}\ \text{s}^{-2}\ [\text{dyne cm}^2\ \text{g}^{-2}]$$
$$= 6.6726 \times 10^{-11}\ \text{m}^3\ \text{kg}^{-1}\ \text{s}^{-2}$$

H atom mass:

$$m_H = 1.6734 \times 10^{-24}\ \text{g} = 1.6734 \times 10^{-27}\ \text{kg}$$

Light speed in vacuum:

$$c = 2.9979 \times 10^{10}\ \text{cm/s} = 2.9979 \times 10^{8}\ \text{m/s}$$

Planck's constant:

$$h = 6.6261 \times 10^{-27}\ \text{erg s} = 6.6261 \times 10^{-34}\ \text{J s}$$

Proton mass:

$$m_p = 1.6726 \times 10^{-24}\ \text{g} = 1.6726 \times 10^{-27}\ \text{kg}$$

Radiation constant:

$$a = 7.5658 \times 10^{-15}\ \text{erg cm}^{-3}\ \text{K}^{-4}$$
$$= 7.5658 \times 10^{-16}\ \text{J m}^{-3}\ \text{K}^{-4}$$

W.J. Maciel, *Astrophysics of the Interstellar Medium*,
DOI 10.1007/978-1-4614-3767-3, © Springer Science+Business Media New York 2013

Rydberg constant:

$R_\infty = 3.2898 \times 10^{-15}$ Hz

Sidereal year = 3.1558×10^7 s

Solar luminosity:

$L_\odot = 3.8458 \times 10^{33}$ erg/s = 3.8458×10^{26} W

Solar mass:

$M_\odot = 1.9891 \times 10^{33}$ g = 1.9891×10^{30} kg

Solar radius:

$R_\odot = 6.9551 \times 10^{10}$ cm = 6.9551×10^8 m

Stefan–Boltzmann constant:

$$\sigma = 5.6704 \times 10^{-5} \text{ erg cm}^{-2}\text{ s}^{-1}\text{ K}^{-4} = 5.6704 \times 10^{-8} \text{ W m}^{-2}\text{ K}^{-4}$$

1 eV = 1.6022×10^{-12} erg = 1.6022×10^{-19} J

1 pc = 3.0857×10^{18} cm = 3.0857×10^{16} m
= 3.2616 light-years = 2.0626×10^5 AU

1 AU = 1.4960×10^{13} cm = 1.4960×10^{11} m

1 atm = 1.0133×10^6 dyne/cm^2 = 1.0133×10^5 N/m^2 = 760 Torr

Index

W.J. Maciel, *Astrophysics of the Interstellar Medium*,
DOI 10.1007/978-1-4614-3767-3, © Springer Science+Business Media New York 2013

Lightning Source UK Ltd.
Milton Keynes UK
UKOW06f1810040516

273569UK00004B/193/P